INSTABILITY IN GEOPHYSICAL FLOWS

Instabilities are present in all natural fluids from rivers to atmospheres. This book considers the physical processes that generate instability from a geophysical perspective. The classical analytical approaches are covered, while emphasizing numerical methods that enable prediction of stability or instability in a system quickly, and with minimal mathematics. The first part of the book describes the normal mode instabilities important to geophysical applications, including convection, shear instability, and baroclinic instability. The second part introduces more advanced ideas, including nonmodal instabilities, the relationships between instability and turbulence, self-organized criticality, and advanced numerical methods. Featuring numerous mathematical and computational exercises, suggestions for projects, and MATLAB coding examples online, it is ideal for advanced students wishing to understand flow instability and apply it in their research, and can be used to teach courses in oceanography, atmospheric science, and environmental science. Also available as Open Access on Cambridge Core at doi.org/10.1017/9781108640084.

WILLIAM D. SMYTH is Professor of Oceanography at Oregon State University. He teaches courses in fluid dynamics, geophysical waves, descriptive oceanography, dynamic meteorology, climate science, and stability of geophysical flows. He studies complex phenomena in nature, especially fluid turbulence. He is working to untangle the relationship between turbulence, waves, and instability in the upper equatorial oceans. He has twice received the Pattullo Award for excellence in teaching. He has been honoured with the Kirby Liang Fellowship from Bangor University and with a Distinguished Visitor Fellowship from Xiamen University, China.

JEFFREY R. CARPENTER is a physical oceanographer at the Institute of Coastal Research, Helmholtz-Zentrum Geesthacht, Germany, where he is the leader of the Small Scale Physics and Turbulence Group. His work focuses on the fluid mechanics of physical process in natural water bodies, and his research interests include turbulent mixing in stable density stratification, shear flows, instability and wave interactions, double-diffusive convection, heat fluxes and eddy formation in the Arctic Ocean, turbulence measurements using ocean gliders, and the impacts of offshore wind farms on the coastal ocean.

INSTABILITY IN GEOPHYSICAL FLOWS

WILLIAM D. SMYTH

Oregon State University

JEFFREY R. CARPENTER

Helmholtz-Zentrum Geesthacht

CAMBRIDGE
UNIVERSITY PRESS

CAMBRIDGE
UNIVERSITY PRESS

University Printing House, Cambridge CB2 8BS, United Kingdom

One Liberty Plaza, 20th Floor, New York, NY 10006, USA

477 Williamstown Road, Port Melbourne, VIC 3207, Australia

314–321, 3rd Floor, Plot 3, Splendor Forum, Jasola District Centre, New Delhi – 110025, India

79 Anson Road, #06–04/06, Singapore 079906

Cambridge University Press is part of the University of Cambridge.

It furthers the University's mission by disseminating knowledge in the pursuit of education, learning and research at the highest international levels of excellence.

www.cambridge.org
Information on this title: www.cambridge.org/9781108703017
DOI: 10.1017/9781108640084

First published 2019

A catalog record for this publication is available from the British Library.

ISBN 978-1-108-70301-7 Paperback

Additional resources for this publication at www.cambridge.org/iigf

While every effort has been made to ensure that the methods and formulae given in this book are accurate, a healthy skepticism on the part of the reader is encouraged. Any statement given here could contain typos or math errors, or could simply be wrong. Don't use a theorem for anything important until you've understood the proof. Neither the authors nor Cambridge University Press assumes legal liability for the results.

Contents

Preface

This book began as lecture notes for an Oregon State University graduate course on instabilities of geophysical fluids, mainly the Earth's atmosphere and oceans. Designed originally for students in physical oceanography, the course is also popular with students of atmospheric dynamics, physics, and mathematics. Besides oceans and atmospheres, flow instabilities are important in rivers, canals, and lakes, hence the course is taken regularly by civil engineers. Additional material has been developed as part of a course at the University of Hamburg.

A defining aspect of the course is its emphasis on the numerical solution of boundary value problems; the student learns techniques whose value extends beyond the present topic. Students develop a collection of software that will allow them to study instabilities and waves in all types of flows, including flows measured observationally. A secondary focus is internal waves, primarily gravity and vorticity waves. These are included for two reasons: first, they arise naturally from the same equations that describe instabilities, and, second, they are often essential parts of the mechanisms that create instability.

In the course and in this book, our main focus is the "big three" instabilities of geophysical flows:

- convection (driven by gravity; Chapter 2),
- shear instability (driven by kinetic energy; Chapter 3), and
- baroclinic instability (controlled by the Earth's rotation; Chapter 8).

The first two of these are foundational: besides describing fundamental mechanisms, which vary and combine to make the assortment of instabilities discussed later, the discussions provide an opportunity to introduce essential ideas and methods in a relatively simple context. We also look at various factors that modify these instabilities such as flow curvature (Chapter 7), viscosity (Chapter 5), and density stratification (Chapters 4 and 6). In the Oregon State course, these main topics are covered in 40 hours of lectures. It is expected that students will devote 80 hours to homework and independent study.

Additional chapters describe specialized mechanisms such as double diffusion (important in the ocean; section 9.2) and conditional convective instability (important in the atmosphere; 9.1), as well as auxiliary topics such as turbulence (Chapter 12) and advanced numerical methods (Chapter 13). These may be included or assigned for independent reading.

Homework exercises are included (Appendix A) and are integral to the course. Much critical information is discovered independently in the course of doing these exercises. Solutions are available from the Cambridge University Press website at www.cambridge.org/iigf, together with example MATLAB codes and authors' contact information. We also include several suggestions for term projects in which a student may explore topics not covered in the main text (Appendix B).

This book is also intended for self-study, with detailed explanations and frequent exercises to confirm understanding. If you take this route, feel free to email the authors with any questions that may arise.

While every effort has been made to ensure that the methods and formulae given in this book are accurate, a healthy skepticism is encouraged. Any statement given here could contain typos, or math errors, or could simply be wrong. Never use a theorem for anything important until you've understood the proof.

Bill Smyth and Jeff Carpenter

Acknowledgments

This book has benefited from the comments and suggestions of many alert students over the years. If you are among them, thank you. Thanks to our colleagues Steve Thorpe, Eyal Heifetz, Alexis Kaminski, Anirban Guha, and Mihir Shete for reading chapters prior to publication. There is yet room for improvement – if you find something wrong or unclear, please let us know.

From Bill Smyth: Grateful thanks to my mentor Dick Peltier for inspiring my interest in flow physics, to my parents, Bill and Barbara Smyth, for their unwavering support, and to my wife Rita Brown for tolerating my nonexistence during this project.

From Jeff Carpenter: Thanks to my parents, Keith and Judy Carpenter, for their patience and for doing such a great job, and to my wife Martina for all the sacrifices she has made for me.

This project was funded by the US National Science Foundation under grant OCE-1537000 to William Smyth and through the Helmholtz Association's PACES II program for Jeffrey Carpenter, who is also grateful for support from the Collaborative Research Centre TRR 181 "Energy Transfers in Atmosphere and Ocean" funded by the German Research Foundation.

Part I
Normal Mode Instabilities

1

Preliminaries

If you would understand anything, observe its beginning.
– Aristotle

Oceanographers think of ocean circulation in terms of the "global conveyer belt," in which cold polar waters sink and then circulate around the ocean basins, eventually being warmed in the tropics. But the truth is that this larger-scale circulation has a typical speed of only a few cm s^{-1}, and it is generally accompanied by variable currents many times faster (1 m s^{-1} is not uncommon), fluctuating on periods ranging from months down to seconds.

The largest such variations are the majestic mesoscale eddies that spin off strong currents like the Gulf Stream (Figure 1.1). Today, much research is focused on the

Figure 1.1 Instability of the Gulf Stream shown in a satellite image. Colors represent sea surface temperature. The darkest red represents warm water flowing northeast along the east coast of the United States. After departing from the coast at Cape Hatteras, the current becomes unstable and breaks down into turbulent eddies. (Image courtesy of the U.S. National Oceanic and Atmospheric Administration, hereafter NOAA.)

next size smaller: the submesoscale eddies. Smaller than this are the gravity waves and, at the smallest scale, three-dimensional turbulence.

One must measure for a year or more in order to average out these various fluctuations and discern the mean "conveyer belt" current. But to think of the variations as something we can average away is to fool ourselves, for it is largely the oscillations that govern the conveyer belt. We can't understand one without the other.

One way to understand such a chaotic profusion of motions is to ask what would happen if, at some initial instant, the ocean was calm, with steady, orderly currents. Would the oscillations develop spontaneously? If so, how? This thought experiment is the essence of instability theory.

For example, Figure 1.1 shows eddies forming in the Gulf Stream. No human mathematician could solve the equations that describe these intricate motions; but, using the methods of linear perturbation theory, we can not only predict their length and time scales but also understand quite a lot about what causes them. The trick is to imagine a fictitious Gulf Stream that is straight and eddy-free, then study what happens in the very first few moments – after the current begins to buckle but before it grows so complex as to be mathematically intractable.

We can think of atmospheric motions in the same way. Imagine a fictitious atmosphere where the winds are purely zonal – mid-latitude westerlies, jet streams, and polar easterlies, all blowing straight in the east-west direction only. It turns out that those winds would be unstable, and as a result would break up into the large vortical structures we know as synoptic weather systems (Figure 1.2). To calculate the details requires a supercomputer, but we can understand the basic mechanics and predict the dominant length and time scales (a few thousand kilometers, a few days) quite easily.

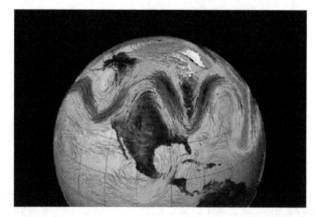

Figure 1.2 The atmospheric jet stream: speed (red = fast) and direction (streaks), showing baroclinic instability. (Visualization courtesy of the US National Aeronautics and Space Administration, hereafter NASA).

The Earth's mantle provides a third example. Suppose the mantle were perfectly motionless. Heating from the radioactive core would lead to the growth of convection cells, exactly as we see reflected in the slow drift of the continents and the attendant seismic and volcanic activity.

In this book we will study instabilities on scales from centimeter to global, controlled mainly by gravity, sheared winds and currents, and the Earth's rotation. While our main focus is the Earth, analogous phenomena are found in atmospheres and magnetospheres of other planets, stellar interiors, and interstellar plasma flows.

1.1 What Is Instability?

Suppose that the emergency brake in your car doesn't work, and you have to park somewhere in hilly country. Where can you park so that your car doesn't roll away (Figure 1.3)? We hope you would park at point (a), the bottom of a valley. But what about point (b), the top of a hill? You could park there in theory, but you would have to park at *exactly* the right spot, and even then any little disturbance would cause your car to roll away.

In mathematical terms, we say that both points (a) and (b) are equilibrium states,[1] i.e., states at which the system can remain steady in time. The difference between (a) and (b) is in what happens when the system is displaced slightly from equilibrium. If you park at the bottom of the valley (a), and if something then pushes the car slightly to the left or the right, gravity will pull it back toward its original location. The car will rock back and forth and eventually come to rest due to friction. In contrast, if you park at the top of the hill (b) and the car is moved even slightly, gravity pulls it further from the equilibrium point. The further the car travels, the steeper the slope and the stronger the pull of gravity. Goodbye car! We say that equilibrium (a) is stable, while (b) is unstable.

The equations that describe geophysical fluid systems are in general far too complicated to solve analytically. One way to get around this problem is to look for equilibria, i.e., solutions that are valid when all time derivatives are set to zero. Flows are often found to be close to such equilibria. For example, the surface of

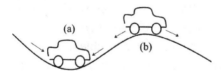

Figure 1.3 No brakes! Where would you park? Arrows show the gravitational force that acts when the car is displaced slightly from equilibrium.

---

[1] Highlighted text is used as an extra level of emphasis for important concepts.

a lake is in equilibrium if it is horizontal. Although this is never exactly true, it is pretty close on average.

Once we have identified an equilibrium state, the next step is to determine its stability. If the equilibrium is stable, disturbances will often have the form of oscillations (e.g., the car in Figure 1.3a), or waves. If the equilibrium is unstable, then small disturbances grow exponentially. Instabilities will be our focus here, though we will find it useful to examine wave phenomena as well.

1.1.1 The Cycle of Instability

Because unstable systems are by their nature ephemeral, you might reasonably wonder why we ever observe them. It is much more usual to see systems close to stable equilibria. For example, the surface of a lake is never perfectly horizontal, but it's usually pretty close, because the horizontal equilibrium state is stable.

But a sufficiently strong wind destabilizes that horizontal equilibrium state, and waves grow as a result.[2] If the waves grow large enough, they fall prey to a second kind of instability as the crests roll over and break (related to convective instability; Chapter 2). The surface then relaxes toward the horizontal state until a new set of waves emerges. Eventually the wind dies down and the horizontal state is once again stable.

The oceans and atmosphere are almost always turbulent, and this cycle of instability is the reason. Forcing by wind, sun, gravity, and planetary rotation tends to push the system toward unstable states. Instability and turbulence then act to relax the system back toward stability. This cyclic instability regime is discussed further in section 12.3.

1.2 Goals

Our exploration of instability will have three main goals.

(i) **Mechanisms:** We aim to understand, on an intuitive level, the basic physical processes that generate instability. In the car example, we've seen how motion away from equilibrium alters the effect of gravity (arrows in Figure 1.3), resulting in oscillations or instability. Geophysical examples will take a bit more work to understand, but we'll do it.

(ii) **Rules of thumb:** We would like to be able to predict the stability or instability of a system quickly with minimal math. In the car example, we are able to predict whether an equilibrium point will be stable or unstable without knowing the details of the shape of the hill or valley. All we need to know

---

[2] The process is similar to shear instability, covered in Chapters 3–5.

is whether the equilibrium is a maximum or a minimum of elevation, i.e., whether the curvature at that point is negative or positive.

We can invent similar rules for most types of geophysical flow instability. These allow us to estimate not only the likelihood of instability, but also the spatial and temporal scales on which it will grow. These can help us identify the particular mechanism through which a geophysical flow becomes unstable. For example, the Gulf Stream eddies shown in Figure 1.1 could be due to different instabilities (which you will learn about later). By comparing their observed length scales, and the time they take to grow, with rules of thumb based on various known instability types, we can take a first guess as to the mechanism.

(iii) **Numerical solution methods:** Sometimes a rule of thumb is not enough. We want to determine quantitative details of an instability, perhaps in a situation where many physical factors interact. In that case we may have to solve a nontrivial set of differential equations. Many advanced analytical methods are available, but in this book we will focus on numerical methods. Since the 1980s, computers have had the capacity to do something unprecedented: *to solve a differential equation whose coefficients are defined using observational measurements.* That capability is now in use in the analysis of oceanographic and atmospheric observation.

1.3 Tools

Below are three topics we'll expect you to have some familiarity with. Under each topic is listed one or more things that you should be able to do.

(i) **Calculus**:

- Solve this boundary value problem:

$$y'' = -y; \quad y(0) = y(\pi) = 0. \tag{1.1}$$

- Derive this Taylor series approximation:

$$\frac{1}{1+x} \approx 1 - x + x^2, \quad \text{for } |x| \ll 1.$$

- Understand the meaning of (though not necessarily solve) a partial differential equation, e.g.,

$$\frac{\partial u}{\partial t} + u\frac{\partial u}{\partial x} = -\frac{\partial \pi}{\partial x}.$$

- Define the Dirac delta function.

(ii) **Linear algebra**:

- Compute the eigenvalues of a 2×2 matrix.

(iii) **Programming**: Homework will be done using the Matlab programming environment[3] or something equivalent.[4] You do not have to be an expert; you will learn as you go. But if you've never used the software at all it would be worth familiarizing yourself with the basic syntax. Try the following:

- Write a function, and a script that calls it.
- Define a matrix and compute its eigenvalues.
- Make a line plot and label the axes.
- Make an image plot.

1.4 Numerical Solution of a Boundary Value Problem

The basic geophysical flow instabilities are analyzed as solutions of two-point boundary value problems. In this section we'll define this class of problems and introduce a simple method for solving them.

1.4.1 Defining the Problem

Let $f(x)$ be the solution to a second-order ordinary differential equation with independent variable x. Complete specification of f requires two pieces of information in addition to the equation itself. These can be either

- values of f and its first-derivative at some initial point, which we'll label as zero, i.e., $f(0)$ and $f'(0)$, or
- values of f at two points, say $f(0)$ and $f(L)$.

The first case is called an initial value problem; the second is called a boundary value problem.

A critical difference between these two classes of problem is that the first generally has a solution while the second generally does not. Here's a simple example:

---

[3] Many universities make the Matlab software available free to students.

[4] Python is another programming environment that we recommend. It is freely available at www.python.org. The commands you need here are found in two packages that will be used over and over. Most of the numerical mathematics and matrix operations come from the `numpy` package, whereas the plotting commands are from the `matplotlib` package. You should start your Python scripts with the following lines:

```
import numpy as np
import matplotlib as plt
```

to load these packages and give them the short cuts `np` and `plt`, respectively. Plotting a line can then be done with the command `plt.plot`, and finding eigenvalues can be accomplished with `np.eig`.

$$f'' = -k^2 f. \tag{1.2}$$

The general solution is

$$f = A \cos kx + B \sin kx, \tag{1.3}$$

where A and B are constants to be determined. Consider first the initial value problem. Suppose we have initial conditions $f(0) = 0$ and $f'(0) = 1$. The solution is then (1.3) with $A = 0$ and $B = 1/k$ (try it). Note that this solution works for *any* value of k.

Now, consider the boundary value problem with conditions

$$f(0) = 0; \quad f(L) = 0. \tag{1.4}$$

The first condition is satisfied if $A = 0$, but the second can then be satisfied only if $k = \pm n\pi/L$, where n is any integer. These special values of k are called the eigenvalues of the problem, and unless k is equal to one of those eigenvalues, the problem has no solution. We also call (1.2–1.4) a differential eigenvalue problem. It is analogous to the more familiar matrix eigenvalue problem, and can in fact be solved numerically using the same methods.

Here's how it works. Suppose that

- $\vec{x}$ is a list of possible values of x arranged as a vector;
- $\vec{f}$ and $\vec{f}^{(2)}$ are vectors composed of the corresponding values of f and its second-derivative, respectively;
- D is a matrix such that $\mathsf{D}\vec{f} = \vec{f}^{(2)}$.

We can now write (1.2) as

$$\mathsf{D}\vec{f} = -k^2 \vec{f}, \tag{1.5}$$

which is a standard matrix eigenvalue problem with eigenvalue $-k^2$. Because the matrix eigenvalue problem can be easily solved using standard numerical routines (e.g., the Matlab function `eig`[5]), this approach suggests a convenient way to solve the differential eigenvalue problem. But first we have to identify this matrix D that transforms a vector into its second-derivative.

1.4.2 Discretization and the Derivative Matrix

We discretize the independent variable x by choosing a vector of values:

$$x_i = x_0 + i\Delta, \quad \text{where } i = 0, 1, 2, \ldots, N+1.$$

---

[5] `Blue text` indicates Matlab syntax. We give coding examples in Matlab, assuming that readers preferring other software environments will substitute the equivalent expressions.

Preliminaries

Figure 1.4 Discretizing the *x* axis.

The first and last values correspond to the boundaries, say $x_0 = 0$ and $x_{N+1} = L$ (Figure 1.4). This requires that

$$\Delta = L/(N+1).$$

Note that the x_i are evenly spaced. This restriction is not necessary, but it simplifies the math. We can now discretize the solution f,

$$f_i = f(x_i),$$

and the kth derivative

$$f_i^{(k)} = \left. \frac{d^k f}{dx^k} \right|_{x=x_i}.$$

The finite difference approximation to the derivative $f^{(k)}$ is a weighted sum of f_i values at neighboring points. A well-known example is:

$$f_i' = \frac{f_{i+1} - f_i}{\Delta}, \tag{1.6}$$

which approximates the first-derivative to arbitrary accuracy as $\Delta \to 0$. In general

$$f_i^{(k)} = \sum_{j=j_1}^{j_2} A_j^{(k)} f_{i+j}.$$

The range of the summation, $j = j_1, \ldots, j_2$, is called the stencil. For example, in (1.6), $k = 1$, $j_1 = 0$, and $j_2 = 1$, and the weights are $A_0^{(1)} = -1/\Delta$ and $A_1^{(1)} = 1/\Delta$.

The weights are computed by means of a Taylor series expansion of f about x_i:

$$f_{i+j} = f(x_i + j\Delta) = f_i + j\Delta f_i^{(1)} + \frac{1}{2}(j\Delta)^2 f_i^{(2)} + \ldots + \frac{1}{k!}(j\Delta)^k f_i^{(k)}. \tag{1.7}$$

For example, suppose we want to approximate the first-derivative using the three-point stencil $j = -1, 0, 1$:

$$\tilde{f}_i' = A f_{i-1} + B f_i + C f_{i+1},$$

where the tilde identifies the approximation.[6] Substituting from (1.7) gives

$$Af_{i-1} + Bf_i + Cf_{i+1} = A\left[f_i - \Delta f_i' + \frac{1}{2}\Delta^2 f_i'' - \frac{1}{6}\Delta^3 f_i''' + \ldots\right] + B[f_i]$$

$$+ C\left[f_i + \Delta f_i' + \frac{1}{2}\Delta^2 f_i'' + \frac{1}{6}\Delta^3 f_i''' + \ldots\right]$$

$$= (A + B + C)f_i + (-A + C)\Delta f_i' + (A + C)\frac{1}{2}\Delta^2 f_i''$$

$$+ (-A + C)\frac{1}{6}\Delta^3 f_i''' + \ldots = f_i'. \qquad (1.8)$$

The final equality expresses our wish that the approximation $\tilde{f}_i'$ equal the true value f_i' (a wish that will not be granted). We try to find values for A, B, and C so that (1.8) is satisfied *for all functions* f, which requires that the final equality hold separately for the terms multiplying each derivative $f^{(k)}$. The terms multiplying f_i, f_i', and f_i'' (colored blue) give:

$$A + B + C = 0$$
$$(-A + C)\Delta = 1$$
$$(A + C) = 0.$$

We now have three equations for three unknowns. Since these are all the equations we can satisfy, we can equate only the blue terms in (1.8); the red term is neglected. The solution is:

$$A = -\frac{1}{2\Delta}, \quad B = 0, \quad C = \frac{1}{2\Delta}, \qquad (1.9)$$

or

$$\boxed{\tilde{f}_i' = \frac{f_{i+1} - f_{i-1}}{2\Delta}} \qquad (1.10)$$

This is called a centered difference owing to its symmetry.

How accurate is this approximation? Recall that, to solve (1.8), we had to ignore the red term. That term is a measure of the error in our approximation. Substituting (1.9) for $A - C$ in (1.8), we have

$$\tilde{f}_i' = \frac{f_{i+1} - f_{i-1}}{2\Delta} = f_i' + \frac{1}{6}\Delta^2 f_i''' + \ldots$$

We can't tell the value of the error term in general because it depends on the function f. What we can do is recognize that, as we shrink the grid spacing Δ to zero for a given f, the error decreases in proportion to Δ^2. We therefore say that the approximation is accurate to second order in Δ.

---

6 The constants A, B, and C are just simpler expressions for A_{-1}, A_0, and A_1.

For comparison, you could derive (1.6) in the same way (try it!), and you'd find that the error is proportional to Δ, i.e., (1.6) is accurate only to first order. We conclude that (1.10) is more accurate than (1.6) in the sense that the error decreases more rapidly as $\Delta \to 0$.

We can represent (1.10) using a matrix:

$$f_i' = D_{ij}^{(1)} f_j.$$

For example, if $N = 4$, then

$$
D = \frac{1}{2\Delta}
\begin{bmatrix}
-1 & 0 & 1 & 0 & 0 & 0 \\
0 & -1 & 0 & 1 & 0 & 0 \\
0 & 0 & -1 & 0 & 1 & 0 \\
0 & 0 & 0 & -1 & 0 & 1
\end{bmatrix}
$$

Note, however, that the matrix is not square. It requires values of f at $x_0, x_1, \ldots, x_{N+1}$, but returns the derivative only at $x_1, \ldots, x_N$. This is unsatisfactory. In particular, for an eigenvalue problem such as (1.5), only a square matrix will do.

One solution is to replace the first and last equations (the top and bottom rows) with expressions that don't depend on f_0 and f_{N+1}. This requires the use of one-sided derivatives, which are derived in the same way as (1.10). The simplest choice is to use (1.6) for the top row and its counterpart for the bottom row:

$$f_1' = \frac{f_2 - f_1}{\Delta}; \quad f_N' = \frac{f_N - f_{N-1}}{\Delta} \tag{1.11}$$

We can represent the result using the matrix

$$
D = \frac{1}{2\Delta}
\begin{bmatrix}
-2 & 2 & 0 & 0 \\
-1 & 0 & 1 & 0 \\
0 & -1 & 0 & 1 \\
0 & 0 & -2 & 2
\end{bmatrix},
$$

which we call a derivative matrix. Recall that (1.6) is only accurate to first order. This is not ideal, since (1.10) is second order. In homework problem 1 in Appendix A, you will derive an alternative to (1.6) that is second order accurate. Another strategy is to use the top and bottom rows to incorporate boundary conditions, as we discuss in the next section.

1.4.3 Incorporating Boundary Conditions

If the derivative matrix is intended for use in solving a boundary value problem, then instead of one-sided derivatives we incorporate the boundary conditions into the top and bottom rows. For example, suppose we have the Dirichlet boundary conditions $f(0) = f_0 = 0$ and $f(L) = f_{N+1} = 0$.[7] Then (1.10) gives, for $i = 1$ and N,

$$f_1' = \frac{f_2}{2\Delta}; \qquad f_N' = \frac{-f_{N-1}}{2\Delta}, \tag{1.12}$$

and the derivative matrix for $N = 4$ becomes

$$D = \frac{1}{2\Delta} \begin{bmatrix} 0 & 1 & 0 & 0 \\ -1 & 0 & 1 & 0 \\ 0 & -1 & 0 & 1 \\ 0 & 0 & -1 & 0 \end{bmatrix}$$

For higher N, of course, the pattern is repeated through the interior of the matrix.

Note that we don't actually calculate the derivative at the boundary points x_0 and x_{N+1}. For that reason they are referred to as "ghost points"; their influence is felt, not seen.

Test your understanding: In homework exercise 2, you will derive the matrix representing the second-derivative, using either one-sided derivatives or boundary conditions $f_0 = f_{N+1} = 0$:

$$D^{(2)} = \frac{1}{\Delta^2} \begin{bmatrix} -2 & 1 & 0 & 0 & \cdots \\ 1 & -2 & 1 & 0 & \cdots \\ 0 & 1 & -2 & 1 & \cdots \\ & & \ddots & & \\ \cdots & 0 & 1 & -2 & 1 \\ \cdots & & & 1 & -2 \end{bmatrix} \tag{1.13}$$

In exercise 3, you will verify that the eigenvalues and eigenvectors of $D^{(2)}$ correspond to the analytical solution of the differential eigenvalue problem (1.2) with the same boundary conditions. Try it with different values of N and see how it affects the accuracy of the results. See if the error decreases in proportion to Δ^2.

1.5 The Equations of Motion

Assume that space is measured by a Cartesian coordinate system $\vec{x} = \{x, y, z\}$, with z directed opposite to gravity.[8] Corresponding unit vectors are $\hat{e}^{(x)}$, $\hat{e}^{(y)}$, and $\hat{e}^{(z)}$. The velocity is $\vec{u} = D\vec{x}/Dt = \{u, v, w\}$. Here D/Dt represents the material derivative, i.e., the time derivative as measured by an observer moving with the flow:

---

[7] In case you don't remember, Dirichlet boundary conditions specify the value of the solution, while Neumann conditions specify the derivative. We will use both kinds in this book.

[8] We'll generalize to other coordinate systems later.

Table 1.1 *Typical terrestrial parameter values. Beware when using these "constants" for general purposes; some of them can vary significantly. Viscosity and diffusivity are assumed to be molecular in origin.*

name	symbol	unit	seawater	air
dynamic viscosity	μ	kg m^{-2}s^{-3}	10^{-3}	1.6×10^{-5}
gravitational acceleration	g	ms^{-2}	9.81	9.81
Coriolis parameter	f_0	s^{-1}	1.458×10^{-4}	1.458×10^{-4}
density	ρ_0	kg m^{-3}	1027	1.2
kinematic viscosity	ν	m^{2}s^{-1}	10^{-6}	1.4×10^{-5}
thermal density coefficient	α	K^{-1}	10^{-4}	3×10^{-3}
saline density coefficient	β	psu^{-1}	7×10^{-4}	
thermal diffusivity	κ_T	m^{2}s^{-1}	1.4×10^{-7}	1.9×10^{-5}
saline diffusivity	κ_S	m^{2}s^{-1}	10^{-9}	

$$\frac{D}{Dt} \equiv \frac{\partial}{\partial t} + \vec{u} \cdot \vec{\nabla}. \tag{1.14}$$

The principle of mass conservation (valid provided that no nuclear reactions are present) requires that the density (mass per unit volume) ρ vary as the velocity field converges or diverges:

$$\frac{D\rho}{Dt} = -\rho \vec{\nabla} \cdot \vec{u}. \tag{1.15}$$

This is also called the continuity equation.

Conservation of momentum (Newton's second law) is represented by the Navier-Stokes equation for velocities measured in a rotating reference frame:

$$\rho \left[\frac{D\vec{u}}{Dt} - \vec{u} \times f\hat{e}^{(z)} \right] = -\vec{\nabla}p - \rho g \hat{e}^{(z)} + \mu \nabla^2 \vec{u}. \tag{1.16}$$

The second term in brackets is the acceleration due to the Coriolis force. On a spherical planet, $f = f_0 \sin \phi$, where ϕ is the latitude and f_0 is twice the planetary rotation rate (see Table 1.1 for values). In this planetary context, we assume that phenomena of interest are small compared with the size of the planet, so that the Cartesian geometry is valid and ϕ can be treated as a constant.

The three terms on the right-hand side represent forces (per unit volume): the pressure gradient force, gravity (with acceleration g), and friction due to viscosity (with dynamic viscosity μ).

1.5.1 Approximate Forms

For most of the book we will use a simplified form of (1.16) based on two assumptions:

(i) The fluid is incompressible:

$$\boxed{\vec{\nabla} \cdot \vec{u} = 0.}$$
(1.17)

(ii) The density remains close to a uniform value ρ_0:

$$\rho = \rho_0 + \rho^*, \quad \text{where} \ |\rho^*| \ll \rho_0.$$
(1.18)

Expanding the pressure as

$$p = p_0 + p^*,$$

where p_0 is in hydrostatic balance with ρ_0 (i.e., $\vec{\nabla} p_0 = -\rho_0 g \hat{e}^{(z)}$; discussed further in section 2.1.1), gives

$$(\rho_0 + \rho^*) \left[\frac{D\vec{u}}{Dt} - \vec{u} \times f \hat{e}^{(z)} \right] = -\vec{\nabla}(\cancel{p_0} + p^*) - (\cancel{\rho_0} + \rho^*) g \hat{e}^{(z)} + \mu \nabla^2 \vec{u}$$

or

$$(\rho_0 + \rho^*) \left[\frac{D\vec{u}}{Dt} - \vec{u} \times f \hat{e}^{(z)} \right] = -\vec{\nabla} p^* - \rho^* g \hat{e}^{(z)} + \mu \nabla^2 \vec{u}.$$

Now, based on (1.18), we replace $(\rho_0 + \rho^*)$ on the left-hand side with ρ_0. Finally, we divide through by ρ_0, resulting in

$$\boxed{\frac{D\vec{u}}{Dt} = -\vec{\nabla} \pi + b \hat{e}^{(z)} + \nu \nabla^2 \vec{u} + \vec{u} \times f \hat{e}^{(z)}.}$$
(1.19)

The accelerations appearing on the right-hand side of (1.19) are

- the pressure gradient, with

$$\pi = \frac{p^*}{\rho_0},$$
(1.20)

- the buoyancy, defined as

$$b = -g \frac{\rho^*}{\rho_0},$$
(1.21)

- the viscosity, with kinematic viscosity

$$\nu = \frac{\mu}{\rho_0}.$$ (1.22)

- the Coriolis acceleration.

Equation (1.19) is commonly known as the Boussinesq approximation to the momentum equations.

The density of seawater is governed by two separate properties: temperature T and salinity S. If the water is close to a uniform state with $T = T_0$ and $S = S_0$, we can use the linearized equation of state:

$$b = \alpha g(T - T_0) - \beta g(S - S_0).$$ (1.23)

Here, α and β are coefficients for thermal and saline buoyancy, taken to be constants. In the absence of sources, T and S obey Fickian diffusion equations with constant diffusivities κ_T and κ_S:

$$\frac{DT}{Dt} = \kappa_T \nabla^2 T; \qquad \frac{DS}{Dt} = \kappa_S \nabla^2 S.$$ (1.24)

In Chapter 9, we will discuss double diffusive instabilities, which depend on the separate effects of temperature and salinity as described by (1.23) and (1.24). In the atmosphere, salinity is obviously not a factor, but buoyancy is affected by humidity in a related way. For most applications, however, the fact that buoyancy has two components is not important, and we will use a single equation

$$\boxed{\frac{Db}{Dt} = \kappa \nabla^2 b.}$$ (1.25)

In these cases, you can think of b as proportional to temperature.

1.5.2 Viscosity and Turbulence

To the human senses, and in most measurements, a fluid appears as a continuous medium. Although we recognize that a fluid is really made of discrete molecules, the science of fluid mechanics is not concerned with such microscopic details. When we talk about, say, the velocity $\vec{u}$ at a point $\vec{x}$, we really mean the average molecular velocity in some volume of space, centered on $\vec{x}$, that is tiny but nonetheless large enough to encompass many molecules. With that understanding, we can think of $\vec{u}$ as a continuous function of $\vec{x}$, and therefore employ the powerful tools of calculus to understand the motion.

Although we are not interested in molecular motions *per se*, we must account for the effect they have on the motions that we *are* interested in. That is where viscosity comes in – it models the frictional effect that molecular interactions exert on the macroscopic motions that we can perceive and measure. The assumption that molecular effects can be represented in this way is called the continuum hypothesis.

In the study of geophysical fluids, the continuum hypothesis is extended to larger scales. We are not only not interested in the motions of individual molecules, but we are also not interested in macroscopic motions smaller than a certain scale. In the study of weather, for example, we do not try to understand every little gust of wind. When we talk about the wind speed at a certain time and place, we usually mean an average that encompasses many gusts.

As with molecular motions, though, we must account for the effect the gusts have on the larger-scale motions that we're trying to understand. Exploiting the obvious analogy, the effect of the gusts is usually represented as an "effective" viscosity, often called turbulent viscosity or eddy viscosity. While this analogy is imperfect,[9] the eddy viscosity concept is a useful first step toward understanding the effect of small-scale turbulence.

The assumption that eddy viscosity is uniform in space and time is, well, better than nothing. In this book, the quantity ν that we call "viscosity" can refer to either molecular or eddy viscosity. Similarly, the diffusivity κ can refer to diffusion either by molecular motions or by small-scale turbulence.

1.6 Further Reading

The derivation of (1.16) is something every student should see at least once. A detailed account is given in Smyth (2017), which has the added virtue of being free. Thorpe (2005) provides an excellent introduction to instability and turbulence in the oceanic context. A theoretical discussion of turbulence in general, with particular attention to eddy viscosity, may be found in chapters 4, 5, and 10 of Pope (2000).

1.7 Appendix: A Closer Look at Perturbation Theory

1.7.1 The Parking Problem Revisited

We now consider the car-parking example from a more rigorous perspective. This example will illustrate some of the fundamental ideas of perturbation theory. Let ℓ be the distance the car rolls along the hill, measured left-to-right from some

---

[9] Eddy viscosity is not a property of the fluid but of the flow, and it can vary greatly in space and time in ways we do not understand. Most properly, eddy viscosity is not even a scalar but is actually a second-rank tensor.

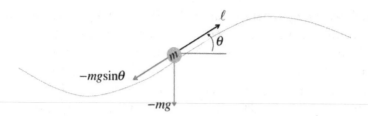

Figure 1.5 Force diagram for the car-parking example.

arbitrary origin (Figure 1.5). The force of gravity is $-mg$, where m is the mass of the car and $g = 9.81 \text{ m s}^{-2}$ at Earth's surface. The component of gravity in the direction of ℓ is $-mg \sin\theta$, where $\theta(\ell)$ is the angle of the road from the horizontal at any point ℓ. Newton's second law is therefore

$$m\frac{d^2\ell}{dt^2} = -mg \sin\theta. \tag{1.26}$$

Abbreviating $\sin\theta$ as $s(\ell)$ and removing the common factor m, we have

$$\frac{d^2\ell}{dt^2} + gs(\ell) = 0. \tag{1.27}$$

A general solution requires specifying the function $s(\ell)$, which, of course, is different for every road. An exact solution would be very difficult since $s(\ell)$ is in general nonlinear. Progress can be made if we identify equilibrium points: values of ℓ at which $s(\ell) = 0$, and concern ourselves only with the behavior close to those points. Equilibria exist when $\theta = 0$, i.e., at the bottom of a valley or the top of a hill (as can be seen by setting all time derivatives to zero in 1.27).

Let $\ell = \ell_0$ be an equilibrium point, and seek a solution

$$\ell(t) = \ell_0 + \epsilon\ell_1(t) + \epsilon^2\ell_2(t) + \dots, \tag{1.28}$$

where ϵ is a parameter measuring the amplitude of the disturbance. Because we'll be assuming that ϵ is small, the early terms in the series are the most important. We now expand the unknown function $s(\ell)$ in a Taylor series about ℓ_0 and substitute (1.28) into the result:

$$s(\ell) = s(\ell_0) + s'(\ell_0)(\ell - \ell_0) + \dots$$
$$= s(\ell_0) + \epsilon s'(\ell_0)\ell_1 + \epsilon^2 G + \dots$$

Here, G stands for some complicated terms whose details don't matter.[10] Now substitute into (1.27):

---

[10] Try it if you like. You should get $G = s'(\ell_0)\ell_2 + \frac{1}{2}s''(\ell_0)\ell_1^2$.

$$\frac{d^2\ell_0}{dt^2} + \epsilon\frac{d^2\ell_1}{dt^2} + \epsilon^2\frac{d^2\ell_2}{dt^2} + \ldots + g\left[s(\ell_0) + \epsilon s'(\ell_0)\ell_1 + \epsilon^2 G + \ldots\right] = 0,$$

or, gathering powers of ϵ,

$$\frac{d^2\ell_0}{dt^2} + gs(\ell_0) + \epsilon\left[\frac{d^2\ell_1}{dt^2} + gs'(\ell_0)\ell_1\right] + \epsilon^2\left[\frac{d^2\ell_2}{dt^2} + gG\right] + \ldots = 0. \quad (1.29)$$

Now here is a subtle but important point. Regardless of values of the quantities in square brackets, we can always find one or more values of ϵ that satisfy (1.29). But that's not what we're looking for. What we want is to take the limit $\epsilon \to 0$, and have the equation be satisfied throughout that limiting process. In other words, the equation has to be true for *every* value of ϵ. That can only be true if the coefficients of the powers of ϵ all vanish individually.[11] That leaves us with an infinite sequence of equations whose solutions are the unknown functions ℓ_0, $\ell_1(t)$, $\ell_2(t)$, etc.:

$$\frac{d^2\ell_0}{dt^2} + gs(\ell_0) = 0, \quad (1.30)$$

$$\frac{d^2\ell_1}{dt^2} + gs'(\ell_0)\ell_1 = 0, \quad (1.31)$$

$$\frac{d^2\ell_2}{dt^2} + gG = 0, \quad (1.32)$$

$$\ldots \text{etc.}$$

Now, let's look again at our putative solution (1.28). We already know the first term, ℓ_0; it's just the equilibrium position (hill or valley). As a result, the first equation (1.30) is satisfied trivially; $s(\ell_0) = 0$ and $d^2\ell_0/dt^2 = 0$.

The next term in (1.28), $\epsilon\ell_1(t)$, is the only one that matters if ϵ is made sufficiently small. We therefore concern ourselves only with the second equation, (1.31), whose solution is ℓ_1. This is a linear ordinary differential equation, and very easy to solve because $s'(\ell_0)$ is a constant. For tidiness we'll abbreviate $s'(\ell_0)$ as s'_0. We'll consider two cases.

If $s'_0 > 0$, we define $gs'_0 = \omega^2$, and the general solution is

$$\ell_1 = A\sin\omega t + B\cos\omega t,$$

where A and B are constants to be determined by the initial conditions. This oscillatory solution describes the car rocking back and forth after being displaced from a stable equilibrium point (a valley).

---

[11] As an analogy, consider a quadratic equation $ax^2 + bx + c = 0$. For given values of a, b, and c, you can easily find solutions for x using the quadratic formula. But what if the equation has to be satisfied for *every* x? That's only possible if $a = b = c = 0$.

If $s_0' < 0$, we define $-gs_0' = \sigma^2$, and the solution is

$$\ell_1 = Ae^{\sigma t} + Be^{-\sigma t}.$$

As long as $A \neq 0$, the first term grows exponentially and will eventually dominate the solution. This describes the unbounded motion of the car away from an unstable equilibrium, i.e., the top of a hill.

Here are some general features of stability analysis that the car-parking problem illustrates:

(i) The equation (1.31) is a linear, homogeneous ordinary differential equation. This is always true. In general, though, the coefficients will not be constant, and the solution will be much more difficult. For that reason we will often resort to numerical methods.

(ii) Solutions can be oscillatory or exponential. Our main interest is in exponential solutions with positive growth rate σ. Often, we can define a simple rule of thumb to tell us which type of solution will be found; here, we only need to know the sign of s_0'.

(iii) Having solved for ℓ_1, it is possible to substitute the result into (1.32) and solve for ℓ_2, and so on to even higher orders. Some brave souls do this, but we won't.

(iv) The solution is valid only if the neglected terms in (1.28) are indeed negligible. How can we tell if this is true? Most commonly, we regard the smallness of the neglected terms as a *hypothesis* whose validity we test by comparing the solution with reality.[12]

1.7.2 A Mechanical Spring-Mass System

Consider a fixed wire frame in the shape of a circle with a bead of mass m that is free to slide, without friction, along it. This is a direct analog of the car-parking problem we just saw. If we were to do a linear stability analysis of this system, we would find that the two equilibrium points correspond to the bottom (stable) and top (unstable) points of the frame. However, we will now modify the problem by adding a spring that connects point Q, at the top of the wire frame, with the mass at point P (as is depicted in Figure 1.6).[13] We wish to determine the stability of the bead in this new system.

As a first step we need to find an equation to describe its motion. To do this, we let the resting length of the spring be l, i.e., equal to the radius of the frame, and measure the position of the bead in terms of the angle, $\theta(t)$. The tension force

---

[12] We don't recommend this for the car-parking problem.
[13] This example comes from the excellent book by Acheson (1997).

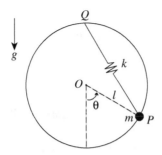

Figure 1.6 Pendulum system with a spring connecting the bead of mass m, at point P, to the top position of the frame at point Q.

exerted by the spring will be proportional to the displacement from the resting length, and given by

$$T = k \times \text{displacement} = k[2l\cos(\beta) - l], \tag{1.33}$$

where k is the spring coefficient with units of N m^{-1}, and β is the angle made by the spring to the vertical. Using some geometry we can express this in terms of θ as $\beta = \theta/2$. The component of this force that is along the wire is then $T\sin(\theta/2)$, and we can write our spring-pendulum equation as

$$ml\frac{d^2\theta}{dt^2} + mg\sin(\theta) - kl[2\cos(\theta/2) - 1]\sin(\theta/2) = 0, \tag{1.34}$$

given that the radial component of the acceleration is $ld^2\theta/dt^2$.

At this point it is helpful to stop and compare this equation with our original nonlinear equation from the car-parking problem (1.26).[14] Since the mass does not drop out of this equation, we have four parameters: m, l, g, and k. In addition we have the extra term that arises from the presence of the spring. We can simplify the problem by measuring time in the units of $(l/g)^{1/2}$, so that the dimensionless time variable becomes $t_\star \equiv t/(l/g)^{1/2}$ and

$$\frac{d^2\theta}{dt_\star^2} = \frac{d^2\theta}{dt^2}\frac{g}{l},$$

We can now write (1.34) as

$$\frac{d^2\theta}{dt_\star^2} + \sin(\theta) - S\sin(\theta/2)[2\cos(\theta/2) - 1] = 0, \tag{1.35}$$

where $S = kl/mg$. By representing our equation in dimensionless form we have achieved a great simplification – we have reduced the number of parameters from four to one.

---

[14] Note that since the path of motion of the mass is circular we can write the left-hand side of (1.26) as $d^2\ell/dt^2 = ld^2\theta/dt^2$.

The parameter S describes the strength of the spring relative to the weight of the bob. Consider two extreme cases. If S is very small, the spring is weak and has little effect, so the bob oscillates about $\theta = 0$ just as in a normal pendulum. If S is large, the spring is very tight, and we expect the bob to be pulled strongly away from the bottom of the hoop. In the latter extreme, is the equilibrium state $\theta = 0$ stable or unstable? Imagine that you started with a loose spring and the bob at $\theta = 0$, then gradually tightened the spring. What would happen? Can you picture, at a critical value of S, the bob suddenly snapping away from $\theta = 0$? Next we will examine that process mathematically.

First, it is necessary to search for equilibrium points of the system by setting the time derivatives to zero in (1.35) and solving for the position of the bead. This leads to

$$\sin(\theta) - S\sin(\theta/2)[2\cos(\theta/2) - 1] = 0,$$

which can be simplified using the identity $\sin(2x) = 2\cos(x)\sin(x)$ with $x = \theta/2$:

$$\sin(\theta/2)[2(S - 1)\cos(\theta/2) - S] = 0.$$

The system is therefore in equilibrium if one of two criteria is satisfied:

$$\sin(\theta/2) = 0, \tag{1.36}$$

or

$$\cos(\theta/2) = S/[2(S - 1)]. \tag{1.37}$$

The former criterion is satisfied if $\theta = 0$ or π. To perform a linear stability analysis about an equilibrium point θ_0 we substitute $\theta(t) = \theta_0 + \epsilon\theta'(t)$, and expand the nonlinear functions in their Taylor series. For the case $\theta_0 = 0$,

$$\sin(\epsilon\theta') = \epsilon\theta' + O(\epsilon^3)$$
$$\cos(\epsilon\theta') = 1 + O(\epsilon^2).$$

Collecting the terms of order ϵ, we have

$$\frac{d^2\theta'}{dt^2} + \left(1 - \frac{S}{2}\right)\theta' = 0.$$

This is analogous to equation (1.31) of the parking problem. As in that case, the result is a linear, homogeneous ordinary differential equation whose solutions are either oscillatory or exponential. Substituting the test solution $\theta(t) \propto e^{\sigma t}$ gives

$$\sigma = \pm\left(\frac{S}{2} - 1\right)^{1/2}.$$

When $S < 2$, σ is imaginary, and the mass oscillates about $\theta = 0$. When $S > 2$, however, σ is real (with one positive and one negative value), and the equilibrium point $\theta = 0$ is therefore *unstable*. The value $S = 2$ represents a critical state. This can be understood physically by interpreting S as the ratio of the destabilizing tension kl, which pulls the mass away from the equilibrium position, to the gravitational force mg that acts to restore the mass to equilibrium.

Something else very interesting happens at the critical value $S = 2$: the system acquires two new equilibrium points. Since $-\pi \leq \theta \leq \pi$, we know that $0 \leq \cos(\theta/2) \leq 1$, and therefore the second criterion for equilibrium, (1.37), can be satisfied only when $S \geq 2$.

Let's recall what the equilibrium points represent. When the time derivatives are set to zero in our equation of motion of the bead, we are essentially saying that the sum of the forces is equal to zero. In other words, the equilibrium points are positions of the mass for which all forces balance. In this example of the spring-pendulum system, we have just shown that the tension from the spring can balance the gravitational force for $\theta \neq 0, \pi$ when the spring force is strong enough, i.e., for $S > 2$. The locations (θ_e) at which this balance occurs are shown in Figure 1.7 for each value of S. We leave it as an exercise for the student to show that these equilibrium points are stable (Figure 1.7).

We have shown that this spring-mass system has either one or two equilibrium states depending on the value of the dimensionless parameter, S. When S exceeds the critical value 2, the bottom position loses stability and the two new stable equilibria appear. The determination of simple "rules of thumb" like this is one of the primary goals of linear stability analysis.

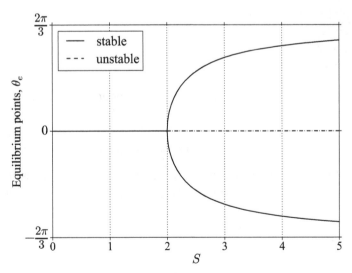

Figure 1.7 Equilibrium points of the spring-pendulum system and their stability. The asymptotic value $2\pi/3$ is reached for large S.

2

Convective Instability

The mechanism of convective instability is intuitively straightforward: dense fluid sinks under the action of gravity, while light fluid rises to take its place. Familiar examples include water heated on a stove, cumulus clouds (Figure 2.1a), and the slow currents in the Earth's mantle that drive continental drift (Figure 2.1b). The mechanism is complicated by the action of viscosity and diffusion, which tend to oppose the instability.

In the simplest convection problems, no differential equations must be solved; the problem is entirely algebraic. In this simple context, we will introduce several important concepts that will be used repeatedly when we address more complicated problems. Watch for the following:

- normal modes,
- the fastest-growing mode,
- isomorphism,
- scaling, and
- critical states.

We will also introduce the use of the Dirac delta function in modeling instabilities.

2.1 The Perturbation Equations

We begin with the equations of motion developed in section 1.5, assuming that the Coriolis effect is negligible. We collect the equations here for convenience:

$$\vec{\nabla} \cdot \vec{u} = 0 \tag{2.1}$$

$$\frac{D\vec{u}}{Dt} = -\vec{\nabla}\pi + b\hat{e}^{(z)} + \nu\nabla^2\vec{u} \tag{2.2}$$

$$\frac{Db}{Dt} = \kappa\nabla^2 b. \tag{2.3}$$

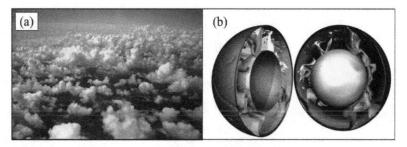

Figure 2.1 Examples of convection in geophysical flows. (a) Atmospheric convection cells revealed by cloud formation in the updrafts (NOAA). (b) Numerical simulation of convection in the Earth's mantle. Gray-red: cold surface plates sinking. Yellow-white: hot mantle plumes rising (courtesy F. Crameri, University of Oslo).

2.1.1 Hydrostatic Equilibrium

Under what conditions may a fluid be, and remain, motionless? In other words, if we assume that (1) $\vec{u} = 0$ and (2) $\partial/\partial t = 0$, can we extract a solution from the equations of motion?

The incompressibility condition (2.1) is satisfied automatically. The momentum equation (2.2) is a bit more complicated. First, when $\partial/\partial t = 0$ and $\vec{u} = 0$, the material derivative of any field vanishes:

$$\frac{D}{Dt} = \frac{\partial}{\partial t} + \vec{u} \cdot \vec{\nabla} = 0.$$

So, the left-hand side of (2.2) is zero. Also, with $\vec{u} = 0$, the viscous term is zero, leaving

$$0 = -\vec{\nabla}\pi + b\hat{e}^{(z)},$$

expressing hydrostatic balance between the pressure and buoyancy fields. The horizontal components give

$$\frac{\partial \pi}{\partial x} = \frac{\partial \pi}{\partial y} = 0.$$

Therefore, pressure is a function of z only. In this book we denote equilibrium fields with uppercase letters, so the hydrostatic pressure is

$$\boxed{\pi = \Pi(z).}\qquad(2.4)$$

From the vertical component we have

$$\frac{\partial \pi}{\partial z} = b,$$

or

$$b = B(z) = \frac{d\Pi}{dz} \qquad (2.5)$$

Finally we have the diffusion equation for buoyancy, (2.3). As with (2.2), the material derivative on the left-hand side is zero. Using (2.5), we obtain

$$\kappa \frac{d^2 B}{dz^2} = 0. \qquad (2.6)$$

We sometimes assume that κ is vanishingly small, in which case (2.6) is satisfied automatically. Otherwise, B must be a linear function of z:

$$B = B_0 + \frac{dB}{dz} z, \quad \text{where} \quad \frac{dB}{dz} \equiv B_z = \text{constant.} \qquad (2.7)$$

The derivative dB/dz is the square of the buoyancy, or Brunt-Väisälä, frequency, often denoted as N^2. What (2.7) tells us is that, for equilibrium, dB/dz must be constant; otherwise, the buoyancy profile will evolve due to diffusion. We abbreviate dB/dz as B_z.

Together with the assumption $\vec{u} = 0$, (2.4), (2.5), and (2.7) define the motionless (or "static") equilibrium state.

2.1.2 Small Departures from Equilibrium: the Linearized Equations

Here we will derive a pair of equations that govern the evolution of small disturbances to a motionless fluid. We begin by assuming that buoyancy, pressure, and velocity are close to hydrostatic equilibrium:

$$b = B(z) + \epsilon b'(\vec{x}, t), \qquad (2.8)$$
$$\pi = \Pi(z) + \epsilon \pi'(\vec{x}, t), \qquad (2.9)$$
$$\vec{u} = 0 + \epsilon \vec{u}'(\vec{x}, t). \qquad (2.10)$$

We refer to B and Π as the "background state" and to the primed quantities as the "perturbations." The evolution of the perturbations is our main focus. We have introduced the ordering parameter ϵ. In perturbation theory, many quantities are assumed to be "small," and ϵ is basically a bookkeeping trick that will help us keep track of them. Any quantity proportional to ϵ is "small"; anything proportional to ϵ^2 is "really small," etc. (See section 1.7 for a more rigorous discussion.)

We now substitute (2.8–2.10) into the equations of motion, (2.1), (2.2), and (2.3). We start with the easiest one: $\vec{\nabla} \cdot \epsilon \vec{u}' = 0$, and therefore

$$\boxed{\vec{\nabla} \cdot \vec{u}' = 0.} \qquad (2.11)$$

The second-simplest equation is (2.3). We work on the left-hand and right-hand sides separately. The left-hand side is:

$$
\begin{aligned}
\frac{Db}{Dt} &= \left(\frac{\partial}{\partial t} + \epsilon \vec{u}' \cdot \vec{\nabla} \right) \left(B(z) + \epsilon b' \right) \\
&= \frac{\partial B}{\partial t} + \epsilon \vec{u}' \cdot \vec{\nabla} B(z) + \epsilon \frac{\partial b'}{\partial t} + \epsilon^2 \vec{u}' \cdot \vec{\nabla} b' \\
&= 0 + \epsilon w' \frac{dB}{dz} + \epsilon \frac{\partial b'}{\partial t} + 0 \\
&= \epsilon \left(\frac{\partial b'}{\partial t} + w' \frac{dB}{dz} \right).
\end{aligned}
$$

The first term, $\partial B / \partial t$, is zero by construction. The term proportional to ϵ^2 (red) is neglected because it is the product of two small quantities and is therefore "really small."

On the right-hand side,

$$
\kappa \nabla^2 b = \kappa \frac{d^2 B}{dz^2} + \epsilon \kappa \nabla^2 b'.
$$

From (2.6) the first term on the right is zero. As a result, (2.3) becomes:

$$
\boxed{\frac{\partial b'}{\partial t} + w' B_z = \kappa \nabla^2 b'.} \tag{2.12}
$$

Recall from (2.6) that either (1) $\kappa \neq 0$ and B_z is a constant, or (2) $\kappa = 0$ and B_z is an arbitrary function of z. The second term on the left describes perturbations to the local buoyancy as a result of vertical motions advecting the "background" buoyancy (i.e., that found in the static equilibrium state).

Two important aspects of (2.12) should be noted.

- (2.12) is a *linear* equation, because the nonlinear term $\epsilon^2 \vec{u}' \cdot \vec{\nabla} b'$ has been discarded.
- The ordering parameter ϵ does not appear.

Finally we address the momentum equation (2.2). The left-hand side is

$$
\frac{D \vec{u}}{Dt} = \left(\frac{\partial}{\partial t} + \epsilon \vec{u}' \cdot \vec{\nabla} \right) \epsilon \vec{u}' = \epsilon \frac{\partial \vec{u}'}{\partial t},
$$

where once again the $O(\epsilon^2)$ term has been discarded. The remaining substitutions are straightforward:

$$\epsilon \frac{\partial \vec{u}'}{\partial t} = -\vec{\nabla}(\Pi + \epsilon \pi') + (B + \epsilon b')\hat{e}^{(z)} + \nu \nabla^2 \epsilon \vec{u}' \tag{2.13}$$

$$= -\frac{d\Pi}{dz}\hat{e}^{(z)} + B\hat{e}^{(z)} - \epsilon \vec{\nabla}\pi' + \epsilon b'\hat{e}^{(z)} + \epsilon \nu \nabla^2 \vec{u}'.$$

From (2.5), we see that the first two terms on the right-hand side add up to zero.[1] The remaining terms are all proportional to ϵ, which therefore cancels, leaving:

$$\boxed{\frac{\partial \vec{u}'}{\partial t} = -\vec{\nabla}\pi' + b'\hat{e}^{(z)} + \nu \nabla^2 \vec{u}'}. \tag{2.14}$$

Once again, the equation is linear and does not involve ϵ. We now have a closed system: (2.11), (2.12), and (2.14) comprise five equations for the five unknowns u', v', w', b', and π'.

Note that (2.12) actually involves only two of these unknowns: w' and b'. Life could be made much easier if we could find another equation involving only those unknowns, for then we would have only two equations to solve. A good place to start is the vertical component of (2.14):

$$\frac{\partial w'}{\partial t} = -\frac{\partial \pi'}{\partial z} + b' + \nu \nabla^2 w'. \tag{2.15}$$

This involves three unknowns, but we can eliminate the pressure as follows. We begin by taking the divergence of (2.14):

$$\vec{\nabla} \cdot \frac{\partial \vec{u}'}{\partial t} = -\vec{\nabla} \cdot \vec{\nabla}\pi' + \vec{\nabla} \cdot (b'\hat{e}^{(z)}) + \vec{\nabla} \cdot \nabla^2 \vec{u}',$$

or,

$$\frac{\partial}{\partial t}\underbrace{(\vec{\nabla} \cdot \vec{u}')}_{=0} = -\nabla^2 \pi' + \frac{\partial b'}{\partial z} + \nabla^2 \underbrace{(\vec{\nabla} \cdot \vec{u}')}_{=0}.$$

The first and last terms vanish by (2.11), and we are left with a Poisson equation for π' in terms of b':

$$\boxed{\nabla^2 \pi' = \frac{\partial b'}{\partial z}}. \tag{2.16}$$

Now take the Laplacian of (2.15):

$$\frac{\partial}{\partial t}\nabla^2 w' = -\frac{\partial}{\partial z}\nabla^2 \pi' + \nabla^2 b' + \nu \nabla^4 w',$$

---

[1] Another way to think about this is to recognize that, other than being $\ll 1$, ϵ can take *any* value. Therefore, the only way that (2.13) can be valid is if the terms proportional to each power of ϵ balance separately. See section 1.7.1 for a fuller discussion.

and substitute from (2.16):

$$\frac{\partial}{\partial t}\nabla^2 w' = -\frac{\partial^2 b'}{\partial z^2} + \nabla^2 b' + \nu\nabla^4 w'$$
$$= \nabla_H^2 b' + \nu\nabla^4 w',$$

where

$$\nabla_H^2 = \frac{\partial^2}{\partial x^2} + \frac{\partial^2}{\partial y^2}.$$

Success! We now have two equations in two unknowns:

$$\frac{\partial}{\partial t}\nabla^2 w' = \nabla_H^2 b' + \nu\nabla^4 w', \tag{2.17}$$

$$\frac{\partial}{\partial t} b' = -B_z w' + \kappa\nabla^2 b'. \tag{2.18}$$

In the following subsections we will look at solutions of these equations in three cases of increasing complexity. We'll begin by neglecting viscosity and diffusion. We'll then restore those effects, and finally we'll add upper and lower boundaries.

2.2 Simple Case: Inviscid, Nondiffusive, Unbounded Fluid

We first investigate the highly simplified case $\nu = \kappa = 0$, in which (2.17) and (2.18) are

$$\frac{\partial}{\partial t}\nabla^2 w' = \nabla_H^2 b'; \qquad \frac{\partial}{\partial t} b' = -B_z w'.$$

Applying $\partial/\partial t$ to the first of these and substituting from the second, we obtain

$$\frac{\partial^2}{\partial t^2}\nabla^2 w' = -B_z\nabla_H^2 w'. \tag{2.19}$$

For now, we will assume that B_z is a constant, but later in this subsection we will consider more general forms of B_z.

2.2.1 The Normal Mode Solution

In this simple context we will introduce a flow structure that will be used repeatedly in this course: the plane-wave, or normal mode perturbation. For the vertical velocity, this has the form

$$W(\vec{x}, t) = \hat{w}\exp\{\iota(kx + \ell y + mz - \omega t)\}, \tag{2.20}$$

Several aspects of (2.20) should be noted:

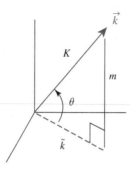

Figure 2.2 Definition sketch for the wavevector $\vec{k}$ and its horizontal and vertical projections $\tilde{k}$ and m.

- $\iota = \sqrt{-1}$.
- The amplitude $\hat{w}$ is a complex constant.
- While W is a complex function, only the real part is physically relevant:

$$w' = W_r.$$

- $\{k, \ell, m\} = \vec{k}$ is the wave vector (Figure 2.2). Its components k, ℓ, and m are the wavenumbers corresponding to the x, y, and z directions, and are assumed to be real. The corresponding periodicity intervals (or wavelengths) are $2\pi/k$, $2\pi/\ell$, and $2\pi/m$, respectively. In MKS units, $\vec{k}$ is measured in m^{-1}.
- We define the magnitude of the wave vector $K = \sqrt{k^2 + \ell^2 + m^2}$ and its horizontal part $\tilde{k} = \sqrt{k^2 + \ell^2}$.
- The angle of elevation, measured from the horizontal plane, is θ, so that $\tilde{k} = K\cos\theta$ and $m = K\sin\theta$. Since $\tilde{k} \geq 0$ by definition, $-\pi/2 < \theta \leq \pi/2$.
- ω is the complex frequency. In MKS units, ω is measured in s^{-1}.
- The real part of ω is interpreted as the angular frequency, measured in radians per second. Dividing by 2π gives the cyclic frequency f, measured in cycles per second.
- If ω has a nonzero imaginary part, it represents exponential growth or decay.

The solution form (2.20) is more general than it may seem, because the equations (2.19 in this case) are linear and therefore obey the *principle of superposition*: two solutions can be added to make a third solution. Using the techniques of Fourier analysis, any perturbation can be expressed as a sum of expressions like (2.20).

When we substitute normal mode solutions such as (2.20) into (2.19), or (2.17), considerable simplification results as derivatives are replaced by multiplications. For example, differentiating (2.20) with respect to x produces the factor ιk but otherwise changes nothing. We now list the most common differential operations and how they apply to (2.20):

$$\frac{\partial}{\partial t} \rightarrow -\iota\omega, \quad \text{or } \frac{\partial}{\partial t} \rightarrow \sigma \qquad (2.21)$$

$$\vec{\nabla} \rightarrow \iota\vec{k}, \quad \text{e.g., } \frac{\partial}{\partial x} \rightarrow \iota k \qquad (2.22)$$

$$\nabla^2 \rightarrow -K^2 \qquad (2.23)$$

$$\nabla_H^2 \rightarrow -\tilde{k}^2. \qquad (2.24)$$

After we substitute these expressions, (2.19) becomes

$$(-\omega^2)(-K^2)W = -B_z(-\tilde{k}^2)W = 0$$

or

$$(\omega^2 K^2 - B_z\tilde{k}^2)W = 0.$$

If there is a nonzero perturbation ($w' \neq 0$), the expression in parentheses must vanish:

$$\omega^2 = B_z\frac{\tilde{k}^2}{K^2} = B_z\cos^2\theta. \qquad (2.25)$$

You may recognize this as the dispersion relation for internal gravity waves. However, the gravity wave solution applies only when $B_z > 0$.

2.2.2 Instabilities and the Fastest-Growing Mode

What if $B_z < 0$, i.e., if dense fluid overlies light fluid? In that case $\omega^2 < 0$. Writing

$$\omega = \iota\sigma,$$

(2.25) gives

$$\sigma = \pm\sqrt{-B_z} \cos\theta. \qquad (2.26)$$

When σ is positive the mode is identified as unstable. Because $e^{-\iota\omega t} = e^{\sigma t}$, we can identify σ as an exponential growth rate. Like ω, σ is measured in s^{-1}.

The Fastest-Growing Mode

Naturally occurring flows are never as simple as the equilibrium state we defined in section 2.1.1; there is almost always an additional component consisting of small-scale, quasi-random fluctuations. According to Fourier's theorem, the fluctuations are equivalent to a superposition of normal modes like (2.20) with a range of ω and $\vec{k}$ (and hence θ) connected by a dispersion relation such as (2.25) or, equivalently, (2.26).

If a given $\vec{k}$ represents a maximum of σ, then that mode will grow the fastest. Because the growth is exponential, even a small difference in σ can make a big difference to the amplitude later on. We usually *assume* that the initial fluctuations are infinitesimal (far too small to be detectable), so that by the time the disturbance grows to visible amplitude, this so-called fastest-growing mode (FGM) will dominate. Based on this assumption, we focus our attention on the FGM.

Fastest-Growing Mode of Convection for an Inviscid, Nondiffusive Fluid

In the present case, (2.26) shows that the FGM has $\cos\theta = 1$, so that $\theta = 0$ and $m = K\sin\theta = 0$. The wave vector is purely horizontal. Motions associated with the fastest-growing mode are purely vertical,[2] which stands to reason because vertical motions are oriented ideally to respond to gravity.

The FGM therefore consists of spatially alternating columns of upward and downward motion, i.e., updrafts and downdrafts (Figure 2.3). The wavelength (the distance between updrafts, say) is $2\pi/\tilde{k}$. The growth rate is $\sqrt{-B_z}$. A property of particular importance in this example is that, while the growth rate depends on the orientation of the wave vector, it does not depend on its magnitude. As a result, the FGM can have any wavelength, including arbitrarily large and small wavelengths. A random noise field can grow to produce a random distribution of updrafts and downdrafts on a wide range of scales. This is called a broadband instability (e.g., Figure 2.4). In section 2.3 we'll re-introduce viscosity and diffusion and see how this can produce an FGM with a single, well-defined wavelength.

Classification of Normal Modes

In the case studied here (perturbations of a motionless, inviscid, nondiffusive fluid with B_z a negative constant), the growth rate has turned out to be purely real

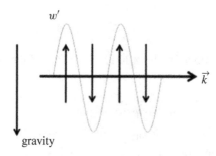

Figure 2.3 Normal mode in an inviscid, nondiffusive, uniformly stratified fluid. The wavevector is horizontal (i.e., $\theta = 0$); hence the motions are vertical. This orientation "feels" gravity most strongly: it exhibits the fastest oscillations when $B_z > 0$ (see equation 2.25), and the fastest exponential growth when $B_z < 0$ (equation 2.26).

---

[2] To verify this, note that the normal mode form of (2.16) is $-K^2\hat{\pi} = im\hat{b}$ and, since $m = 0$ and $K^2 \neq 0$ for the FGM, there is no pressure perturbation. Examination of the horizontal components of (2.14) then shows that, with $\pi' = 0$, there is no horizontal velocity perturbation.

Figure 2.4 Filaments formed by Rayleigh-Taylor convective instability (section 2.2.4) in a supernova, the Crab Nebula. A rapidly expanding shell of dense ejecta is gradually decelerated as it accumulates circumstellar material, leaving it convectively unstable. Photographic mosaic created by the Hubble Space Telescope, completed January 2000, courtesy of the US National Aeronautics and Space Administration (NASA).

[recall 2.26]. We will soon see that this is true for arbitrary B_z as well. In general, though, σ also has an imaginary part, which represents an oscillatory component of the motion:

$$\sigma = \sigma_r + \iota\sigma_i.$$

Some terminology:

- When $\sigma_r > 0$, the solution grows exponentially and is called an unstable mode, or an instability. (Conversely, solutions with $\sigma_r < 0$ are called decaying modes. Those are of less interest here.)
- We define the e-folding time $1/\sigma_r$, the time needed for an unstable disturbance to grow by a factor of $e = 2.71828$.
- When $\sigma_i = 0$ [as is the case in (2.26) with $B_z < 0$, for example], we have stationary instability.
- If σ_i were *nonzero* (and $\sigma_r > 0$) the instability would be called oscillatory, and would have the form of an exponentially growing oscillation. We will see examples of this later.

- If $\sigma_r = 0$ and $\sigma_i \neq 0$, the solution represents a wave that propagates without change of amplitude. The internal gravity wave described by (2.26) with $B_z > 0$ is an example.
- The normal mode form (2.20) can be written in terms of σ:

$$W(\vec{x}, t) = \hat{w}\, e^{\sigma t + \iota (kx + \ell y + mz)}.\qquad(2.27)$$

2.2.3 Instabilities of an Arbitrary Buoyancy Profile

In the foregoing sections 2.2.1 and 2.2.2, we explored normal mode solutions for the simple case in which B_z is a constant. However, in the nondiffusive fluid that we consider in this section, any buoyancy profile $B(z)$ can remain steady. Therefore, the theory may be applied to the wide range of buoyancy profiles found in geophysical fluids if diffusion is considered negligible.

Suppose that B_z is an arbitrary function of z. Equation (2.19) is still valid, but it now has coefficients that vary with z, and therefore (2.20) is no longer a solution. Instead, we use the more general normal mode form

$$W(\vec{x}, t) = \hat{w}(z)e^{\iota (kx + \ell y) + \sigma t},\qquad(2.28)$$

where the complex amplitude $\hat{w}(z)$ is a function of height whose form is yet to be determined. The wave vector $\vec{k} = (k, \ell)$ is now directed entirely in the horizontal. The vertical wavenumber component, m, the three-dimensional magnitude, K, and the angle of elevation, θ, are no longer relevant. As before, only the real part is retained: $w' = W_r$.

Substituting in (2.19), we obtain

$$\sigma^2 \left(\frac{d^2}{dz^2} - \tilde{k}^2 \right) \hat{w} = B_z \tilde{k}^2 \hat{w}.\qquad(2.29)$$

This is not an algebraic equation, as in sections 2.2.1 and 2.2.2, but rather is a second-order ordinary differential equation. In general, (2.29) must be solved numerically. Numerical solution methods will be discussed in the next chapter. For now, we'll derive a general theorem that tells us something important about the instability.

Suppose that the fluid is confined between two horizontal planes $z = z_1$ and $z = z_2$ where $\hat{w} = 0$. These may represent impermeable boundaries, or boundaries at infinity where the amplitude goes to zero. We now multiply (2.29) by the complex conjugate $\hat{w}^*$ and integrate the result over the vertical domain:

$$\sigma^2 \int_{z_1}^{z_2} \left(\hat{w}^* \frac{d^2 \hat{w}}{dz^2} - \tilde{k}^2 \hat{w}^* \hat{w} \right) dz = \int_{z_1}^{z_2} B_z \tilde{k}^2 \hat{w}^* \hat{w} \, dz.\qquad(2.30)$$

The first term in parentheses can be integrated by parts:

$$\int_{z_1}^{z_2} \hat{w}^* \frac{d^2\hat{w}}{dz^2} dz = \hat{w}^* \frac{d\hat{w}}{dz}\Big|_{z_1}^{z_2} - \int_{z_1}^{z_2} \frac{d\hat{w}^*}{dz}\frac{d\hat{w}}{dz} dz = -\int_{z_1}^{z_2} \Big|\frac{d\hat{w}}{dz}\Big|^2 dz. \qquad (2.31)$$

Substituting in (2.30), we have

$$\sigma^2 \int_{z_1}^{z_2} \Big(-\Big|\frac{d\hat{w}}{dz}\Big|^2 - \tilde{k}^2|\hat{w}|^2\Big) dz = \int_{z_1}^{z_2} B_z \tilde{k}^2 |\hat{w}|^2 dz. \qquad (2.32)$$

Close inspection of (2.32) yields three useful observations.

- While σ^2 may be complex in general, in this case it is purely real [since every other quantity in (2.32) is real]. As a result, solutions separate cleanly into two categories:

  (i) $\sigma^2 > 0$, a growing mode and a decaying mode
  (ii) $\sigma^2 < 0$, a pair of waves propagating in opposite directions.

- Suppose that B_z is positive throughout $z_1 < z < z_2$. In that case the integral on the right-hand side of (2.32) must be positive. On the left-hand side, the quantity in parentheses is negative, and hence $\sigma^2 < 0$. The solution represents a wave, not an instability, i.e., category 1 above.
- The solution represents instability ($\sigma^2 > 0$, category 1) if the right-hand side is negative. This requires, at least, that the minimum value of B_z in $z_1 < z < z_2$ be negative. The latter is a *necessary* condition for $\sigma^2 > 0$, but it is not *sufficient*; the sufficient condition is that (1) there be some z where $B_z < 0$ and (2) the vertical motions be concentrated in that region enough to make the integral $\int_{z_1}^{z_2} B_z|\hat{w}|^2 dz$ negative.

With a bit of rearranging we can place an upper bound on the growth rate. After moving the second term to the right-hand side, (2.32) becomes

$$\sigma^2 \int_{z_1}^{z_2} \Big|\frac{d\hat{w}}{dz}\Big|^2 dz = -\int_{z_1}^{z_2} \Big(\sigma^2 + B_z\Big)\tilde{k}^2|\hat{w}|^2 dz. \qquad (2.33)$$

For $\sigma^2 > 0$, the left-hand side, and hence the right-hand side, must be positive. This means that the integrand on the right must be negative somewhere in $z_1 < z < z_2$, and therefore the *minimum* value of B_z must not only be negative (as we already know) but must also be less than $-\sigma^2$. Rearranging this inequality, we have

$$\sigma < \sqrt{-\min_z B_z}. \qquad (2.34)$$

You can check this by comparing with (2.26) and with homework problems 6 and 16.

We will also find that the upper bound (2.34) is actually reached (to arbitrary precision), provided only that $B_z > 0$ for some z. In that case

$$\lim_{\tilde{k} \to \infty} \sigma = \sqrt{-\min_z B_z}. \tag{2.35}$$

You can learn more about this important result in section 7.8.1 and project B.7.

2.2.4 Convection at an Interface – Rayleigh-Taylor Instability

We now examine instabilities that arise from convection between two layers of inviscid, nondiffusive fluid with different buoyancy by solving (2.29) with the appropriate function $B(z)$. We place the interface at $z = 0$ and define buoyancy with respect to the lower layer, so that the background buoyancy profile has the step function form:

$$B(z) = \begin{cases} b_0, & z > 0 \\ 0, & z < 0 \end{cases} \tag{2.36}$$

When dealing with discontinuous profiles like (2.36), it will be helpful to use a general notation to represent the jump in a quantity, f say, occurring at the level $z = z_i$:

$$[[f]]_{z_i} \equiv f(z_i^+) - f(z_i^-). \tag{2.37}$$

Obviously $[[f]] = 0$ for a continuous function. But in (2.36), the jump in buoyancy at $z = 0$ is $[[B]]_0 = b_0$.

The buoyancy gradient, B_z, is expressed in terms of the Dirac delta function: $B_z = b_0 \delta(z)$. If the delta function is unfamiliar to you, spend some time looking at Figure 2.5.

For $z \neq 0$, $B_z = 0$ (see property 2 in Figure 2.5). Therefore, in both the upper and lower layers, (2.29) becomes

$$\frac{d^2 \hat{w}}{dz^2} - \tilde{k}^2 \hat{w} = 0. \tag{2.38}$$

This ordinary differential equation must be solved to determine $\hat{w}(z)$ in each layer. The general solution is

$$\hat{w} = \begin{cases} A_1 e^{-\tilde{k}z} + A_2 e^{\tilde{k}z}, & \text{for } z > 0 \\ A_3 e^{-\tilde{k}z} + A_4 e^{\tilde{k}z}, & \text{for } z < 0, \end{cases}$$

where A_1, A_2, A_3, and A_4 are constants to be determined. We assume that the domain is vertically unbounded, and that $\hat{w}$ is continuous and finite for all z. Those

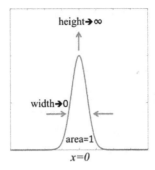

Figure 2.5 The Dirac delta function $\delta(x)$ can be thought of as a peak centered at $x = 0$, with zero width, infinite height, and unit area. It has the following properties:

1. $\delta(0) = \infty$.
2. $\delta(x) = 0$, for $x \neq 0$.
3. $\int_{-\infty}^{\infty} \delta(x)dz = 1$.
4. $\int_{-\infty}^{\infty} f(x)\delta(x)dz = f(0)$.
5. $\lim_{\epsilon \to 0} \int_{-\epsilon}^{\epsilon} f(x)\delta(x)dz = f(0)$.
6. $\lim_{\epsilon \to 0} \int_{x_0-\epsilon}^{x_0+\epsilon} f(x)\delta(x - x_0)dz = f(x_0)$.

conditions require $A_2 = 0$, $A_3 = 0$, and $A_4 = A_1$. (Show this.) Our solution can then be written compactly as

$$\hat{w}(z) = A_1 e^{-\tilde{k}|z|}. \tag{2.39}$$

Note that the largest amplitude of vertical velocity is found at the interface level, and that the motions decay with distance from the interface over a length scale proportional to the wavelength of the disturbance, $\tilde{k}^{-1}$.

As we have stated, $\hat{w}$ is continuous. In fact, this is always true. (If it were not, voids would appear within the fluid, and we do not observe this.) What about its derivative? If the coefficients of our differential equation were nice, smooth functions with no infinities, $\hat{w}_z$ would be continuous also, and this would furnish an additional constraint on the solution. But at $z = 0$, because $B_z \propto \delta(z)$, $\hat{w}_z$ is discontinuous, as you can also see by examination of (2.39). To complete the solution, we use the properties of $\delta(z)$ to establish the change in $\hat{w}_z$, then require that the derivative change by that amount.

The function $\delta(x)$ has several properties (Figure 2.5) that allow us to remove it by means of integration. Here, we take advantage of property 5 by integrating (2.29) over a layer extending vertically from $z = -\epsilon$ to $z = \epsilon$, i.e.,

$$\sigma^2 \int_{-\epsilon}^{\epsilon} \left(\frac{d^2 \hat{w}}{dz^2} - \tilde{k}^2 \hat{w} \right) dz = b_0 \tilde{k}^2 \int_{-\epsilon}^{\epsilon} \delta(z) \hat{w}(z) dz, \tag{2.40}$$

and take ϵ to be vanishingly small. Integration of the first term is trivial. The second term will vanish along with ϵ, given that $\hat{w}$ is finite. The integral on the right-hand side is simplified using property 5 of the delta function (Figure 2.5), resulting in

$$\boxed{\sigma^2 \left[\!\left[\frac{d\hat{w}}{dz} \right]\!\right]_0 = b_0 \tilde{k}^2 \hat{w}(0).} \tag{2.41}$$

This is called a jump condition, and the technique is one we'll use repeatedly in this book. For the solution (2.39),

$$\hat{w}(0) = A_1, \quad \text{and} \quad \left[\!\left[\frac{d\hat{w}}{dz} \right]\!\right]_0 = -2\tilde{k} A_1. \tag{2.42}$$

Substituting, we can solve for the growth rate:

$$\sigma = \pm\left(-\frac{b_0 \tilde{k}}{2} \right)^{1/2}. \tag{2.43}$$

Recall that for a top-heavy buoyancy profile we have $b_0 < 0$, so that instability is always present. The largest growth rates occur at the smallest scales (i.e., where $\tilde{k}$ is large). This situation is called ultraviolet catastrophe. It arises here because of our neglect of viscosity and diffusion, which would damp the instability at small scales.[3] We will consider these effects in the next section.

A beautiful laboratory demonstration of convective instability arising at the interface between two fluids is shown in Figure 2.6. The experimental setup consists of a rectangular container that initially separates the two fluids with a gate. In order to minimize the disturbances caused to the interface as the gate is removed, a fabric covering was used over the gate, which "peeled," rather than "slid," away from the interface. The removal of this gate can be seen at the left of the top panel.

If $b_0 > 0$, σ is imaginary, and the solution represents a pair of oppositely propagating gravity waves with phase speeds

$$c = \pm\left(\frac{b_0}{2\tilde{k}} \right)^{1/2}. \tag{2.44}$$

We'll encounter these waves again in Chapter 4.

2.3 Viscous and Diffusive Effects

We now return our attention to the viscous, diffusive case described by (2.17) and (2.18) with the equilibrium condition (2.7). Remember that equilibrium requires

---

[3] In fact, due to the highly simplified setup of this problem, there are only three characteristic scales present in our problem: σ, $\tilde{k}$, and b_0, and only two fundamental dimensions: length and time. This means that, before doing any stability analysis, we could have predicted from the dimensions of these parameters alone that the growth rate must scale as $\sigma \propto (b_0 \tilde{k})^{1/2}$.

Figure 2.6 Dye visualization of convective instability at an interface (Davies Wykes, and Dalziel, 2014). Upper panel: removal of barrier (at left) 4 seconds after initial gate opening. Lower panel: 11 seconds after gate opening, disturbance is larger at the left due to peeling back of the barrier.

uniform B_z. Assuming a normal mode solution, we make the transformations (2.21–2.24) in (2.17), resulting in

$$-\sigma K^2 \hat{w} = -\tilde{k}^2 \hat{b} + \nu(-K^2)^2 \hat{w},$$

or

$$\sigma \hat{w} = \cos^2 \theta \, \hat{b} - \nu K^2 \hat{w},$$

and in (2.18):

$$\sigma \hat{b} = -B_z \, \hat{w} - \kappa K^2 \hat{b}.$$

These can be written as a 2×2 matrix eigenvalue equation, with eigenvalue σ and eigenvector $[\hat{w} \, \hat{b}]^T$:

$$\sigma \begin{bmatrix} \hat{w} \\ \hat{b} \end{bmatrix} = \begin{bmatrix} -\nu K^2 & \cos^2 \theta \\ -B_z & -\kappa K^2 \end{bmatrix} \begin{bmatrix} \hat{w} \\ \hat{b} \end{bmatrix}. \tag{2.45}$$

The eigenvalues are determined by the characteristic equation:

$$\begin{vmatrix} -\nu K^2 - \sigma & \cos^2 \theta \\ -B_z & -\kappa K^2 - \sigma \end{vmatrix} = 0,$$

or

$$\boxed{\sigma^2 + (\nu + \kappa)K^2 \sigma + \nu \kappa K^4 + B_z \cos^2 \theta = 0.} \tag{2.46}$$

Note that, in the limit $K^2 \to 0$, we recover the inviscid, nondiffusive result (2.25). This tells us that *viscosity and diffusivity have negligible influence on disturbances with sufficiently large spatial scale*, a common result in fluid mechanics.

In general, the solutions are

$$\sigma = \frac{1}{2} \left[-(\nu + \kappa)K^2 \pm \sqrt{\mathcal{D}} \right], \tag{2.47}$$

where the discriminant is

$$\mathcal{D} = (\nu + \kappa)^2 K^4 - 4(\nu\kappa K^4 + B_z \cos^2 \theta).$$

There are three classes of solutions based on $\mathcal{D}$:

- If $\mathcal{D} < 0$, both values of σ are complex. Their real parts are the same:

$$\sigma_r = -\frac{(\nu + \kappa)K^2}{2},$$

  and both are negative, so the flow is stable. The imaginary parts have opposite signs and thus represent waves propagating in opposite directions, decaying as they go.
- If $0 < \mathcal{D} < (\nu + \kappa)^2 K^4$, then both solutions are real and negative, so the disturbance simply decays in place and the flow is classified as stable.
- If $\mathcal{D} > (\nu + \kappa)^2 K^4$, both solutions are real and one has $\sigma_r > 0$, i.e., the flow is unstable.

The condition for instability $\mathcal{D} > (\nu + \kappa)^2 K^4$ is equivalent to

$$\nu\kappa K^4 + B_z \cos^2 \theta < 0,$$

which can only happen if $B_z < 0$. Solving for K^2 gives

$$K^2 < \sqrt{\frac{-B_z \cos^2 \theta}{\nu\kappa}}.$$

This condition is illustrated in Figure 2.7. Differentiating (2.46) with respect to K^2, one gets

$$2\sigma \frac{\partial \sigma}{\partial K^2} + (\nu + \kappa)K^2 \frac{\partial \sigma}{\partial K^2} + (\nu + \kappa)\sigma + 2\nu\kappa K^2 = 0,$$

or

$$\frac{\partial \sigma}{\partial K^2} = -\frac{(\nu + \kappa)\sigma + 2\nu\kappa K^2}{2\sigma + (\nu + \kappa)K^2} < 0,$$

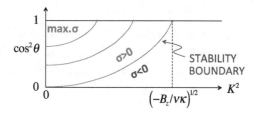

Figure 2.7 Growth rate for convection in an unbounded fluid.

which is negative in the unstable regime because σ is real and positive. In the same way, you can show that

$$\frac{\partial \sigma}{\partial \cos^2 \theta} > 0.$$

Note that

- Short waves are stabilized by viscosity and diffusion, and the effect increases as K^2 increases.
- The fastest-growing instability is found in the limit $K^2 \to 0$ and has purely vertical motions ($\cos^2 \theta = 1$) just as in the inviscid, nondiffusive case.
- In the special case $B_z = 0$ (either the fluid is homogeneous or gravity is negligible), (2.46) has two decaying solutions: $\sigma = -\nu K^2$ and $\sigma = -\kappa K^2$. Again, the smallest scales decay the fastest.
- If $B_z < 0$, the flow is unstable, i.e., there will always be some values of K and θ such that the growth rate is positive.

The fact that growth rate increases monotonically with wavelength tells us that convective instability prefers the largest possible scale, and therefore that boundary conditions are important. We'll address that issue in the next section.

2.4 Boundary Effects: the Rayleigh-Benard Problem

Consider a layer of fluid in motionless equilibrium bounded by frictionless, horizontal plates at $z = 0$ and $z = H$ (Figure 2.8). At these boundaries, w' must vanish. To satisfy that condition, we use a variant of the normal mode solution (2.20):

$$W(\vec{x}, t) = \hat{w} \sin(mz) \exp\{\iota(kx + \ell y) + \sigma t\}, \quad \text{where } m = n\frac{\pi}{H}; \; n = 1, 2, \ldots$$

$$(2.48)$$

where once again $w' = W_r$. We assume in addition that the buoyancy at the boundaries is fixed, so that $b' = 0$ and the normal mode form for buoyancy is therefore the same as (2.48). These modes have a form very different from the planar motions described by (2.20). Counter-rotating cells are arranged so that there is no vertical motion at the boundaries (Figure 2.9). The vertical scale decreases with increasing mode number n.

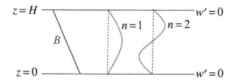

Figure 2.8 Definition sketch for Rayleigh-Benard convection.

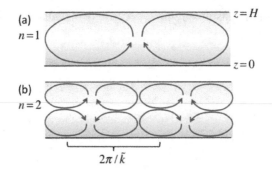

Figure 2.9 Normal modes (2.48) for the Rayleigh-Benard convection problem.

Substituting (2.48) in (2.17) and (2.18), we once again obtain (2.45); the only difference is the quantization condition $m = n\pi/H$. The characteristic equation is again (2.46), but we will now find it more convenient to write it in terms of $\tilde{k}$ rather than $\cos\theta$:

$$\sigma^2 + (\nu + \kappa)K^2\,\sigma + \nu\kappa K^4 + B_z \frac{\tilde{k}^2}{K^2} = 0. \tag{2.49}$$

Because of the quantization of m, the squared magnitude of the wave vector is now:

$$K^2 = k^2 + \ell^2 + m^2 = \tilde{k}^2 + \frac{n^2\pi^2}{H^2}; \quad n = 1, 2, \ldots. \tag{2.50}$$

2.4.1 Diffusion Scaling

Note that the solution of (2.49, 2.50) depends on six parameters:

$$\sigma = \mathcal{F}(\tilde{k}, n; B_z, \nu, \kappa, H). \tag{2.51}$$

The semicolon separates parameters describing the disturbance from those describing the environment it evolves in. Even though the solution for σ is relatively simple, the exploration of its properties in a six-dimensional parameter space is inefficient. We can reduce the dimensionality of the parameter space by means of *scaling*, a concept that we will use repeatedly. To set the stage for scaling, we define the concept of isomorphism.

Isomorphism

Two mathematical problems are isomorphic if they have, literally, the same form. For example, these two quadratic equations are isomorphic:

$$Ax^2 + Bx + C = 0 \tag{2.52}$$

$$\alpha y^2 + \beta y + \gamma = 0. \tag{2.53}$$

We write this relationship as (2.52) $\leftrightarrow$ (2.53).

Now suppose we have a solution algorithm for (2.52). This could be an analytical formula or a subroutine. In the present example, our solution algorithm could be the quadratic formula, as implemented in the following Matlab subroutine:

```
function x = quad(A,B,C)
D = B∧2-4*A*C;
x(1) = -B/2+sqrt(D)/2;
x(2) = -B/2-sqrt(D)/2;
return
```

We can use the *same* subroutine to solve (2.53); we simply substitute the appropriate variables:

y=quad(α, β, γ).

To sum up, if two problems are isomorphic, we can solve them using the same solution algorithm with different variables.[4] In this case we would write:

$$A \to \alpha$$
$$B \to \beta$$
$$C \to \gamma$$
$$x \to y.$$

To apply this idea to (2.51), we define a length scale and a time scale, which we choose to be H and H^2/κ, respectively. We then use these scales to define the nondimensional variables $\sigma^\star$, $\tilde{k}^\star$, and $K^\star$:

$$\sigma = \sigma^\star \frac{\kappa}{H^2} \tag{2.54}$$

$$\tilde{k} = \frac{\tilde{k}^\star}{H} \tag{2.55}$$

$$K = \frac{K^\star}{H}. \tag{2.56}$$

Substituting these into (2.49), we have

$$\sigma^{\star 2} \frac{\kappa^2}{H^4} + (\nu + \kappa) \frac{K^{\star 2}}{H^2} \sigma^\star \frac{\kappa}{H^2} + \nu\kappa \frac{K^{\star 4}}{H^4} + B_z \frac{\tilde{k}^{\star 2}/H^2}{K^{\star 2}/H^2} = 0.$$

Is this isomorphic to (2.49)? We can make the first term look the same by multiplying through by H^4/κ^2:

---

[4] Mathematicians have a more exacting definition of "isomorphism." Here, the essential characteristic defining isomorphic equations is that a single algorithm solves both.

$$\sigma^{*2} + (\frac{\nu}{\kappa} + 1)K^{*2}\,\sigma^{*} + \frac{\nu}{\kappa}K^{*4} + \frac{B_z H^4}{\kappa^2}\,\frac{\tilde{k}^{*2}}{K^{*2}} = 0. \tag{2.57}$$

This form of the equation contains two new parameters: the Prandtl number

$$Pr = \frac{\nu}{\kappa},$$

and the Rayleigh number,

$$Ra = \frac{-B_z H^4}{\nu\kappa}.$$

The Prandtl number depends only on the chemical makeup of the fluid. For air, $Pr \approx 1$. For water, $Pr \approx 7$ when the stratification is thermal, 700 when the stratification is saline. The Rayleigh number quantifies the relative importance of gravity and viscosity/diffusion.[5]

Substituting the definitions of Pr and Ra in (2.57), we have

$$\sigma^{*2} + (Pr + 1)K^{*2}\,\sigma^{*} + Pr K^{*4} - Ra\,Pr\,\frac{\tilde{k}^{*2}}{K^{*2}} = 0. \tag{2.58}$$

We can also write a nondimensional version of the quantization condition (2.50):

$$K^{*2} = \tilde{k}^{*2} + n^2\pi^2. \tag{2.59}$$

Note that (2.58) and (2.59) are isomorphic to (2.49) and (2.50). Now suppose that, in (2.51), $\mathcal{F}$ stands for some solution algorithm for (2.49, 2.50). The same solution algorithm will then work for (2.58, 2.59), after we make the appropriate substitutions:

$$\sigma \to \sigma^{*}$$
$$\nu \to Pr$$
$$K \to K^{*}$$
$$\kappa \to 1$$
$$B_z \to -Ra\,Pr$$
$$\tilde{k} \to \tilde{k}^{*}$$
$$H \to 1,$$

resulting in

$$\sigma^{*} = \mathcal{F}(\tilde{k}^{*}, n; -Ra\,Pr, Pr, 1, 1). \tag{2.60}$$

---

[5] To see this, recall that the maximum growth rate due to gravity alone (in the absence of viscosity and diffusion; see 2.26) is $\sqrt{-B_z}$. A disturbance with length H is damped by viscosity at a rate $\sim \nu/H^2$, and by diffusion at a rate $\sim \kappa/H^2$. The Rayleigh number is therefore a nondimensional combination of rates: the gravitational growth rate (squared) divided by the product of viscous and diffusive decay rates.

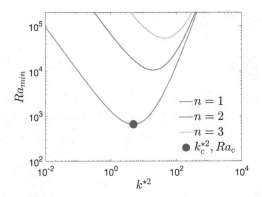

Figure 2.10 Minimum Rayleigh number for instability versus squared scaled wavenumber $\tilde{k}^{\star 2}$. The bullet shows the critical point $\tilde{k}^{\star 2} = \pi^2/2$, $Ra_c = 657.5$.

The number of independent parameters is now reduced to four, $\tilde{k}_\star$, n, Ra, and Pr, greatly simplifying the subsequent analysis.

In this book, we will explore both numerical and analytical solutions to many instability problems. When developing subroutines, we will use dimensional variables, as in (2.51). But when exploring solutions in multidimensional parameter spaces, we will simplify the task by calling the subroutines using appropriately scaled variables, as in (2.60).

2.4.2 The Critical State for the Onset of Convective Instability

The solution of (2.58) is

$$\sigma^\star = \frac{1}{2}\left[-(Pr+1)K^{\star 2} \pm \sqrt{\mathcal{D}}\right],$$

where

$$\mathcal{D} = (Pr+1)^2 K^{\star 4} - 4Pr\left(K^{\star 4} - Ra\frac{\tilde{k}^{\star 2}}{K^{\star 2}}\right).$$

This is just (2.47) with the diffusion scaling applied, but because of the boundaries we must also impose the quantization condition (2.59).

The solution is unstable if and only if

$$\mathcal{D} > (Pr+1)^2 K^{\star 4},$$

or

$$K^{\star 4} - Ra\frac{\tilde{k}^{\star 2}}{K^{\star 2}} < 0,$$

Figure 2.11 Turbulent convection in a numerical simulation. Blue (red) indicates buoyant (dense) fluid. From Cabot and Cook (2006).

or

$$Ra > \frac{(\tilde{k}^{\star 2} + n^2\pi^2)^3}{\tilde{k}^{\star 2}} \tag{2.61}$$

(see Figure 2.11).

This illustrates something interesting about the physics of convection in a bounded fluid: *it is not enough that dense fluid overlie light fluid*, i.e., $B_z < 0$, or $Ra > 0$, because disturbances amplified by gravity are simultaneously damped by viscosity and diffusion. As we saw previously in the unbounded cases (section 2.3), the relative strength of gravity and viscosity/diffusion is determined by two factors: the spatial scale of the normal mode, and the orientation of the motions. Specifically:

- Large-scale motions are relatively immune to damping by viscosity and diffusion. In the absence of boundaries, as long as $B_z < 0$, there will always be a normal mode large enough that viscosity and diffusion are too weak to prevent its growth. Boundaries limit the spatial scale, however, so that a sufficiently large wavelength may not fit.
- Gravity works most effectively on motions that are purely vertical (section 2.2.2). For a fixed value of m, as $\tilde{k}$ decreases (i.e., for wider convection cells, see Figure 2.9), the motion becomes mainly horizontal, so the gravitational acceleration is weaker.

Because of this competition, there is an *optimal horizontal scale for growth*: a scale at which viscosity and diffusion are small but gravity is still effective. There is also a minimum value of Ra below which even the mode with the optimum horizontal scale will not grow.

We now quantify these intuitive ideas by analyzing (2.61). Differentiating the right-hand side with respect to $\tilde{k}^{*2}$ and setting the result to zero, we find that the minimum Ra for instability is smallest, for a given n, when

$$\tilde{k}^* = n\pi/\sqrt{2}.$$

With $\tilde{k}^*$ set to this optimal value, the minimum Ra for instability is $27/4 \times n^4\pi^4$. The higher the vertical mode number n, the higher Ra has to be (because higher n means stronger damping by viscosity/diffusion). The minimum Ra is therefore lowest when $n = 1$. The critical Ra for instability is therefore

$$\boxed{Ra_c = \frac{27}{4}\pi^4 \approx 657.5.}$$

The horizontal wavelength $2\pi/\tilde{k}$ is $2\sqrt{2}H$. Since a wavelength comprises two convection cells, each cell has width about 1.4 times the thickness of the convecting layer.

So, suppose that B_z starts out at zero but gradually becomes more negative, as happens when fluid is heated from below. When B_z becomes negative enough that Ra exceeds 657.5, convective motions begin. These have vertical mode number $n = 1$ (i.e., a single row of convection cells that occupy the entire vertical domain) and $\tilde{k}^* = \pi/\sqrt{2}$. Therefore, the spacing between updrafts (or between downdrafts) is about three times the layer thickness.[6]

Note that the parameters of this critical state do not depend on the Prandtl number. As a result, they apply equally well to air, seawater, magma, or any other convecting fluid.

2.5 Nonlinear Effects

While the growth rate depends on the magnitude of the horizontal wavenumber $\tilde{k}$, the *direction* of the wave vector (in the horizontal) is of no consequence. Therefore, a random initial perturbation will produce convection rolls with all possible horizontal orientations. By Fourier's theorem, these can sum to make a variety of planforms. Nonlinear analyses predict that the dominant planform is hexagonal (Schluter et al., 1965), and laboratory experiments have confirmed this (e.g., Figure 2.12a, also Figure 9.1). In a much more energetic environment such as the Sun's surface (Figure 2.12b), the planform is less regular.

---

[6] The foregoing solution applies to a fluid with frictionless boundaries. If one assumes frictional boundaries, at which the horizontal velocity must vanish, the critical state has $Ra = 1708$.

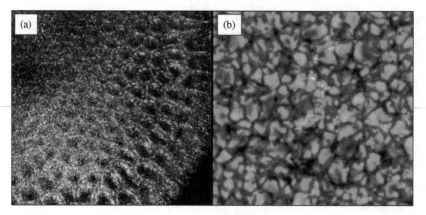

Figure 2.12 (a) Convection cells at centimeter scale in a lab experiment. A mixture of silicone oil and fine aluminum powder is heated from below. Cells form with fluid rising in the center and sinking at the edges (NOAA). (b) Convection cells at 1000 km scale in the solar photosphere. Bright (dark) regions are rising (sinking). (NASA Hinode Solar Optical Telescope, Hinode JAXA/NASA/PPARC).

In most geophysical examples of convection, Ra is many orders of magnitude greater than the critical value. Convection is turbulent, with chaotic motions over a broad range of scales (e.g., Figures 2.1 and 2.11). The vertical motion is dominated by thin plumes. The spacing of the strongest plumes is generally of the same order of magnitude as the thickness of the convecting layer, reflecting the underlying linear instability.

Linden (2000) gives a clear and detailed description of environmental convection, including the nonlinear regime.

2.6 Summary

The main "rules of thumb" about convection that one should take away from this chapter are:

- In a layer with frictionless boundaries, convection requires that $Ra > 657.5$.
- The exact value of Ra that must be exceeded for convection depends on the boundary conditions, so a more general rule of thumb is $Ra \gtrsim 10^3$.
- The spacing between updrafts (or between downdrafts) is about three times the layer thickness.

2.7 Appendix: Waves and Convection in a Compressible Fluid

Compressibility is a property of all fluids, though it is more important in some fluids (e.g., the atmosphere) and less so in others (e.g., lakes and rivers).

Compressibility mainly affects the meaning and measurement of the buoyancy gradient B_z, which is of course critical to the prediction of gravity waves and convective instability. Here we will examine the stability characteristics of a compressible fluid using the parcel method, a less-precise, but often more-intuitive alternative to the normal mode method. After introducing the parcel method in an incompressible fluid, we'll conduct a thought experiment aimed at understanding convection and waves in a compressible fluid. To highlight compressibility effects, we will neglect viscosity and diffusion.

2.7.1 Thought Experiment: Assessing Stability via the Parcel Method

Consider a layer of inviscid, incompressible fluid that is stably stratified, so that its vertical density gradient ρ_z is negative as shown in Figure 2.13a. Now imagine that a parcel of fluid is lifted up a distance η above its equilibrium height. If the parcel is sufficiently small, the resulting pressure change is negligible and the parcel's motion is governed only by the buoyancy force. If the displacement is rapid, there will be no exchange of heat with the surrounding fluid, and therefore no change in the density of the parcel. This is called an adiabatic displacement.

At its new elevation, the parcel is denser than the surrounding fluid and is therefore accelerated downward (upper blue arrow in Figure 2.13a). Conversely, a parcel displaced below its equilibrium height will be accelerated upward. Buoyancy therefore creates a restoring force that can support oscillatory motion.

Let's make this observation a bit more quantitative. If the vertical displacement is small, then the buoyant acceleration is $-g\Delta\rho/\rho$, where $\Delta\rho$ is the density of the parcel minus that of the surrounding fluid. To first order in η, we can approximate $\Delta\rho$ as $-\rho_z\eta$ (show this), and therefore

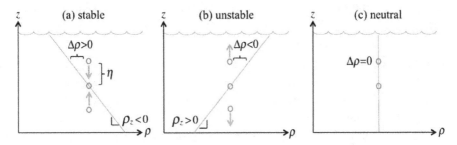

Figure 2.13 The parcel method applied to gravity waves (a), convection (b), and neutral stability (c). Circles represent small fluid parcels displaced vertically from their equilibrium heights. The density differential $\Delta\rho$ is the parcel's density minus that of the surrounding fluid. The fluid is assumed to be incompressible and the motion adiabatic, so the parcel's density remains constant as it moves.

$$\frac{d^2\eta}{dt^2} = \frac{g}{\rho}\rho_z\eta.$$ (2.62)

With $\rho_z < 0$, the solution is $\eta \sim e^{\pm i\omega t}$, with

$$\omega = \sqrt{-\frac{g}{\rho}\rho_z} = \sqrt{B_z}\,.$$ (2.63)

You will recognize this as the *buoyancy frequency* discussed previously in section 2.1.1. It corresponds to the frequency of an internal gravity wave, in uniform stratification, in which the motion is purely vertical, i.e., $\theta = 0$ in (2.25).

Now, what happens if the density gradient is reversed, so that $\rho_z > 0$ (Figure 2.13b)? Displaced above its equilibrium height, our fluid parcel finds itself *lighter* than its surroundings and therefore continues to rise. Conversely, a parcel displaced downward continues to sink. This is the parcel version of convective instability. The solution of (2.62) is now $\eta \sim e^{\sigma t}$, with the growth rate σ given by

$$\sigma = \sqrt{\frac{g}{\rho}\rho_z} = \sqrt{-B_z}\,,$$

exactly as found in section 2.2 using normal modes (cf. 2.26 with $\theta = 0$).

Finally, suppose that the fluid is homogeneous (Figure 2.13c). In that case the displaced parcel has the same density as its surroundings, and it therefore feels no buoyancy force. This is the state of *neutral* stratification.

2.7.2 Another Thought Experiment: Effects of Compressibility

We'll now repeat the thought experiment described above, but our working fluid will be an imaginary substance whose compressibility we can alter at will. Consider a layer of inviscid, incompressible fluid. Beginning with the simplest case, we assume that the density is *uniform*, as in Figure 2.13c or 2.14a. Now we flip a magic switch and the fluid becomes compressible (Figure 2.14a). Picture what happens next: the surface descends slightly as the fluid is compressed under its own weight. Compression is greatest at the bottom, where the pressure is greatest. In fact, the density increase is nearly proportional to the pressure, so after compression the density gradient ρ_z will equal $-\gamma$, where γ is a constant that quantifies the degree of compressibility.[7]

Based on this negative density gradient, you might think that the fluid is now stably stratified, but you'd be wrong. Suppose that a fluid parcel is displaced upward, as in Figure 2.14a. In contrast to the incompressible case, the parcel does *not* retain

---

[7] The density profile will actually be exponential, but linearity is a fine approximation for small changes.

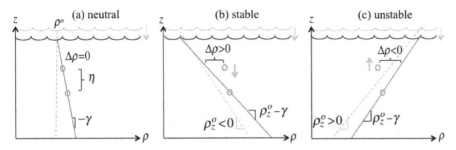

Figure 2.14 Density profiles in a fluid with (solid) and without (dashed) compression effects. Circles illustrate the change in density as a fluid parcel is raised above its equilibrium position.

its original density, but rather expands as the ambient pressure decreases. Its density decreases at exactly the same rate as that of the surrounding fluid, namely $-\gamma$, and it therefore feels *no buoyancy force*. The fluid is neutrally stable, just as when it was incompressible. Put another way, the state of neutral stability in a compressible fluid is defined not by $\rho_z = 0$ but by $\rho_z = -\gamma$.

Next, suppose that our original fluid layer is already stably stratified, with density gradient $\rho_z^o < 0$. When the fluid is allowed to compress (Figure 2.14b), its density will once again increase in proportion to pressure. The density gradient now decreases from its original, negative value ρ_z^o to the *more* negative value $\rho_z^o - \gamma$.

At a glance it looks like the fluid has become even more stable, but what happens to a lifted fluid parcel? Again, the parcel expands as the pressure is released, so that its density decreases with height at the rate $-\gamma$. The difference between the density of the parcel and its surroundings is $-\rho_z^o \eta$, and the parcel oscillates with frequency $\omega = \sqrt{-(g/\rho)\rho_z^o}$. This is superficially similar to the incompressible result (2.63), the difference being that ρ^o represents the *original* density profile, before the fluid was allowed to compress.

As a final example, suppose that the original stratification is unstable, i.e., $\rho_z^o > 0$, so that a displaced parcel will accelerate away from its equilibrium height (Figure 2.14c). The result of compression is to reduce the density gradient by adding $-\gamma$. However, the compressibility of the parcel compensates for that change, so that the displacement grows exponentially with rate $\sigma = \sqrt{(g/\rho)\rho_z^o}$.

Now here is a critical point: *the density gradient we actually observe in a compressible fluid* includes *the compressive part,* $-\gamma$. The original, "uncompressed" state described in our thought experiment is actually a nonexistent fiction, albeit an important one because it determines the stability of the fluid. To assess the stability of a compressible fluid like the atmosphere, we first measure the net density gradient, $\rho_z^o - \gamma$, then remove the compressive part to recover ρ_z^o. Another way to say this is that the squared buoyancy frequency (or buoyancy gradient) B_z is given by the *observed* density gradient, ρ_z, minus the compressive part, which works out to

$$B_z = -\frac{g}{\rho}(\rho_z + \gamma). \qquad (2.64)$$

So in summary, a compressible fluid is less stable (or more unstable) than it looks. The difference is the compressibility parameter γ.

2.7.3 A Quantitative View

We can quantify γ as follows:

$$\gamma = -\frac{d\rho}{dz} = -\frac{\partial\rho}{\partial p}\frac{\partial p}{\partial z}.$$

For adiabatic displacements, the partial derivative $\partial\rho/\partial p$ is the positive quantity $1/c_s^2$. We won't prove it here, but c_s turns out to be the speed of sound in the fluid. Moreover, since the ambient fluid is in equilibrium, its pressure is hydrostatic: $\partial p/\partial z = -\rho g$. Combining these results, we have

$$\gamma = \frac{\rho g}{c_s^2}. \qquad (2.65)$$

In the incompressible limit, the sound speed goes to infinity, and hence $\gamma \to 0$ and $N^2 \to B_z$ (cf. 2.64) as we have found before.

In meteorology, stability is usually discussed in terms of temperature rather than density. Those who have studied such things will recognize that γ is related to the adiabatic lapse rate of temperature, commonly called Γ, via $\gamma = \alpha\rho\Gamma$, α being the thermal expansion coefficient. The Earth's troposphere is often found near a state of neutral stability: its temperature decreases with height at the adiabatic lapse rate, in that case typically 10°C/km.

The density in our imaginary "uncompressed" state, $\rho^o(z)$, is called potential density. Potential density is defined as the density a fluid parcel would have if moved adiabatically to some reference height. In our thought experiment, the reference height is the surface of the fluid layer. For a simple example, refer to Figure 2.14a. There, the uncompressed density profile (dashed line) is just a constant. In the compressed state, lifting a parcel adiabatically from any height moves it along the solid line to the surface, where its density becomes that same constant value ρ^o.

3

Instabilities of a Parallel Shear Flow

A parallel shear flow is a flow that moves in a single direction, and whose velocity changes in a perpendicular direction, e.g., $\vec{u} = U(z)\hat{e}^{(x)}$. This class of flows includes wakes, jets, boundary layers and shear layers (e.g., Figures 3.2 and 3.3). Parallel shear flows are the simplest class of flows exhibiting the phenomenon of shear instability, in which a fluid is unstable because of spatial variations in its velocity.

Here we will look at shear instability in its most basic form, free of complications due to viscosity, diffusion, buoyancy, or planetary rotation. We'll need to solve an ordinary differential equation, the Rayleigh equation. This can be done analytically for some simple examples (section 3.3), but we'll also make frequent use of the numerical methods introduced in Chapter 1.

Not every parallel shear flow is unstable. We'll prove a simple theorem that often allows us to identify stable cases without solving any equations. Even in an unstable flow, not all types of disturbances grow, so we'll prove two additional results that allow us to distinguish those that do.

Besides developing the needed mathematical theory, we will seek an intuitive understanding of the mechanism of shear instability. This latter goal is considerably more challenging than in the previous case of convection.

3.1 The Perturbation Equations

We assume that

- the flow is inviscid: $\nu = 0$;
- the flow is homogeneous: $\rho = \rho_0$, or $b = 0$;
- Coriolis effects are negligible: $f = 0$.

With these assumptions, the equations of motion (1.17, 1.19) become

$$\vec{\nabla} \cdot \vec{u} = 0 \tag{3.1}$$

Figure 3.1 Shear instability at the edge of an altocumulus layer photographed from the International Space Station (NASA).

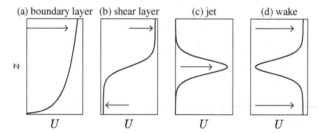

Figure 3.2 Common models of parallel shear flow. (a) Atmospheric or oceanic bottom boundary layer. (b) Shear layer, e.g., Figure 3.3. (c) Jet, e.g., jet stream or Gulf Stream (Figure 1.1). (d) Wake, e.g., island wake in the atmosphere, Figure 3.13.

and

$$\frac{D\vec{u}}{Dt} = -\vec{\nabla}\pi. \tag{3.2}$$

3.1.1 The Equilibrium State

We first seek equilibrium solutions of (3.1, 3.2) having the form of parallel shear flows

$$\vec{u} = U(z)\hat{e}^{(x)}.$$

Note that, while the coordinate z traditionally indicates the vertical direction in geophysical problems, gravity is irrelevant here and z may therefore represent any

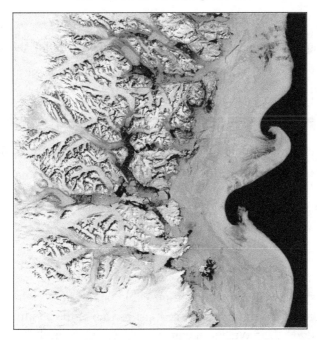

Figure 3.3 Shear instability in a Greenland coastal current made visible by floating glacial ice. Landsat 7 satellite photo courtesy United States Geological Survey (hereafter USGS) and NASA.

direction. For example, the instabilities shown in Figures 3.1, 3.3, and 3.13 grow on horizontally sheared flows. In the analysis of those instabilities, z would be directed horizontally across the mean flow.

The continuity equation (3.1) is clearly satisfied:

$$\vec{\nabla} \cdot \vec{u} = \vec{\nabla} \cdot [U(z)\hat{e}^{(x)}] = \frac{\partial}{\partial x} U(z) = 0.$$

The left-hand side of the momentum equation (3.2) is

$$\frac{D\vec{u}}{Dt} = [\frac{\partial}{\partial t} + \vec{u} \cdot \vec{\nabla}]\vec{u}$$
$$= [\frac{\partial}{\partial t} + U(z) \underbrace{\hat{e}^{(x)} \cdot \vec{\nabla}}_{=\partial/\partial x}]U(z)\hat{e}^{(x)} = 0,$$

and therefore the right-hand side, $-\vec{\nabla}\pi$ must also be zero. This tells us that any parallel shear flow $U(z)\hat{e}^{(x)}$ is an equilibrium state, provided that the pressure is uniform.

3.1.2 Perturbations from Equilibrium

We next assume that the velocity field consists of a parallel shear flow plus a perturbation:

$$\vec{u} = U(z)\hat{e}^{(x)} + \epsilon\vec{u}'(\vec{x}, t), \tag{3.3}$$

$$\pi = \Pi + \epsilon\pi'(\vec{x}, t), \tag{3.4}$$

where Π is an arbitrary constant as determined in section 3.1.1. Substituting in (3.1), we have

$$\vec{\nabla} \cdot \vec{u} = \vec{\nabla} \cdot \left[U(z)\hat{e}^{(x)} + \epsilon\vec{u}'\right] = \frac{\partial}{\partial x}\left[U(z) + \epsilon u'\right] + \frac{\partial}{\partial y}\epsilon v' + \frac{\partial}{\partial z}\epsilon w' = 0,$$

or

$$\boxed{\vec{\nabla} \cdot \vec{u}' = 0.} \tag{3.5}$$

(In fact, this is always true in an incompressible fluid. The background state must be nondivergent, and therefore the same must be true of the perturbation.)

We next address the momentum equation, (3.2). We begin by substituting (3.3) into the material derivative (1.14):

$$\frac{D}{Dt} \equiv \frac{\partial}{\partial t} + \vec{u} \cdot \vec{\nabla}$$

$$= \frac{\partial}{\partial t} + \left[U(z) + \epsilon u'\right]\frac{\partial}{\partial x} + \epsilon v'\frac{\partial}{\partial y} + \epsilon w'\frac{\partial}{\partial z}$$

$$= \frac{\partial}{\partial t} + U(z)\frac{\partial}{\partial x} + \epsilon\vec{u}' \cdot \vec{\nabla}. \tag{3.6}$$

Applying this material derivative to $\vec{u}$ gives the left-hand side of (3.2) in perturbation form:

$$\frac{D\vec{u}}{Dt} = \left(\frac{\partial}{\partial t} + U(z)\frac{\partial}{\partial x} + \epsilon\vec{u}' \cdot \vec{\nabla}\right)\left[U(z)\hat{e}^{(x)} + \epsilon\vec{u}'\right]$$

$$= \underbrace{\left(\frac{\partial}{\partial t} + U(z)\frac{\partial}{\partial x}\right)U(z)\hat{e}^{(x)}}_{=0} + \underbrace{\epsilon\vec{u}' \cdot \vec{\nabla}U(z)\hat{e}^{(x)}}_{=w'dU/dz}$$

$$+ \epsilon\left(\frac{\partial}{\partial t} + U(z)\frac{\partial}{\partial x}\right)\vec{u}' + \underbrace{\epsilon^2[\vec{u}' \cdot \vec{\nabla}]\vec{u}'}_{\approx 0}$$

$$= \epsilon w'\frac{dU}{dz}\hat{e}^{(x)} + \epsilon\left(\frac{\partial}{\partial t} + U(z)\frac{\partial}{\partial x}\right)\vec{u}'.$$

As usual, the $O(\epsilon^2)$ term is assumed to be negligible. The right-hand side of (3.2) is $-\epsilon\vec{\nabla}\pi'$ since $\vec{\nabla}\Pi = 0$.

Equating the left- and right-hand sides and cancelling the common factor ϵ, the momentum equation for the perturbations is:

$$\boxed{\frac{\partial \vec{u}'}{\partial t} + U(z)\frac{\partial \vec{u}'}{\partial x} + w'\frac{dU}{dz}\hat{e}^{(x)} = -\vec{\nabla}\pi'}. \qquad (3.7)$$

The second term on the left-hand side describes advection of velocity perturbations by the background flow. The third term describes the reverse: the vertical velocity perturbation w' advects the background shear dU/dz to produce perturbations in the x-velocity.

For later convenience, we split into components:

$$\frac{\partial u'}{\partial t} + U\frac{\partial u'}{\partial x} = -\frac{\partial \pi'}{\partial x} - \frac{dU}{dz}w' \qquad (3.8)$$

$$\frac{\partial v'}{\partial t} + U\frac{\partial v'}{\partial x} = -\frac{\partial \pi'}{\partial y} \qquad (3.9)$$

$$\frac{\partial w'}{\partial t} + U\frac{\partial w'}{\partial x} = -\frac{\partial \pi'}{\partial z}. \qquad (3.10)$$

The perturbation equations (3.5) and (3.7) comprise four equations in four unknowns, u', v', w', and π'. As in our previous study of convection, we try to reduce these to a single equation for w'. We begin by deriving the Poisson equation for the pressure perturbation (cf. derivation of 2.16). To do this, we take the divergence of (3.7). The first term is easy:

$$\vec{\nabla} \cdot \frac{\partial \vec{u}'}{\partial t} = \frac{\partial}{\partial t}\vec{\nabla} \cdot \vec{u}' = 0,$$

due to (3.5). The second term is

$$\vec{\nabla} \cdot \left(U(z)\frac{\partial \vec{u}'}{\partial x}\right) = \vec{\nabla}U(z) \cdot \frac{\partial \vec{u}'}{\partial x} + U\frac{\partial}{\partial x}\underbrace{\vec{\nabla} \cdot \vec{u}'}_{=0}$$

$$= U_z\hat{e}^{(z)} \cdot \frac{\partial \vec{u}'}{\partial x}$$

$$= U_z\frac{\partial w'}{\partial x},$$

where the abbreviation U_z has been adopted for the total derivative dU/dz.[1] For the third term we have

$$\vec{\nabla} \cdot \left(w'\frac{dU}{dz}\hat{e}^{(x)}\right) = \frac{\partial}{\partial x}\left(w'\frac{dU}{dz}\right) = U_z\frac{\partial w'}{\partial x}$$

(again!).

---

[1] We will do this frequently to simplify complicated expressions. We must take care, though, because subscripts r and i are also used to denote real and imaginary parts of a complex quantity, and must not be confused with derivatives.

Finally, the right-hand side is

$$-\vec{\nabla} \cdot \vec{\nabla}\pi' = -\nabla^2 \pi.$$

We can now assemble the desired equation:

$$\nabla^2 \pi = -2U_z \frac{\partial w'}{\partial x}. \tag{3.11}$$

Together with (3.10), this gives us two equations in the two unknowns w' and π'. To eliminate π', we take the Laplacian of (3.10) and substitute:

$$\nabla^2 \frac{\partial w'}{\partial t} + \vec{\nabla} \cdot \vec{\nabla}\left(U \frac{\partial w'}{\partial x}\right) = -\nabla^2 \frac{\partial \pi'}{\partial z},$$

$$\frac{\partial}{\partial t}\nabla^2 w' + \vec{\nabla} \cdot \left(U_z \hat{e}^{(z)} \frac{\partial w'}{\partial x} + U\vec{\nabla}\frac{\partial w'}{\partial x}\right) = -\frac{\partial}{\partial z}\nabla^2 \pi',$$

$$\frac{\partial}{\partial t}\nabla^2 w' + \frac{\partial}{\partial z}\left(U_z \frac{\partial w'}{\partial x}\right) + \vec{\nabla}U \cdot \vec{\nabla}\frac{\partial w'}{\partial x} + U\nabla^2 \frac{\partial w'}{\partial x} = \frac{\partial}{\partial z}\left(2U_z \frac{\partial w'}{\partial x}\right),$$

$$\frac{\partial}{\partial t}\nabla^2 w' + U_z \frac{\partial}{\partial z}\frac{\partial w'}{\partial x} + U\nabla^2 \frac{\partial w'}{\partial x} = U_z \frac{\partial}{\partial z}\frac{\partial w'}{\partial x} + U_{zz} \frac{\partial w'}{\partial x},$$

and finally we have a single equation for w':

$$\frac{\partial}{\partial t}\nabla^2 w' + U\frac{\partial}{\partial x}\nabla^2 w' = U_{zz} \frac{\partial w'}{\partial x}. \tag{3.12}$$

It is instructive to compare this with the corresponding equation for the convective case, (2.17). Again we have the time derivative of $\nabla^2 w'$. But the buoyancy and viscosity terms are now neglected, and instead we have two new terms describing interactions between the perturbation and the parallel shear flow.

3.2 Rayleigh's Equation

3.2.1 Normal Modes in a Shear Flow

As in section 2.2.3, we cannot use the simplest normal mode solution (2.20), because (3.12) does not have constant coefficients; U and U_{zz} are in general functions of z. To allow for this z-dependence, we use the more general normal mode form (2.28), reproduced here for convenience:

$$W(\vec{x}, t) = \hat{w}(z)e^{i(kx+\ell y)+\sigma t}. \tag{3.13}$$

As in the convective case, we define the complex function W for analytical convenience; when the time comes to interpret the solution physically, we will retain only the real part:

$$w' = W_r.$$

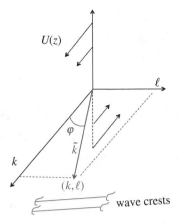

Figure 3.4 Structure of the wave vector for a parallel shear flow, the counterpart of Figure 2.2. The wave vector is horizontal, with components (k, ℓ) and magnitude $\tilde{k}$. The angle of obliquity $\varphi = \cos^{-1} k/\tilde{k}$ is restricted to $-\pi/2 \leq \varphi \leq \pi/2$.

We note a few more properties of this normal mode form:

- The wave vector $\vec{k} = (k, \ell)$ is directed horizontally as shown in Figure 3.4, with k corresponding to the streamwise (x) direction and ℓ the cross-stream (y) direction. The magnitude of the wavevector is $\tilde{k}$, defined as before: $\tilde{k} = \sqrt{k^2 + \ell^2}$.
- The angle of obliquity, φ, is the angle between the wavevector and the direction of the background flow (x), and is in the range $-\pi/2 \leq \varphi \leq \pi/2$. Modes are categorized as
  - two-dimensional (2D) if $\varphi = 0$ or, equivalently $\ell = 0$,
  - oblique if $\varphi \neq 0$.

  2D modes have crests perpendicular to the background flow (Figure 3.4).

The normal mode solution (3.13) can also be written in terms of the frequency $\omega = -\iota\sigma$:

$$W = \hat{w}(z)e^{\iota(kx+\ell y-\omega t)}.$$ (3.14)

A third alternative is to express the solution in terms of the streamwise phase speed $c = \omega/k$:

$$W = \hat{w}(z)e^{\iota k(x-ct)+\iota\ell y}.$$ (3.15)

This is the speed at which a fixed phase of the wave (e.g., a crest or a trough) moves in the x direction (i.e., in a plane $y = $ constant). The phase speed is related to the growth rate by $\sigma = -\iota kc$, or $c = \iota\sigma/k$.

The background flow U is also referred to as the mean flow. Averaging a perturbation of the form (3.13), (3.14), or (3.15) over an integer number of wavelengths in x, y, or both gives zero. Therefore $\bar{u} = \overline{U + u'} = \bar{U} = U$.

3.2.2 Three Forms of Rayleigh's Equation

Returning to the perturbation equation (3.12) and substituting (3.13), the normal mode form written in terms of σ, we obtain a second-order, ordinary differential equation for $\hat{w}(z)$:

$$\sigma \nabla^2 \hat{w} = -\imath k U \nabla^2 \hat{w} + \imath k \frac{d^2 U}{dz^2} \hat{w}, \tag{3.16}$$

where

$$\nabla^2 = \frac{d^2}{dz^2} - \tilde{k}^2. \tag{3.17}$$

This is called Rayleigh's equation after Lord Rayleigh, the inventor of normal modes (Rayleigh, 1880). Together with boundary conditions (often $\hat{w} = 0$ at upper and lower limits of z), these form a differential eigenvalue problem which we will soon convert into an algebraic eigenvalue problem.

Another useful form of Rayleigh's equation results from expanding ∇^2 using (3.17) and rearranging:

$$(\sigma + \imath k U)\left(\frac{d^2}{dz^2} - \tilde{k}^2\right)\hat{w} = \imath k \frac{d^2 U}{dz^2}\hat{w}. \tag{3.18}$$

Rayleigh's equation can also be written in terms of the phase speed. Substituting $\sigma = -\imath k c$ into (3.18), we have

$$\hat{w}_{zz} = \left(\frac{U_{zz}}{U - c} + \tilde{k}^2\right)\hat{w}. \tag{3.19}$$

All three forms of Rayleigh's equation are useful.

3.2.3 Polarization Relations

Normal mode expressions like (3.14) also describe the remaining variables, each with its own vertical structure function: the horizontal velocity components $\hat{u}(z)$ and $\hat{v}(z)$, and the pressure $\hat{\pi}(z)$. Once we know $\hat{w}$, we can calculate the other eigenfunctions. Substituting the normal mode expressions into (3.5), (3.8), and (3.9) gives:

$$\iota(k\hat{u} + \ell\hat{v}) + \hat{w}_z = 0;$$
$$\iota k(U - c)\hat{u} = \iota k\hat{\pi} - U_z\hat{w};$$
$$\iota k(U - c)\hat{v} = -\iota\ell\hat{\pi}.$$

These can be solved algebraically for $\hat{u}$, $\hat{v}$, and $\hat{\pi}$ in terms of $\hat{w}$:

$$\hat{u} = \iota\frac{k}{\tilde{k}^2}\left[\frac{\ell^2}{k^2}\frac{U_z}{U-c}\hat{w} + \hat{w}_z\right], \tag{3.20}$$

$$\hat{v} = \iota\frac{l}{\tilde{k}^2}\left[-\frac{U_z}{U-c}\hat{w} + \hat{w}_z\right], \tag{3.21}$$

$$\hat{\pi} = \iota\frac{k}{\tilde{k}^2}\left[U_z\hat{w} - (U-c)\hat{w}_z\right]. \tag{3.22}$$

Owing to their use in the context of electromagnetic waves, these are referred to as polarization relations.

3.3 Analytical Example: the Piecewise-Linear Shear Layer

The shear layer is a ubiquitous flow shape: it's just a region where the velocity changes from one uniform value to another. Although the mechanism is not as obvious as in the convection case, a shear layer is inherently unstable, and that instability is the main reason naturally occurring flows are almost always turbulent.

In the simplest model of a shear layer, the velocity changes linearly from one value to another. We choose coordinates so that the velocities are $\pm u_0$ and the shear layer boundaries are $z = \pm h$ (Figure 3.5). The velocity profile is then

$$U = u_0 \begin{cases} 1, & z \geq h \\ z/h, & -h \leq z \leq h \\ -1, & z \leq -h \end{cases} \tag{3.23}$$

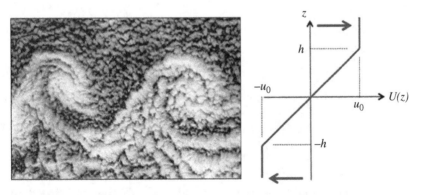

Figure 3.5 Velocity profile (blue) for a linear shear layer with thickness $2h$ and velocity change $2u_0$ (see 3.23). To the left is a disturbance that might grow on such a flow (NASA).

We now use $U(z)$ as input to the Rayleigh equation, which we write in the form (3.19):

$$\hat{w}_{zz} = \left(\frac{U_{zz}}{U - c} + \tilde{k}^2 \right) \hat{w}.$$

Note that U_z is discontinuous at $z = \pm h$, and therefore U_{zz} is infinite at those heights. Specifically

$$U_{zz} = -\frac{u_0}{h} \delta(z + h) + \frac{u_0}{h} \delta(z - h).$$

If the delta function δ is unfamiliar, review section 2.2.4.

3.3.1 Computing the Dispersion Relation

We will solve this problem using the method introduced in section 2.2.4 for convection at an interface, but with two interfaces instead of one. Except at $z = \pm h$, $U_{zz} = 0$, so (3.19) reduces to $\hat{w}_{zz} - \tilde{k}^2 \hat{w} = 0$, and the solution is very simple:

$$\hat{w} = \begin{cases} A_1 e^{\tilde{k}z} + A_2 e^{-\tilde{k}z}, & z \geq h \\ A_3 e^{\tilde{k}z} + A_4 e^{-\tilde{k}z}, & -h \leq z \leq h \\ A_5 e^{\tilde{k}z} + A_6 e^{-\tilde{k}z}, & z \leq -h \end{cases} \tag{3.24}$$

There are six undetermined constants, so we need six constraints to specify the solution. The first two are obvious: the solution cannot blow up as $z \to \pm \infty$, so

$$A_1 = 0 \quad \text{and} \quad A_6 = 0. \tag{3.25}$$

Another pair of constraints expresses the continuity of $\hat{w}$:

$$[\![\hat{w}]\!]_{\pm h} = 0. \tag{3.26}$$

Applying the constraints (3.25) and (3.26) to (3.24), we have

$$A_2 e^{-\tilde{k}h} = A_3 e^{\tilde{k}h} + A_4 e^{-\tilde{k}h}$$

and

$$A_3 e^{-\tilde{k}h} + A_4 e^{\tilde{k}h} = A_5 e^{-\tilde{k}h}.$$

Substituting these relations into (3.24) and redefining the constants A_3 and A_4 as $B_2 e^{-\tilde{k}h}$ and $B_1 e^{-\tilde{k}h}$, respectively, we can replace (3.24) with the compact form:

$$\hat{w}(z) = B_1 e^{-\tilde{k}|z+h|} + B_2 e^{-\tilde{k}|z-h|}. \tag{3.27}$$

The solution is thus a superposition of two functions, each peaked at one edge of the shear layer, a fact that will be of central importance later. To determine B_1 and B_2 we require jump conditions at $z = h$ and $z = -h$. We will first derive the *general* jump condition for a velocity kink, then apply it to the present case.

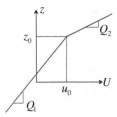

Figure 3.6 Velocity profile (3.28) with a single vorticity interface.

General Jump Condition at a Velocity Kink

Consider an arbitrary piecewise-linear velocity profile with a single kink. The velocity is continuous, but the shear (or vorticity) changes from Q_1 to Q_2 at $z = z_0$:

$$U(z) = u_0 + \begin{cases} Q_2(z - z_0), & z \geq z_0 \\ Q_1(z - z_0), & z \leq z_0 \end{cases} \tag{3.28}$$

The second-derivative is therefore

$$U_{zz} = (Q_2 - Q_1)\, \delta(z - z_0). \tag{3.29}$$

To capture the effect of the kink, we integrate the Rayleigh equation (3.19) over a thin layer from $z_0 - \epsilon$ to $z_0 + \epsilon$, then take the limit as $\epsilon \to 0$:

$$\lim_{\epsilon \to 0} \int_{z_0 - \epsilon}^{z_0 + \epsilon} (\hat{w}_{zz} - \tilde{k}^2 \hat{w})\,dz = \lim_{\epsilon \to 0} \int_{z_0 - \epsilon}^{z_0 + \epsilon} (Q_2 - Q_1)\, \delta(z - z_0)\, \frac{\hat{w}}{U - c}\,dz.$$

On the left-hand side, the first term integrates trivially; the result is the change in $\hat{w}_z$ between just above and just below z_0, a difference we write as

$$\lim_{\epsilon \to 0} \hat{w}_z \Big|_{z_0 - \epsilon}^{z_0 + \epsilon} \equiv [\![\hat{w}_z]\!]_{z_0}.$$

The second term is $\tilde{k}^2$ times the integral of $\hat{w}$ over a vanishingly small interval. Given that $\hat{w}$ is finite, this integral can only be zero.

On the right-hand side, the combination $\hat{w}(z)/[U(z) - c]$ is integrated with the delta function, picking out its value at $z = z_0$ (see property 6 of the delta function, listed on Figure 2.5).

Finally, noting that $Q_2 - Q_1 \equiv [\![Q]\!]_{z_0}$, the general jump condition is

$$\boxed{[U(z_0) - c]\,[\![\hat{w}_z]\!]_{z_0} = [\![Q]\!]_{z_0}\,\hat{w}(z_0).} \tag{3.30}$$

Exercise: Review section 2.2.4, the analysis of a density interface. Note the similarities between the solution (2.39) and the present (3.27), and also between the jump conditions (2.41) and (3.30).

Exercise: Convince yourself that, for the shear layer profile (3.23), $[[Q]]_{\pm h} = \mp u_0/h$. Now apply the jump conditions (3.30) to the solution (3.27).

Shear Layer Dispersion Relation

The two equations that determine B_1 and B_2 are:

$$\left[\frac{u_0}{h}e^{-2\tilde{k}h}\right]B_1 + \left[-2\tilde{k}(u_0 - c) + \frac{u_0}{h}\right]B_2 = 0;$$

$$\left[2\tilde{k}(u_0 + c) - \frac{u_0}{h}\right]B_1 + \left[-\frac{u_0}{h}e^{-2\tilde{k}h}\right]B_2 = 0. \tag{3.31}$$

The set is homogeneous and therefore its determinant must vanish. Solving the resulting equation for c gives the dispersion relation:

$$\boxed{\frac{c^2}{u_0^2} = \left(1 - \frac{1}{2\tilde{k}h}\right)^2 - \frac{e^{-4\tilde{k}h}}{4\tilde{k}^2h^2}.} \tag{3.32}$$

3.3.2 Interpreting the Results

Each $\tilde{k}h$ in (3.32) gives two solutions for c. Since the equation involves c^2 and not c, we know that either both values are real or both are imaginary (Figure 3.7a).

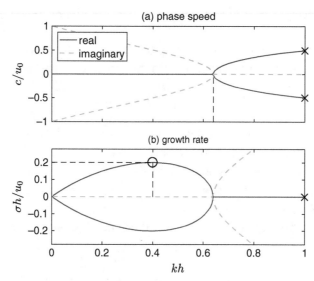

Figure 3.7 Nondimensional phase speed (a) and growth rate (b) for the linear shear layer as given by (3.32, 3.34). Crosses indicate the waves shown on Figure 3.8b,c (red and yellow curves). Circles represent the fastest-growing mode, located at $kh = 0.40$, $\ell h = 0$, $\sigma h/u_0 = 0.20$ (blue curve on Figure 3.8b,c).

Waves

If $c^2 > 0$, c can take either of two real values that are additive inverses. These represent neutrally stable waves with equal but opposite phase speeds. For example, in the limit $\tilde{k}h \to \infty$, (3.32) gives $c/u_0 \to \pm 1$. The positive value gives $B_1 = 0$, i.e., it corresponds to the first term in (3.27),[2] describing a function peaked at $z = +h$. Similarly, $c/u_0 = -1$ gives $B_2 = 0$, corresponding to the peak at $z = -h$.

If $\tilde{k}h$ is finite but sufficiently large, the second term on the right-hand side of (3.32) is negligible (because the exponential function goes to zero faster than any power of its argument) and

$$\frac{c}{u_0} \approx \pm\left(1 - \frac{1}{2\tilde{k}h}\right). \tag{3.33}$$

So, if we start at the right edge of Figure 3.7 and move to lower $\tilde{k}h$, the phase speeds of the oppositely propagating waves converge toward zero. The corresponding eigenfunctions are now combinations of the two terms in (3.27), as shown by the red and yellow curves on Figure 3.8b. Note that each of those functions is dominated by one peak but shows a slight contribution from the other.

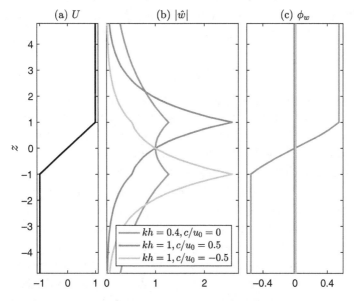

Figure 3.8 Eigenfunction $\hat{w}$ for the piecewise-linear shear layer defined by (3.24) and subsequent constraints. (a) Background velocity profile for reference. (b) Eigenfunction magnitudes for left- and right-going modes with $kh = 1$ (crosses on Figure 3.7) and the fastest-growing mode (circles on Figure 3.7). (c) Eigenfunction phases.

---

[2] To see this, divide (3.31) through by $\tilde{k}$ and set $\tilde{k}h$ to ∞.

Foreshadowing

An interesting interpretation of (3.33) is found by considering the upper and lower edges of the shear layer in isolation. First, suppose that there is no lower edge, i.e., the shear layer extends downward to $-\infty$. In that case, the bounded, continuous solution of the Rayleigh equation is given by the first term of (3.27) alone:

$$\hat{w}(z) = B_2 e^{-\tilde{k}|z-h|}.$$

Applying the jump condition (3.30) leads to the neutral wave solution

$$\frac{c}{u_0} = 1 - \frac{1}{2\tilde{k}h}.$$

Doing the same at the lower edge yields another wave with opposite phase speed:

$$\frac{c}{u_0} = -1 + \frac{1}{2\tilde{k}h}.$$

This pair of solutions is equivalent to (3.33), or to (3.32) without its final term. Therefore, the first term on the right-hand side of (3.32) describes neutral waves that would propagate on each edge of the shear layer if the other edge were not present. The second term, then, can be interpreted as describing the interaction of those two waves. We will have considerably more to say about this in section 3.12.

Instabilities

For $\tilde{k}h < 0.64$ the phase speeds are zero, and the wavelike solutions are replaced by a pair of exponential solutions with imaginary c but real σ (Figure 3.7b). Like the wave solutions, the exponential solutions occur in pairs, now with equal and opposite growth rate.[3] Both the growing mode ($\sigma_r > 0$) and the decaying mode ($\sigma_r < 0$) are classified as stationary – the pattern does not move to the left or the right in a coordinate frame fixed at the center of the shear layer. A mode that is not stationary is called "oscillatory" (cf. discussion in section 2.2.2), because a measurement made at a fixed position oscillates as the disturbance propagates past.

The eigenfunction for the fastest-growing mode (FGM; see section 2.2.2) is shown in Figure 3.8 by the blue curves. In contrast to the wave modes, the eigenfunction is symmetric about $z = 0$, having peaks of equal amplitude on the upper and lower edges of the shear layer. Also in contrast to the wave modes, the unstable mode has vertically variable phase (Figure 3.8c). The significance of this will be explored later in section 3.11.4.

---

[3] This is actually a general property of the Rayleigh equation (3.19), as we will see in section 3.4.

To better understand the factors that govern instability, we now rewrite (3.32) in terms of the growth rate by substituting $c = i\sigma/k$, resulting in

$$\sigma = \frac{k}{\tilde{k}} \frac{u_0}{h} f(\tilde{k}h), \quad \text{where } f(x) = \sqrt{\frac{e^{-4x}}{4} - \left(x - \frac{1}{2}\right)^2}. \qquad (3.34)$$

This expression for σ has three factors: $k/\tilde{k}$, u_0/h, and $f(\tilde{k}h)$, each of which represents an important influence on growth. We'll discuss these in turn.

(i) Recall that the angle of obliquity φ is the angle between the wave vector and the mean flow (Figure 3.4). For 2D modes (those having $\varphi = 0$), the wave crests are perpendicular to the mean flow. The first factor in (3.34) is

$$\frac{k}{\tilde{k}} = \cos \varphi.$$

This factor tells us that growth is optimized when $\varphi = 0$, i.e., for 2D modes.
(ii) The factor u_0/h tells us that the growth rate is proportional to the shear.
(iii) The function $f(\tilde{k}h)$ is positive in the range $0 < \tilde{k}h < 0.64$ with a single peak $f(0.40) = 0.20$ (Figure 3.7b). This tells us that the shear layer is always unstable, i.e., there is always some value of (k, ℓ) for which $\sigma_r > 0$.

3.3.3 *"Rules of Thumb" and the Critical State*

Based on these results, we can state four rules of thumb regarding the fastest-growing instability of a piecewise-linear shear layer:

(i) The angle of obliquity is zero, i.e., the wave crests are perpendicular to the mean flow.
(ii) The growth rate is proportional to the shear: $\sigma = 0.20u_0/h$.
(iii) The nondimensional wavenumber is $kh = 0.40$ or, equivalently, the wavelength $\lambda = 2\pi/k = 15.7h$, or about 8 times the thickness of the shear layer.
(iv) The disturbance travels with the speed of the background flow at the center of the shear layer.

The "critical state" for this flow is just $u_0 = 0$. Because there is no mechanism to damp the instability (e.g., viscosity and diffusion, as in the Rayleigh-Benard problem), (3.23) is unstable for any $u_0 \neq 0$.

The piecewise-linear shear layer examined in this section is only one example of the infinite variety of parallel shear flows that are important in nature. In upcoming sections, we will explore analogs of the "rules of thumb" listed above, and more, that apply to *all* parallel shear flows.

3.4 Solution Types for Rayleigh's Equation

We have seen that the dispersion relation (3.32) for the piecewise-linear shear layer (3.23) admits two solution types: real c and oppositely signed pairs of imaginary c. In fact, values of c occur in complex conjugate pairs regardless of the form of $U(z)$. To see this, note that the complex conjugate of (3.19) is

$$\hat{w}^*_{zz} + \left\{ \frac{U_{zz}}{U - c^*} + k^2 \right\} \hat{w}^* = 0$$

(assuming that the wavenumber is real). This is equivalent to (3.19) with c and $\hat{w}$ replaced by their complex conjugates. Therefore, if $[c, \hat{w}]$ is a solution of the Rayleigh equation, then $[c^*, \hat{w}^*]$ is also a solution, and as a result we will obtain either wavelike solutions with c purely real or pairs of solutions in which one grows and the other decays. *In neither case is the flow actually stable*, in the sense that the perturbed flow returns to its equilibrium state. If $c_i \neq 0$, the flow is unstable; if $c_i = 0$, the disturbance oscillates with constant amplitude, i.e., it is neutrally stable.

3.5 Numerical Solution of Rayleigh's Equation

Rayleigh's equation may be solved analytically for certain very simple cases, like the piecewise-linear shear layer of the previous section. In general, though, it must be solved numerically. This requirement is most obvious when $U(z)$ is a velocity profile derived from direct measurements and can therefore be almost arbitrarily complicated (e.g., Figure 3.9). Here we will convert Rayleigh's equation to discretized form, then discuss its numerical solution.

3.5.1 Discretization and the Generalized Eigenvalue Problem

As in section 1.4.2, we place grid points at

$$z_i = i\Delta; \quad i = 0, 1, 2, \ldots, N, N+1,$$

with z_0 and z_{N+1} located at the boundaries. At the interior points $z_1, z_2, \ldots, z_N$ the solution is

$$\hat{w}_i = \hat{w}(z_i); \quad i = 1, 2, \ldots, N.$$

We convert (3.16) into an algebraic equation by discretizing the derivatives. In this case the only derivative is d^2/dz^2, and we discretize it using the second-order derivative matrix $\mathsf{D}^{(2)}$ as defined in (1.13). The derivative matrix may incorporate the impermeable boundary conditions $\hat{w}_0 = \hat{w}_{N+1} = 0$ as in section 1.4.3. Other choices for the boundary conditions will be discussed later (section 3.5.3). With

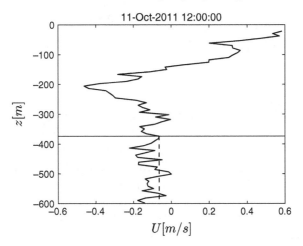

Figure 3.9 Velocity profile measured in the Indian Ocean at the equator at 80.5E longitude (south of Sri Lanka). Total ocean depth is about 4000 m. The horizontal line indicates a plausible depth for a virtual boundary. The fine print: a realistic analysis of this oceanic regime would require inclusion of stratification, viscosity and diffusion (Chapter 6); it is included here only to illustrate a virtual boundary. Data courtesy of Jim Moum, Oregon State University.

the derivative matrix defined, the Laplacian operator (3.17) becomes a matrix that we'll call A:

$$\nabla^2 \;\rightarrow\; D_{ij}^{(2)} - \tilde{k}^2 I_{ij} \;=\; A_{ij}. \qquad (3.35)$$

The symbol I represents the identity matrix (sometimes called the Kronecker delta).

Next we discretize the velocity profile and its second-derivative to form the vectors $\vec{U}$ and $\vec{U}''$:

$$U_i \;=\; U(z_i); \quad U_i'' \;=\; \left.\frac{d^2 U}{dz^2}\right|_{z=z_i}.$$

We can now write the Rayleigh equation in the form (3.16) as

$$\sigma A_{ij}\hat{w}_j \;=\; -\,\iota k U_i A_{ij}\hat{w}_j + \iota k U_i'' I_{ij}\hat{w}_j \quad \text{(with no sum on } i\text{)}.$$

Defining a second matrix B as

$$B_{ij} \;=\; -\,\iota k U_i A_{ij} + \iota k U_i'' I_{ij}, \quad \text{(no sum on } i\text{)}, \qquad (3.36)$$

the equation becomes

$$\sigma A_{ij}\hat{w}_j \;=\; B_{ij}\hat{w}_j. \qquad (3.37)$$

This is a generalized eigenvalue problem, with eigenvalue σ and eigenvector components $\hat{w}_j$.

3.5.2 Digital Implementation

The eigenvalue problem (3.37) is easily solved using the Matlab function *eig*:

$$[\hat{w}, \sigma] = \text{eig(B,A)}. \tag{3.38}$$

or the equivalent in any other programming environment.

The most practical approach is to write a subroutine that assembles the matrices A and B and then solves the eigenvalue problem as in (3.38). In due course you will be shown explicitly how to do this, but your understanding will be much deeper if you try it yourself first. Here are some coding hints.

- The routine should have the following inputs: `z, U, k, l`.
- Define the second-derivative matrix `D2` using the subroutine you developed earlier: `ddz2(z)`. For later convenience, design that subroutine using one-sided derivatives for the top and bottom rows. That gives you an easy way to differentiate functions that do not obey boundary conditions, for example: `Uzz=ddz2(z)*U`. When you are ready to incorporate boundary conditions (section 1.4.3), define `D2=ddz2(z)` then replace the top and bottom rows of `D2` according to the boundary conditions you have chosen (more on this in section 3.5.3).
- Define the identity matrix using `I=eye(N)`.
- Compute $\tilde{k}$ as `kt=sqrt(k^2+l^2)`.Then compute the Laplacian matrix `A=D2-kt^2*I`.
- The multiplications in (3.36) are a bit unusual. To compute $U_i A_{ij}$ with no sum on i, each row of A is multiplied by the corresponding element of $\vec{U}$. As a simple example:

$$\vec{U} \cdot \mathsf{A} = \begin{bmatrix} U_1 A_{11} & U_1 A_{12} & U_1 A_{13} \\ U_2 A_{21} & U_2 A_{22} & U_1 A_{23} \\ U_3 A_{31} & U_3 A_{32} & U_3 A_{33} \end{bmatrix}.$$

This can be written as a standard matrix multiplication:

$$\vec{U} \cdot \mathsf{A} = \begin{bmatrix} U_1 & 0 & 0 \\ 0 & U_2 & 0 \\ 0 & 0 & U_3 \end{bmatrix} \begin{bmatrix} A_{11} & A_{12} & A_{13} \\ A_{21} & A_{22} & A_{23} \\ A_{31} & A_{32} & A_{33} \end{bmatrix}.$$

The diagonal matrix can be formed using the Matlab function *diag*: `diag(U)*A`. Similarly, $U_i'' I_{ij}$ can be coded as `diag(Uzz)*I`, or just `diag(Uzz)`.

3.5.3 Boundary Conditions: Impermeable and Asymptotic

Rayleigh's equation is often solved with the impermeability condition imposed at horizontal upper and lower boundaries. For a discretized normal mode solution, that boundary condition is just

$$\hat{w}_0 = \hat{w}_{N+1} = 0. \tag{3.39}$$

Approximating the second-derivative using the second-order difference formula

$$\hat{w}_i'' = \frac{\hat{w}_{i-1} - 2\hat{w}_i + \hat{w}_{i+1}}{\Delta^2}, \tag{3.40}$$

the first and last cases are

$$\hat{w}_1'' = \frac{-2\hat{w}_1 + \hat{w}_2}{\Delta^2}; \quad \text{and} \quad \hat{w}_N'' = \frac{\hat{w}_{N-1} - 2\hat{w}_N}{\Delta^2}.$$

The expressions define the top and bottom rows of the second-derivative matrix $\mathsf{D}^{(2)}$.

Finding eigenvalues numerically can be very time-consuming; the time needed to analyze an $N \times N$ matrix is typically proportional to N^2. This can make matrix stability analysis numerically intractable if N exceeds a few hundred. It is therefore important to avoid using large values of N. We next discuss one useful strategy: the application of asymptotic boundary conditions at virtual boundaries. Further strategies are discussed in Chapter 13.

Example: the Ocean Surface Mixed Layer and the Bottom Boundary

In nature, shear instability often occurs in a localized layer far from any boundary, and boundary effects are therefore likely to be negligible. For example, measurements in the upper ocean usually reveal flow features on vertical scales of 10 m or less (e.g., Figure 3.9). The bottom boundary may lie 4000 m or more below this. To include the entire ocean depth in a numerical calculation is impractical. To resolve the shear layer would require $\Delta \sim 1m$ or less, and we would therefore need $N \sim 4000$ grid points, i.e., we would have to calculate the eigenvalues of a 4000×4000 matrix – a very slow process!

One alternative is to place a fictitious lower boundary well below the shear layer. If that boundary is far enough from the shear layer that the vertical velocity perturbation associated with any instability that may emerge is negligible, then the boundary should have negligible effect on the solution. In the ocean example, one could easily imagine that a boundary placed, say, 1000 m below the shear layer would have negligible effect on the results, reducing N to ~ 1000. The assumption can be tested by repeating the calculation with boundaries placed successively farther from shear layer and checking that the results converge.

Another strategy is to assume that the boundary is actually at infinity. In that case, the impermeable boundary condition is replaced by the requirement that the solution remain bounded as $z \to \pm\infty$. This is just what we did in the analytical example of the piecewise-linear shear layer (section 3.3). If U is assumed to approach a constant value far from the layer of interest (e.g., below the horizontal line in Figure 3.9), then the solution to the Rayleigh equation decays with depth in proportion to $e^{\tilde{k}z}$ (cf. 3.24 and 3.25). In a numerical solution, this requirement can be enforced by imposing the boundary condition

$$\hat{w}_z = \tilde{k}\hat{w}$$

at a fictitious boundary chosen reasonably far from the region of interest. This is called an asymptotic boundary condition. It generally has less impact on the solution than an impermeable boundary, and it therefore allows us to use a smaller domain and hence smaller N.

Implementation

Suppose that $U(z)$ varies only in a limited range of z, say $z_B \leq z \leq z_T$, and is constant in the semi-infinite regions $z < z_B$ and $z > z_T$. In either of those outer regions, $U_{zz} = 0$ and the Rayleigh equation (3.19) becomes

$$\hat{w}_{zz} - \tilde{k}^2 \hat{w} = 0,$$

with general solution

$$\hat{w} = A e^{\tilde{k}z} + B e^{-\tilde{k}z}.$$

In the upper region $z > z_T$, the solution is unbounded unless $A = 0$, hence

$$\hat{w} = B e^{-\tilde{k}z}, \quad \text{for } z > z_T. \tag{3.41}$$

We can ensure that our computed solution matches smoothly with (3.41) by imposing the condition

$$\boxed{\hat{w}_z = -\tilde{k}\hat{w} \quad \text{at } z = z_T.} \tag{3.42}$$

Similarly, in the lower layer $z < z_B$, we match to the bounded solution $\hat{w} = A e^{\tilde{k}z}$ by requiring

$$\boxed{\hat{w}_z = \tilde{k}\hat{w} \quad \text{at } z = z_B.} \tag{3.43}$$

To implement asymptotic boundary conditions in a numerical calculation, we approximate (3.42) and (3.43) to second order in Δ by

$$\hat{w}_1' = \frac{\hat{w}_2 - \hat{w}_0}{2\Delta} = \tilde{k}\hat{w}_1 \, ; \quad \hat{w}_N' = \frac{\hat{w}_{N+1} - \hat{w}_{N-1}}{2\Delta} = -\tilde{k}\hat{w}_N,$$

and therefore

$$\hat{w}_0 = \hat{w}_2 - 2\Delta\tilde{k}\hat{w}_1 ; \quad \hat{w}_{N+1} = \hat{w}_{N-1} - 2\Delta\tilde{k}\hat{w}_N. \tag{3.44}$$

Now approximate the second-derivative using the second-order difference formula (3.40) and substitute (3.44) to define the top and bottom rows of the derivative matrix:

$$\mathsf{D}^{(2)} = \frac{1}{\Delta^2} \begin{bmatrix} -2-2\Delta\tilde{k} & 2 & 0 & 0 & \cdots \\ 1 & -2 & 1 & 0 & \cdots \\ 0 & 1 & -2 & 1 & \cdots \\ & & \ddots & & \\ \cdots & 0 & 1 & -2 & 1 \\ \cdots & & 2 & -2-2\Delta\tilde{k} \end{bmatrix} \tag{3.45}$$

The best way to code the boundary conditions is to begin with a subroutine that calculates the second-derivative matrix $\mathsf{D}^{(2)}$ using one-sided derivatives for the top and bottom rows. After that subroutine is called, replace the top and bottom rows of $\mathsf{D}^{(2)}$ with appropriate values for the boundary conditions you have chosen. In the case of the asymptotic boundary conditions, the Matlab code would look something like this:

```
D2 = ddz2(z);   %2nd derivative matrix with one-sided
derivatives at the boundaries
del = z(2)-z(1);
D2(1,:)=0;  ...
D2(1,:)=0; D2(1,1)= -2*(1+del*kt) / del^2; D2(1,2)=
2 / del^2;
D2(N,:)=0; D2(N,N)= -2*(1+del*kt) / del^2; D2(N,N-1)=
2 / del^2.
```

3.6 Shear Scaling

Scaling allows us to investigate the stability of an infinite class of flows all at once. In Chapter 2, we used diffusive scaling, together with the concepts of isomorphic equations and solution algorithms, to arrive at some very general conclusions about convective instability (see section 2.4.1 for a detailed description of diffusive scaling). Here, we will describe a different scaling that is useful for parallel shear flows.

Consider a class of parallel shear flows of the form

$$U(z) = u_0 f(z/h), \tag{3.46}$$

where u_0 and h are constants and f is an arbitrary function. We can represent $U(z)$ in the scaled form:

$$U^* = f(z^*),$$

where

$$U^* = \frac{U}{u_0}; \quad z^* = \frac{z}{h}. \tag{3.47}$$

For example, the piecewise-linear shear layer (3.23) would become

$$U^* = \begin{cases} 1, & z^* \geq 1 \\ z^*, & -1 < z^* < 1 \\ -1, & z^* \leq -1 \end{cases} \tag{3.48}$$

Note that this scaling (like the diffusive scaling) yields *nondimensional* forms of U and z.

Now suppose we want to analyze the stability of some class of profiles of the form (3.46) using the Rayleigh equation. We'll use the form (3.18):

$$(\sigma + \iota k U) \left(\frac{d^2}{dz^2} - \tilde{k}^2 \right) \hat{w} = \iota k \frac{d^2 U}{dz^2} \hat{w} \tag{3.49}$$

and assume that we have a solution algorithm

$$\sigma = \mathcal{F}(z, U; k, \ell). \tag{3.50}$$

Our goal is to express (3.49) in a scaled form. We already have scaled forms for z and U (3.47); we now define scaled versions of the other variables appearing in (3.49) using the same velocity and length scales:

$$\sigma = \sigma^* \frac{u_0}{h}$$
$$\hat{w} = \hat{w}^* u_0$$
$$\{\tilde{k}, k, \ell\} = \{\tilde{k}^*, k^*, \ell^*\}/h$$
$$\frac{d}{dz} = \frac{1}{h} \frac{d}{dz^*}; \quad \nabla^2 = \frac{1}{h^2} \left(\frac{d^2}{dz^{*2}} - \tilde{k}^{*2} \right). \tag{3.51}$$

Substituting the scaling transformations (3.47) and (3.51) and multiplying by h^3/u_0^2 yields

$$(\sigma^* + \iota k^* U^*) \left(\frac{d^2}{dz^{*2}} - \tilde{k}^{*2} \right) \hat{w}^* = \iota k^* \frac{d^2 U^*}{dz^{*2}} \hat{w}^*. \tag{3.52}$$

Comparison of (3.52) and (3.49) shows that (3.52) $\leftrightarrow$ (3.49), and the solution algorithm is therefore the same:

$$\sigma^* = \mathcal{F}(z^*, U^*; k^*, \ell^*). \tag{3.53}$$

The shear scaling is both a labor-saving device (you can analyze a whole class of flows at once) and a source of insight. For any class of velocity profiles like (3.46), the fastest-growing mode has a scaled growth rate $\sigma^\star$ and a scaled wave vector $(k^\star, \ell^\star)$, both of which can be calculated using $\mathcal{F}$. For example, in the case of the piecewise-linear shear layer (section 3.3), $\sigma^\star = 0.20$ and $(k^\star, \ell^\star) = (0.40, 0)$.

In dimensional terms, we can draw the following general conclusions from (3.51):

- For every velocity profile of the form (3.46), the fastest growth rate is proportional to the "characteristic shear" u_0/h, with proportionality constant $\sigma^\star$.
- The wavelength $2\pi/\tilde{k}$ of the FGM is proportional to h, with proportionality constant $2\pi/\tilde{k}^\star$.

Admonition: Suppose that the solution algorithm $\mathcal{F}$ is the subroutine described in section 3.5.2. For practical applications, avoid the temptation to write this subroutine in terms of scaled variables. The reason is that you will want to use other scalings in the future. In preparation for that, use the original, dimensional form of the variables, so that the subroutine has the form (3.50). Then, if you want to use shear scaling, call the subroutine using the scaled input variables as in (3.53), and keep in mind that the output will be in scaled form.

Re-dimensionalization

If the solution algorithm described above is used with scaled variables as in (3.53), the results can then be re-dimensionalized to apply to specific situations using (3.47) and (3.51). For example, the piecewise-linear shear layer (3.48) yields instability with scaled growth rate $\sigma^\star = 0.20$ at wavenumber $k^\star = 0.40$ (section 3.9.1). Suppose we want to apply this to a shear layer with half-thickness $h = 2$ m and half velocity change $u_0 = 0.5$ m/s. We would predict a growth rate of $\sigma = \sigma^\star u_0/h = 0.20 \times 0.5 \text{ ms}^{-1}/2 \text{ m} = 0.05 \text{ s}^{-1}$. The e-folding time σ^{-1} is 20 s. The wavelength becomes $2\pi/h = (2\pi/0.40) \times 2 \text{ m} = 31$ m.

3.7 Oblique Modes and Squire Transformations

In Chapter 2, we found that the growth rate of convective instability does not depend on k and ℓ individually, but only on their combination $\tilde{k}$. That is *not* the case for parallel shear flows; the direction of the wave vector (k, ℓ) relative to the mean flow matters a lot. Happily, we can draw some general conclusions about this dependence that will spare us from considering every combination of k and ℓ separately.

In the example of the piecewise-linear shear layer (3.23), we found that the fastest-growing mode is 2D. According to (3.34) and the discussion that follows it, if you take any 2D mode and rotate the wave vector to an angle φ from the mean flow, the growth rate is reduced by a factor $\cos \varphi$. We will now show that this

behavior is not peculiar to the linear shear layer; *it is true for any velocity profile.*
To demonstrate this powerful result, we introduce Squire transformations.

Recall Rayleigh's equation (3.49, or 3.18):

$$(\sigma + \imath k U)\left(\frac{d^2}{dz^2} - \tilde{k}^2\right)\hat{w} = \imath k U_{zz}\hat{w}, \tag{3D}$$

and suppose again that we have a solution algorithm

$$\sigma = \mathcal{F}(z, U; \, k, \ell). \tag{σ3D}$$

The special case of a 2D mode is defined by setting $\ell = 0$ and therefore $\tilde{k} = k$:

$$(\sigma + \imath k U)\left(\frac{d^2}{dz^2} - k^2\right)\hat{w} = \imath k U_{zz}\hat{w}. \tag{2D}$$

We can find σ for this class of modes by setting $\ell = 0$ in the solution algorithm:

$$\sigma_{2D} = \mathcal{F}(z, U; \, k, 0). \tag{σ2D}$$

Now, go back to (3D) and substitute the Squire transformations $k = \tilde{k}\cos\varphi$ and $\sigma = \tilde{\sigma}\cos\varphi$. Dividing out the common factor $\cos\varphi$, we have

$$(\tilde{\sigma} + \imath\tilde{k}U)\left(\frac{d^2}{dz^2} - \tilde{k}^2\right)\hat{w} = \imath\tilde{k}U_{zz}\hat{w}. \tag{$\widetilde{3D}$}$$

The form $(\widetilde{3D})$ is valid for a general 3D mode, but it is also *isomorphic* to the special case (2D):

$$(\widetilde{3D}) \leftrightarrow (2D), \text{ under } \tilde{\sigma} \to \sigma, \tilde{k} \to k.$$

This means that we can use the same solution algorithm as for the 2D case (σ2D):

$$\tilde{\sigma} = \mathcal{F}(z, U; \, \tilde{k}, 0). \tag{$\sigma\widetilde{3D}$}$$

or

$$\sigma = \cos\varphi \, \mathcal{F}(z, U; \, \tilde{k}, 0).$$

So, suppose we have a 2D mode with growth rate σ_{2D}. Now rotate the wave vector by an angle φ. The resulting oblique mode will have growth rate $\sigma_{2D}\cos\varphi$. Conversely, *for every oblique mode with wave vector (k, ℓ) and growth rate σ, there is a corresponding 2D mode $(\tilde{k}, 0)$ with higher growth rate $\tilde{\sigma} = \sigma/\cos\varphi$ (Figure 3.10)*. As a consequence, the fastest-growing mode is always 2D.

Like the shear scaling discussed in the previous section, the Squire transformation is a labor saver. If we have some arbitrary flow profile $U(z)$, and we want to know the growth rate for all k, ℓ, we need only calculate the growth rate for the 2D cases, i.e., $\ell = 0$, then for any $\ell \neq 0$ simply multiply the result by $\cos\varphi$. Better still, if we just want to find the FGM, we need only search the 2D cases.

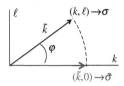

Figure 3.10 Black: wave vector (k, ℓ) of an oblique mode with growth rate σ. Blue: wave vector $(\tilde{k}, 0)$ of the corresponding 2D mode, with growth rate $\tilde{\sigma} = \sigma / \cos \varphi \geq \sigma$.

3.8 Rules of Thumb for a General Shear Instability

Based on sections 3.6 and 3.7, we can now list three rules that apply to the fastest-growing instability of *every* parallel shear flow:

(i) The fastest-growing mode has wave vector parallel to the mean flow.
(ii) The growth rate is proportional to u_0/h.
(iii) The wavelength is proportional to h.

Rules (i–iii) for the piecewise-linear shear layer (section 3.3.3) are a special case of these.

3.9 Numerical Examples

Here we look at two model shear flows (Figure 3.11) for which U is a smooth function of z and the Rayleigh equation must be solved numerically.

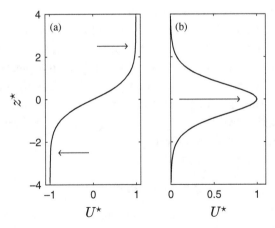

Figure 3.11 Background flow profiles for (a) the hyperbolic tangent shear layer, $U^\star = \tanh z^\star$ and (b) the Bickley jet, $U^\star = \operatorname{sech}^2 z^\star$.

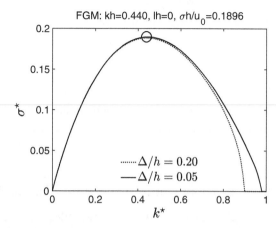

Figure 3.12 Growth rate versus wavenumber for $U^* = \tanh z^*$ using two levels of grid resolution. Asymptotic boundary conditions are employed at $z^* = \pm 4$. In the exact solution, the growth rate drops to zero at $k^* = 0$ and $k^* = 1$. Circle shows the fastest-growing mode.

3.9.1 The Hyperbolic Tangent Shear Layer

We consider a smoother version of the piecewise-linear shear layer (Figure 3.11a, cf. Figure 3.5). The velocity profile is modeled by a hyperbolic tangent function $U^* = \tanh z^*$. There is no analytical solution, but it can be shown that the growth rate is nonzero for $0 < k^* < 1$ and $\ell^* = 0$ (e.g., section 4.4).

The growth rate is computed numerically for $\ell^* = 0$ and a range of k^* (Figure 3.12). To test for numerical convergence, the computation is repeated at two values of the grid spacing Δ^* (solid and dotted curves). The difference is greatest as $k^* \to 1$. In the coarsely resolved calculation (dotted), the growth rate drops to zero around $k^* = 0.9$, in contrast to the exact value $k^* = 1$. The finely resolved calculation (solid) approximates the exact result more closely. If only the fastest-growing mode is needed, you might decide that the coarser resolution is sufficient.

The result shown in Figure 3.12 is comparable to the piecewise-linear shear layer: the growth rate rises to a peak around $k^* = 0.44$, and the maximum value is $\sigma^* = 0.19$. Recall that, for the piecewise-linear shear layer, we got 0.40 and 0.20, respectively (Figure 3.7b). The ratio of wavelength to shear layer thickness for the hyperbolic tangent shear layer is about $(2\pi/0.44)/2 = 7$.

3.9.2 The Bickley Jet

The Bickley jet, $U^* = \mathrm{sech}^2 z^*$, is a common model for both jets and wakes (Figure 3.11b). An example is the atmospheric island wake shown in Figure 3.13. The fastest-growing mode has scaled wavenumber $k^* = 0.9$ and growth rate

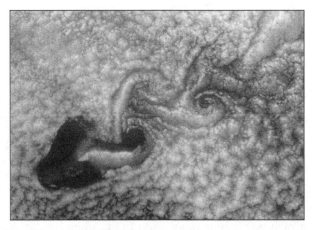

Figure 3.13 Vortex street in the lee of Guadalupe Island in the eastern Pacific. The vortices are formed by the same shear instability that causes a flag to flutter in the wind, the sinuous instability of a plane jet. (NASA)

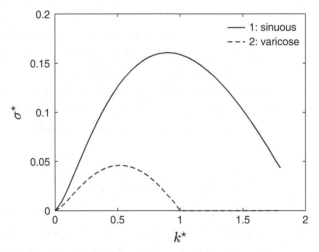

Figure 3.14 Growth rate versus wavenumber for the Bickley jet $U^\star = \mathrm{sech}^2 z^\star$. Asymptotic boundary conditions are employed at $z^\star = \pm 4$. Solid and dashed curves show the sinuous and varicose modes, respectively.

$\sigma^\star = 0.16$ (Figure 3.14). The wavelength is usefully expressed as an aspect ratio, more specifically as a multiple of the jet width $2h$:

$$\frac{\lambda}{2h} = \frac{(2\pi/k^\star)h}{2h} = \frac{\pi}{0.9} = 3.5.$$

This aspect ratio can be compared with Figure 3.13: take the jet width to be the width of the island and λ to be the wavelength of the instability. It is left to the reader to judge whether that ratio compares favorably with our theoretical value.

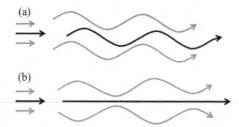

Figure 3.15 Streamlines distinguishing the sinuous (a) and varicose (b) modes of a plane jet. For the sinuous mode (a), the alternating regions of widely separated streamlines correspond to the alternating vortices in the island wake shown in Figure 3.13.

The scaled growth rate 0.16 is quite close to the value for the shear layer, 0.19. This is not an accident; the jet may be thought of as a pair of shear layers set back to back. The instability is then similar to a pair of shear layer instabilities staggered so as to form a vortex street as in Figure 3.13. This jet instability is called the "sinuous mode" (Figure 3.15a).

The dashed curve in Figure 3.14 indicates a second mode of instability called the "varicose mode." This disturbance is also similar to a pair of shear layer instabilities, but in this case the resulting vortices are not staggered but are instead arranged side by side so that the streamlines alternately bulge and constrict (Figure 3.15b). Upper and lower motions oppose one another, so that the varicose mode grows much more slowly than its sinuous counterpart and is seldom seen in naturally occurring flows.

3.10 Perturbation Energetics

The perturbation analyses that we have described so far can tell us whether or not a given flow is unstable and allow us to calculate the length and time scales of the unstable modes. They do not, however, tell us in any intuitive, physical sense how the instability works. To understand instability on this intuitive level, several kinds of auxiliary analysis are helpful. For example, it is almost always enlightening to look at the processes that control the energy of the instability. Here, we will do this for the case of kinetic energy.

A parallel shear flow has kinetic energy. An unstable perturbation grows by accessing that energy and converting it to perturbation kinetic energy. Our goals in this section are to (1) understand the processes by which that energy conversion happens, and (2) learn to quantify those processes via numerical calculations.

3.10.1 *Kinetic Energy Evolution in a General Disturbance*

For a general physical system, the evolution equation for the kinetic energy is obtained by dotting the velocity vector onto Newton's second law. Here, we apply this formalism to explore the growth mechanisms for perturbations to a homogeneous, inviscid, parallel shear flow $U(z)$.

The perturbation momentum equation is

$$\frac{\partial \vec{u}'}{\partial t} + U(z)\frac{\partial \vec{u}'}{\partial x} + w'\frac{dU}{dz}\hat{e}^{(x)} = -\vec{\nabla}\pi'$$

[reproducing (3.7)]. Dotting with $\vec{u}'$, we have

$$\vec{u}' \cdot \frac{\partial \vec{u}'}{\partial t} + U(z)\vec{u}' \cdot \frac{\partial \vec{u}'}{\partial x} + w'\frac{dU}{dz}\vec{u}' \cdot \hat{e}^{(x)} = -\vec{u}' \cdot \vec{\nabla}\pi',$$

or

$$\frac{\partial}{\partial t}\left(\frac{\vec{u}' \cdot \vec{u}'}{2}\right) + U(z)\frac{\partial}{\partial x}\left(\frac{\vec{u}' \cdot \vec{u}'}{2}\right) + u'w'\frac{dU}{dz} = -\vec{\nabla} \cdot (\vec{u}'\pi'). \qquad (3.54)$$

[To compute the right-hand side of (3.54), we have used both the product rule for the divergence of the vector-scalar product $\vec{u}'\pi'$ and the fact that $\vec{\nabla} \cdot \vec{u}' = 0$.]

Now we apply a horizontal average, to be denoted by an overbar. For this discussion, we assume that there is no dependence on y, so $v' = 0$ and $\partial/\partial y = 0$. (The analysis is easily extended to include y-dependence, the only expense being more complicated algebra; see section 5.9.) Therefore, the horizontal average is taken over x only. Note that $\partial/\partial x$ of any averaged quantity will be zero.

After averaging, the first term on the left-hand side of (3.54) is $\partial K/\partial t$, where

$$K(z, t) = \frac{1}{2}\overline{\vec{u}' \cdot \vec{u}'} = \frac{1}{2}(\overline{u'u'} + \overline{w'w'}) \qquad (3.55)$$

is the horizontally averaged perturbation kinetic energy per unit mass. The second term is

$$U\frac{\partial K}{\partial x} = 0.$$

The third term does not simplify; it's just

$$\overline{u'w'}\frac{dU}{dz}.$$

Finally, the right-hand side becomes

$$-\frac{\partial}{\partial z}\overline{w'\pi'}.$$

The result is an equation describing the evolution of the horizontally averaged perturbation kinetic energy:

$$\boxed{\frac{\partial K}{\partial t} = SP - \frac{\partial}{\partial z}EF,} \qquad (3.56)$$

where

$$SP = -\frac{dU}{dz}\overline{u'w'} \tag{3.57}$$

represents the shear production, the rate at which kinetic energy is transferred from the mean flow to the disturbance, and

$$EF = \overline{w'\pi'} \tag{3.58}$$

represents a vertical flux of kinetic energy. The energy flux transports kinetic energy in the vertical, and its convergence (or divergence) at a given z causes energy to accumulate (or be depleted) at that height.

Equation (3.56) shows that the evolution of perturbation kinetic energy at a given height z is driven by the combination of (1) production by the shear production term and (2) the convergence or divergence of EF. Note that the flux vanishes at the boundaries because $w' = 0$ there. (This is true for impermeable boundaries and also for boundaries at infinity.) Therefore, the vertical integral of $\partial(EF)/\partial z$ over the entire domain is zero, and

$$\frac{d}{dt}\int K dz = \int SP dz. \tag{3.59}$$

This tells us that only shear production actually creates perturbation kinetic energy; the energy flux just moves it around. The shear production, as defined in (3.57), is revealed as the critical quantity for determining shear instability. We'll take advantage of this fact later.

3.10.2 Kinetic Energy Evolution in a Normal Mode Instability

By plotting the various budget quantities as functions of z, we may gain insight into the processes that drive instability growth. To calculate these quantities, we first substitute the normal mode form (3.14):

$$w' = \{\hat{w}(z)e^{\sigma t+\iota kx}\}_r, \tag{3.60}$$

where the real part has been specified explicitly. Normal mode expressions like (3.60) also describe the remaining variables u' and π'. The vertical structure functions are obtained in terms of $\hat{w}$ by simplifying (3.20) and (3.22) for the special case $\ell = 0$:

$$\hat{u} = \frac{\iota}{k}\frac{d\hat{w}}{dz}, \tag{3.61}$$

$$\hat{\pi} = -(\sigma + \iota kU)\frac{1}{k^2}\frac{d\hat{w}}{dz} + \frac{\iota}{k}\frac{dU}{dz}\hat{w}. \tag{3.62}$$

We're now ready to evaluate the terms in (3.56). We begin with $\overline{u'w'}$. Recall that the real part of any complex quantity a can be written as $(a + a^*)/2$, where the asterisk represents the complex conjugate. Applying this to w' and u' and averaging, we have:

$$\overline{u'w'} = \frac{k}{2\pi} \int_0^{2\pi/k} \frac{1}{2} \left\{ \hat{u} e^{\sigma t} e^{\imath kx} + \hat{u}^* e^{\sigma^* t} e^{-\imath kx} \right\} \times \frac{1}{2} \left\{ \hat{w} e^{\sigma t} e^{\imath kx} + \hat{w}^* e^{\sigma^* t} e^{-\imath kx} \right\} dx$$

$$= \frac{k}{8\pi} \int_0^{2\pi/k} \left\{ \hat{u} \hat{w} e^{2\sigma t} e^{2\imath kx} + \hat{u} \hat{w}^* e^{(\sigma + \sigma^*)t} + \hat{u}^* \hat{w} e^{(\sigma + \sigma^*)t} + \hat{u}^* \hat{w}^* e^{2\sigma^* t} e^{-2\imath kx} \right\} dx.$$

In the final integral, the first and last terms integrate to zero. The second and third terms are complex conjugates and do not depend on x, so

$$\overline{u'w'} = \frac{k}{8\pi} \int_0^{2\pi/k} \left\{ \hat{u} \hat{w}^* e^{(\sigma + \sigma^*)t} + \hat{u}^* \hat{w} e^{(\sigma^* + \sigma)t} \right\} dx$$

$$= \frac{1}{4} \left\{ \hat{u} \hat{w}^* e^{(\sigma + \sigma^*)t} + \hat{u}^* \hat{w} e^{(\sigma^* + \sigma)t} \right\} = \frac{1}{2} \left\{ \hat{u} \hat{w}^* e^{(\sigma + \sigma^*)t} \right\}_r$$

$$= \frac{1}{2} \left\{ \hat{u} \hat{w}^* \right\}_r e^{2\sigma_r t},$$

We can now generalize this result to give the horizontal average of a product of *any two perturbation quantities*:

$$\boxed{ \overline{a'b'} = \frac{1}{2} \left\{ \hat{a} \hat{b}^* \right\}_r e^{2\sigma_r t}. } \tag{3.63}$$

This formula may be used in the evaluation of K, SP, or EF. When plotting these quantities, one normally suppresses the time dependence (because it is trivial) by setting $t = 0$. For example, the momentum flux that appears in SP is computed from $\hat{w}$ as:

$$\overline{u'w'} = \frac{1}{2} \left\{ \hat{u} \hat{w}^* \right\}_r$$

where the subscript "r" denotes the real part. In Matlab, this expression is easily computed using the functions `conj` and `imag` and your subroutine `ddz` that forms the first-derivative matrix.

In (3.63), note that the time derivative of $\overline{a'b'}$ is just $2\sigma_r \, \overline{a'b'}$. Applying this result to the perturbation kinetic energy (3.55), the left-hand side of (3.56) becomes $2\sigma_r \, K$. The kinetic energy equation in normal mode form is therefore

$$\boxed{ 2\sigma_r K = SP - \frac{d}{dz} EF, } \tag{3.64}$$

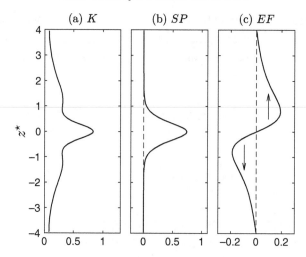

Figure 3.16 Perturbation kinetic energy budget (3.56) for the instability of a shear layer $U^\star = \tanh z^\star$. Kinetic energy (a) is extracted from the mean flow within the shear layer via shear production (b), then transported vertically away from the shear layer by the energy flux (c).

where

$$K = \frac{|\hat{u}|^2 + |\hat{v}|^2 + |\hat{w}|^2}{4},$$ (3.65)

$$SP = -\frac{1}{2}\frac{dU}{dz}\,(\hat{u}^\ast\hat{w})_r,$$ (3.66)

$$EF = \frac{(\hat{w}^\ast\hat{\pi})_r}{2}.$$ (3.67)

Figure 3.16 shows sample results for the instability of a hyperbolic tangent shear layer $U^\star = \tanh z^\star$ (section 3.9.1). Kinetic energy is transferred from the mean flow to the perturbation near the center of the layer by SP (Figure 3.16b), then fluxed outward by EF (Figure 3.16c). Both of these processes are reflected in the shape of the K profile. K is sharply peaked at $z = 0$ because SP is concentrated there, but also shows significant amplitude outside that peak because EF carries some of the energy vertically and deposits it beyond the shear layer.

3.11 Necessary Conditions for Instability

For a particular mode to grow on a particular mean flow, both mode and mean flow must satisfy certain criteria. These are useful to know, because they often allow us to rule out instability without having to do the stability analysis explicitly.

3.11.1 Instability Requires an Inflection Point

We know from (3.59) that growth requires positive net shear production:

$$\int SP \, dz > 0,$$

where the integral covers the whole vertical domain. This means that SP has to be positive for some z. But SP can't be positive for all z, because $\hat{w} = 0$, and therefore $SP = 0$, at the boundaries. Therefore SP must have at least one positive local maximum somewhere in the interior of the flow (e.g., Figure 3.16b). What conditions must the mean flow satisfy for this to be true?

Combining (3.57) and (3.63), we write the shear production as

$$SP = -\frac{1}{2} U_z \left(\hat{u} \hat{w}^* \right)_r. \tag{3.68}$$

(Don't be confused: The subscript z indicates a derivative, while the subscripts r and i specify the real and imaginary parts, respectively.) Because the fastest-growing mode is invariably two-dimensional (section 3.7), we restrict our attention to 2D modes ($\ell = 0$), in which case

$$\hat{u} = \frac{\iota}{k} \hat{w}_z.$$

With that substitution,

$$\boxed{SP = \frac{U_z}{2k} \left(\hat{w}_z \hat{w}^* \right)_i.} \tag{3.69}$$

If SP is a maximum, its derivative must be zero. Differentiating, we obtain

$$SP_z = \underbrace{\frac{U_{zz}}{2k} \left(\hat{w}_z \hat{w}^* \right)_i}_{(1)} + \frac{U_z}{2k} \Big[\underbrace{\left(\hat{w}_{zz} \hat{w}^* \right)_i}_{(2)} + \underbrace{\left(\hat{w}_z \hat{w}_z^* \right)_i}_{(3)} \Big].$$

We next simplify the terms (1), (2), and (3) individually.

- (3) is the easiest; it's the imaginary part of an absolute value and is therefore zero.
- (1) can be written as

$$(1) = SP \frac{U_{zz}}{U_z}.$$

- (2) is a bit more involved. We begin by writing the Rayleigh equation (3.19) for a 2D disturbance:

$$\hat{w}_{zz} = M \hat{w}, \quad \text{where } M = \frac{U_{zz}}{U - c} + k^2.$$

Now

$$(2) = (\hat{w}_{zz}\hat{w}^*)_i = (M\hat{w}\hat{w}^*)_i = |\hat{w}|^2 M_i.$$

The imaginary part of M is

$$M_i = \left(\frac{U_{zz}}{U-c}\frac{U-c^*}{U-c^*}+k^2\right)_i = \frac{U_{zz}}{|U-c|^2}c_i.$$

Recalling that $c_i = \sigma_r/k$, we can now write (2) as

$$(2) = |\hat{w}|^2\frac{U_{zz}}{|U-c|^2}\frac{\sigma_r}{2k^2}.$$

Combining these results we have

$$
\begin{aligned}
SP_z &= SP\frac{U_{zz}}{U_z} + |\hat{w}|^2\frac{U_zU_{zz}}{|U-c|^2}\frac{\sigma_r}{2k^2} \\
&= U_zU_{zz}\left(\frac{SP}{U_z^2}+\left|\frac{\hat{w}}{U-c}\right|^2\frac{\sigma_r}{2k^2}\right).
\end{aligned}
\tag{3.70}
$$

This expression has three factors, at least one of which must be zero if $SP_z = 0$. Because we need a positive maximum of SP, U_z cannot be zero. The term in brackets is positive definite because $SP > 0$ and $\sigma_r > 0$. Therefore, $U_{zz} = 0$. This result includes

Rayleigh's inflection point theorem: *For an inviscid, homogeneous parallel shear flow, a necessary condition for instability is that there exist an inflection point somewhere in the flow.* In addition, the local maximum of SP (where perturbation kinetic energy is produced) coincides with the inflection point.

3.11.2 The Inflection Point Must Be a Shear Maximum

The inflection point specified above (section 3.11.2) may represent a concentration of shear, as in the center of a shear layer, but it may also indicate a layer of reduced shear (compare Figures 3.17a and b). Here we'll show that the former case *may* be unstable, but the latter is definitely *not*.

Note first that the first two factors on the right-hand side of (3.70) can be written as

$$U_zU_{zz} = \frac{1}{2}(U_z^2)_z.\tag{3.71}$$

Therefore, the inflection point is also an extremum of the squared shear: $(U_z^2)_z = 0$, and we can rewrite (3.70) as

$$SP_z = \frac{1}{2}(U_z^2)_z\left(\frac{SP}{U_z^2}+\left|\frac{\hat{w}}{U-c}\right|^2\frac{\sigma_r}{2k^2}\right).$$

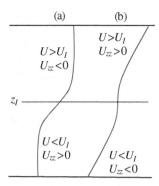

Figure 3.17 Inflectional velocity profiles. $U_I = U(z_I)$ is the velocity at the inflection point z_I. According to the shear production theorem (which includes the Rayleigh and Fjørtoft theorems), (a) may be unstable, whereas (b) is definitely not.

Differentiating, we obtain

$$SP_{zz} = \underbrace{\frac{1}{2} \left(U_z^2 \right)_{zz} \left(\frac{SP}{U_z^2} + \left| \frac{\hat{w}}{U - c} \right|^2 \frac{\sigma_r}{2k^2} \right)}_{>0} + \frac{1}{2} \underbrace{\left(U_z^2 \right)_z}_{=0} \left(\frac{SP}{U_z^2} + \left| \frac{\hat{w}}{U - c} \right|^2 \frac{\sigma_r}{2k^2} \right)_z .$$

Now if our extremum of SP is a maximum, then $SP_{zz} < 0$. Because the quantity in parentheses is positive and the second term is zero,

$$\frac{1}{2} \left(U_z^2 \right)_{zz} < 0.$$

Therefore, the maximum of shear production must also be a maximum of the squared shear.

Alternatively, one can draw the same conclusion in terms of the absolute shear $|U_z|$. Instead of (3.71), start with

$$U_z U_{zz} = |U_z| |U_z|_z,$$

and follow the analysis through in the same way (try it!). The result is that the maximum of shear production must also be a maximum of the *absolute* shear. In visual terms, this means that *the velocity profile around the inflection point must look like Figure 3.17a, not 3.17b*, if SP is to be a maximum. This statement is equivalent to:

Fjørtoft's theorem: *A necessary condition for instability is that $U_{zz}(U - U_I) < 0$ somewhere within the domain of flow, where $U_{zz}(z_I) = 0$, and $U_I = U(z_I)$ is the velocity at the inflection point.*

The standard derivations of the Rayleigh and Fjørtoft theorems are given in the Appendix to this chapter. Also, both results are explained in mechanistic terms in section 3.12.

The results from this and the previous subsection can be summarized as follows:

Shear production theorem: *In an unstable normal mode growing on an inviscid, homogeneous, parallel shear flow, the production of perturbation kinetic energy occurs at a local maximum of the absolute (or squared) shear. If there is no such local maximum, the flow is stable.*

Test your understanding: Is the converse true, i.e., does the presence of a shear maximum guarantee instability?[4]

3.11.3 Instability Requires a Critical Level

In the previous section we found conditions that the *mean flow* must satisfy to be unstable. In this section and the next, we'll look at two conditions that *the perturbation* must satisfy in order to grow.

A glance at Rayleigh's equation in the form (3.19) shows that something special will happen at any height where $U(z) = c$. Since $U - c$ appears in the denominator, such a height is a singularity of (3.19). For unstable modes, c is complex and therefore singularities are located at complex values of z. Near a stability boundary, singularities approach the real z axis. This can become a problem for numerical analysis, since the solution varies rapidly near a singularity and is therefore hard to resolve (Figure 3.18).

On the real z axis, the nearest point to the singularity is that where $U(z) - c_r = 0$. This is the height where the phase speed, relative to the background flow, is zero, i.e., the disturbance is effectively standing still. We call this the critical level.

We now show that every unstable mode must have a critical level. To begin with, write the Rayleigh equation (3.89) with the change of variables

$$\hat{w} = \iota k(U - c)\hat{\eta}. \tag{3.72}$$

Here, $\hat{\eta}$ represents the eigenfunction of the vertical displacement, defined so that $w = D\eta/Dt$. After some algebra (try it, cf. homework problem 9), this leads to

$$[(U - c)^2 \hat{\eta}_z]_z = \tilde{k}^2 (U - c)^2 \hat{\eta}. \tag{3.73}$$

We now multiply by $\hat{\eta}*$ and integrate in the vertical across the domain from z_B to z_T, which may be at $\pm\infty$. We assume that $w' = 0$ at z_B and z_T, which may mean

---

[4] Answer: No. A shear maximum is a necessary, but not a sufficient, condition for instability. A counterexample is derived in homework problem 12 in Appendix A.

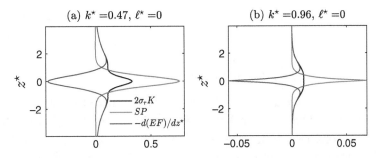

Figure 3.18 Kinetic energy budget terms for two modes of a hyperbolic tangent shear layer $U^* = \tanh z^*$. Two cases are compared: (a) far from the stability boundary, and (b) close to the stability boundary. The calculation is done with asymptotic boundary conditions and $\Delta^* = 0.05$. Case (b) is more difficult to resolve due to the sharp gradients near the critical level $z_c = 0$.

that the boundaries are impermeable or that the disturbance is localized sufficiently to vanish at the boundaries. Integrating the left-hand side by parts and recognizing that $\hat{\eta} = 0$ at the boundaries for all unstable modes,[5] we obtain

$$\int_{z_B}^{z_T} \hat{\eta}^* [(U-c)^2 \hat{\eta}_z]_z dz = -\int_{z_B}^{z_T} \hat{\eta}_z^* (U-c)^2 \hat{\eta}_z dz.$$

Applying the same operations to the right-hand side of (3.73) yields

$$\tilde{k}^2 \int_{z_B}^{z_T} \hat{\eta}^* (U-c)^2 \hat{\eta} dz.$$

Combining, we have

$$\int_{z_B}^{z_T} (U-c)^2 (|\hat{\eta}_z|^2 + \tilde{k}^2 |\hat{\eta}|^2) dz = 0.$$

Now take the imaginary part:

$$c_i \int_{z_B}^{z_T} (U-c_r)(|\hat{\eta}_z|^2 + \tilde{k}^2 |\hat{\eta}|^2) dz = 0.$$

For a growing (or decaying) mode, $c_i \neq 0$, and therefore the integral must be zero. Because the second factor in the integrand is positive definite, the first factor, $U - c_r$, must change sign somewhere in the range of z. In other words:

Critical level theorem: *For an unstable normal mode of a homogeneous, inviscid, parallel shear flow, the phase speed c_r must lie within the range of the background flow.*

---

[5] Referring to (3.72), assume $\hat{w} = 0$ at the boundaries and note that $U - c$ cannot be zero if $c_i \neq 0$.

This result is a corollary of Howard's semicircle theorem, which will be proved in section 4.8.

3.11.4 Shear Production and Phase Tilt

How does one recognize a growing instability observationally? In its early stages, an instability looks like an internal wave, except that it's growing in time. If all you have is a single observation at some point in time, you can't see the growth. A good indicator, then, is the vertical phase structure. At progressively deeper locations, the phase of the wave may remain constant or it may shift in one direction or the other, and that shift can tell us whether the wave is configured properly for growth.

The interpretation of the phase shift depends on the quantity being measured. The easiest choice to handle theoretically is the vertical velocity, so we'll discuss that one here. Suppose that, at some fixed time that we'll call $t = 0$, the vertical velocity has the normal mode form

$$w' = \{\hat{w}(z)e^{\iota kx}\}_r. \tag{3.74}$$

Now let's write the complex amplitude $\hat{w}(z)$ in polar form:

$$\hat{w} = r(z)e^{\iota \phi(z)}. \tag{3.75}$$

Combining (3.74) and (3.75), we have

$$w' = \{r(z)e^{\iota[kx+\phi(z)]}\}_r. \tag{3.76}$$

Any particular phase of the wave (a crest, say, or a trough) is a curve in the $x - z$ plane on which the phase function $\Phi(x, z) = kx + \phi(z)$ has a fixed value (e.g., black dashed line on Figure 3.19). Along such a curve, constancy of $\Phi(x, z)$ requires

$$\Delta\Phi = k\Delta x + \phi_z\Delta z = 0, \quad \text{or} \quad \left(\frac{dz}{dx}\right)_\Phi = -\frac{k}{\phi_z}.$$

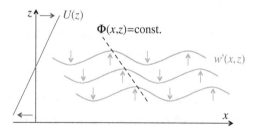

Figure 3.19 Schematic of a vertical velocity field of the form (3.76) with the amplitude r fixed. Note how the phase lines (black) tilt against the sheared background flow (blue).

Now recall that the shear production must be positive for instability. Substituting (3.75) in (3.69),

$$
\begin{aligned}
SP &= \frac{U_z}{2k}\, (\hat{w}_z \hat{w}^*)_i \;=\; \frac{U_z}{2k}\left\{ \left(r_z e^{\iota\phi} + \iota\phi_z r e^{\iota\phi} \right) r e^{-\iota\phi} \right\}_i \\
&= \frac{U_z}{2k}\{ (r_z + \iota\phi_z r)\, r \}_i \;=\; \frac{U_z}{2k}\, r^2 \phi_z \;=\; -\,\frac{U_z}{2}\,\frac{r^2}{(dz/dx)_\Phi}.
\end{aligned}
$$

Therefore $SP > 0$ implies that U_z and $(dz/dx)_\Phi$ have opposite signs. In other words, shear production is positive if and only if the phase lines of w' tilt **against the shear**.

3.11.5 Summary: Conditions for Instability

The following conditions, derived previously in this section, must be satisfied in order for an unstable mode to grow. Conditions (i) and (ii) relate to the mean flow, while conditions (iii) and (iv) relate to the perturbation.

 (i) The mean velocity profile must have an inflection point.
 (ii) The inflection point must represent a maximum (not a minimum) of the absolute shear.
 (iii) The mode must have a critical level.
 (iv) The phase lines of w' must tilt against the mean shear.

Each of these is a necessary (not a sufficient) condition for growth.

3.12 The Wave Resonance Mechanism of Shear Instability

How does shear instability work? We can solve equations that demonstrate its existence, but is there some intuitive explanation? In the case of convective instability, the answer is obvious: light fluid rises, dense fluid falls. Several explanations have been proposed for shear instability, none as simple as that for convection. Here we will describe a mechanistic picture that has turned out to be quite powerful, not just for shear instability but also for more complex processes such as the baroclinic instability that we will look at later (Chapter 8).

3.12.1 Thought Experiment: Wave Resonance

Consider a piecewise-linear shear layer as in Figure 3.20a, and suppose that the upper edge of the shear layer is somehow deformed to make a sinusoidal wave (upper black curve). Suppose further that this wavelike deformation is held stationary. (Don't be concerned about how this could actually happen; we'll get to that in

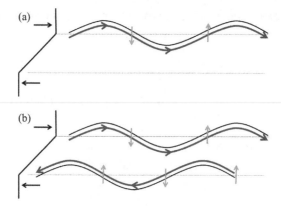

Figure 3.20 Stationary wavelike disturbances in a shear flow. (a) The wavy black curve represents artificially imposed corrugations of the upper edge of a shear layer. The blue arrow shows the path of the flow past the corrugations and green arrows its vertical component. (b) Both edges are corrugated. Mutual amplification results when upward (downward) motions line up with crests (troughs).

the next subsection. For now, just picture inserting a corrugated plastic sheet into the flow.) The flow along the sinusoid is generally from left to right (blue curve), but because of the corrugations it is not purely horizontal; it has a vertical component that alternates in sign, with amplitude greatest at the nodes (green arrows).

Now suppose that the lower edge of the shear layer is also deformed into a stationary sinusoid (Figure 3.20b). The flow along this deformation is from right to left (lower blue curve), again with the vertical component maximized at the nodes.

Finally, suppose that the phase relationship between the two sinusoids is such that the *upward* motions along each sinusoid coincide with the *crests* of the other, and the *downward* motions coincide with the *troughs*, as shown in Figure 3.20b. The result is positive feedback: the bigger the upper disturbance gets, the more it amplifies the lower disturbance, and vice versa. Positive feedback results in exponential growth.

But what kind of waves are these? And how is it that they are able to remain stationary? We'll address the second question first. Two effects combine to allow waves like this to remain stationary. First, Doppler-shifting by the sheared background flow creates the possibility that the phase speeds will be the same (i.e., if each wave propagates *against* the background flow at its own height). The second effect is more subtle.

In Figure 3.21, the solid curves once again show two waves whose phase relationship is optimal for mutual amplification. The solid vertical arrow on the lower wave (marked "1") represents the upward component of the associated motion, extended to emphasize its alignment with the crest of the upper wave.

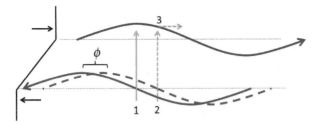

Figure 3.21 Flow along stationary corrugations of the edges of a shear layer as in Figure 3.20. Blue arrows show the path of the flow; green arrows 1 and 2 represent its maximum upward component. The phase relationship shown by the solid blue lines is optimal for mutual amplification, while that with the dashed blue is suboptimal. The interaction accelerates the upper wave to the right (green arrow 3), shifting the phase relationship toward optimal.

Now suppose that the lower wave is shifted to the right by a small amount ϕ (blue dashed curve). The vertical motion associated with the lower wave (arrow 2) is now directed slightly to the *right* of the crest of the upper wave. While this vertical motion still acts to amplify the upper wave, it also tends to shift it to the right (arrow 3), reducing the phase difference ϕ. The same is true of all nodes of both waves. The result is that the upper wave is shifted to the right and the lower wave to the left, reducing ϕ in both cases.

So as long as ϕ is not too large (details in section 3.13.3), the waves tend not only to amplify each other but also to hold each other in place so that amplification can continue. When two waves maintain the same phase velocity, we say that they are phase locked. In the next subsection we will show how waves like this can occur naturally in a shear flow.

3.12.2 Vorticity Waves

We now describe a type of wave whose propagation is driven by a change in vorticity, such as the upper or lower edge of the shear layer in Figure 3.20. Vorticity is a vector field equal to the curl of the velocity. In a two-dimensional flow, vorticity has only a single nonzero component, directed perpendicular to the plane of the flow. That component can be treated as a scalar. For flow in the $x - z$ plane, the scalar vorticity is $q(x, z, t) = u_z - w_x$. For two-dimensional flow in a homogeneous, inviscid fluid,

$$\frac{Dq}{Dt} = 0, \tag{3.77}$$

i.e., a fluid particle's vorticity does not change as it moves through the flow.[6]

---

[6] You can confirm this by writing the equations of motion (1.17, 1.19) for the 2D case:

$$u_x + w_z = 0$$

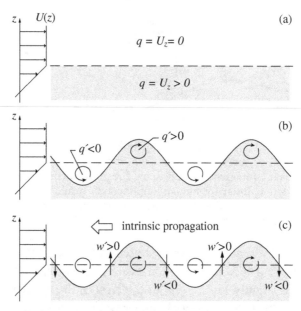

Figure 3.22 Schematic of a vorticity wave propagating on the upper edge of a piecewise-linear shear layer (cf. Figure 3.5). (a) The undisturbed flow, with vorticity positive (clockwise) in the shear layer and zero above. (b) Upward displacements of the interface carry the clockwise vorticity of the shear layer (gray), so the vorticity anomaly in the crests is positive. Downward displacements carry zero vorticity, so the anomaly is negative in the troughs. (c) Counter-rotating vorticity perturbations induce alternately upward and downward motion, causing the pattern to propagate leftward relative to the background flow.

Now, consider the upper edge of a piecewise-linear shear layer (Figure 3.22a, compare with Figure 3.5). Within the undisturbed shear layer, the vorticity is uniform and is given by $q = U_z$. As drawn, U_z is positive, so the sense of the vorticity is clockwise. Outside the shear layer, the vorticity is zero. We will refer to these kinks in the velocity profile, where the vorticity changes abruptly, as vorticity interfaces.

Next let us ask what becomes of a sinusoidal disturbance at the upper edge of the shear layer (Figure 3.22b). Consider a fluid parcel carried upward at a wave crest, i.e., into the region where the vorticity was originally zero. The parcel brings with it

and

$$u_t + uu_x + wu_z = -\pi_x \tag{3.78}$$
$$w_t + uw_x + ww_z = -\pi_z, \tag{3.79}$$

where subscripts indicate partial derivatives. Now differentiate (3.78) and (3.79) with respect to z and x, respectively, and subtract:

$$q_t + uq_x + wq_z = \frac{Dq}{Dt} = 0.$$

its positive vorticity, and the change in vorticity at that location is therefore positive (clockwise circular arrows in Figure 3.22b). Conversely, in regions of downward displacement, the positive ambient vorticity is replaced by zero vorticity from outside the shear layer, so the change is negative. The result is a vorticity perturbation consisting of a row of counter-rotating vortices.

Between each pair of vortices, a region of vertical flow is induced (Figure 3.22c). The direction of this vertical flow alternates, so that the interface moves alternately upward and downward with maximum vertical velocity at its nodes. This causes the whole pattern to move to the left. Note that this leftward motion is opposite, or upstream, relative to the background flow. At the lower edge of the shear layer, the same process produces a rightward propagating wave. Again, propagation is opposite to the mean flow.

Vorticity waves can propagate wherever there is a change in the background vorticity.[7] When the vorticity change takes the form of a sharp kink in the velocity profile, as in this example, the intrinsic propagation is toward the *concave* side of the kink.

So, can vorticity waves at the upper and lower edges of a shear layer be stationary, as we imagined in our thought experiment (section 3.12.1)? The mechanism described above suggests that it is possible, since each wave propagates oppositely to the mean flow at its own elevation. In fact, it can happen rather easily, as will be shown quantitatively in section 3.13.3.

3.12.3 Resonance and the Conditions for Instability

A virtue of the wave resonance model for shear instability is that it allows us to understand, in a visual, mechanistic way, the necessary conditions for instability that we have previously only been able to derive mathematically (section 3.11.5). You will now demonstrate this for yourself by repeating the graphical construction in Figure 3.23 for three other background velocity profiles.

For each of the velocity profiles shown in Figure 3.24, sketch upper and lower waves as follows:

 (i) Draw the *upper wave* at an arbitrary horizontal position.
 (ii) Determine the sense of the vorticity anomaly (clockwise or counterclockwise) in each crest and trough.
(iii) Show the resulting vertical velocity perturbations at the nodes.

---

[7] A well-known example of vorticity waves are the planetary-scale Rossby waves, driven by the gradient in the Coriolis effect between the equator and the poles (Gill, 1982; Pedlosky, 1987). Rossby waves propagate westward relative to the mean flow in the same way that the vorticity waves sketched in Figure 3.22 propagate to the left.

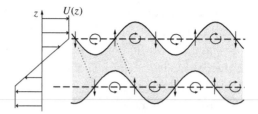

Figure 3.23 Resonant vorticity waves on a shear layer. The mean flow is shown at the left. The upper wave corresponds to Figure 3.22c. Curving arrows represent the vorticity anomalies in the crests and troughs of each wave. Vertical arrows represent the induced vertical motions. Each wave is stationary with respect to the other. Dotted lines are phase lines of w'.

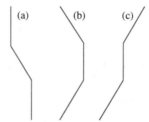

Figure 3.24 Schematics of three piecewise-linear background velocity profiles for use with exercise 3.12.3.

(iv) Indicate the direction of propagation relative to the background flow.
 (v) Sketch the *lower wave* so that its crests and troughs are amplified by the upper wave.
(vi) Now determine the vorticity anomalies, the vertical velocities, and the propagation direction for the lower wave.

Determine whether the following criteria for resonant growth are satisfied.

 (i) The vertical velocity perturbations of each wave amplify the crests and troughs of the other, creating resonance.
(ii) The propagation directions and mean flow allow for the waves to be stationary relative to each other (i.e., phase-locked), so that the resonance can be sustained over time.

If so, observe how each of the four conditions for growth summarized in section 3.11.5 is satisfied. If not, why not? Identify the condition that is violated.

Results

Admonition: Do not read this until you have tried it yourself!

- For the profile shown in Figure 3.24(a), you should basically reproduce Figure 3.23 with the horizontal direction reversed. Note that the phase lines (lines along which the vertical velocity is constant, i.e., upward arrows or downward arrows) tilt against the shear. If the tilt was opposite, mutual amplification would not occur. Note also that the right-going wave travels against the rightward background flow, and likewise the left-going wave moves against the leftward background flow. If this was not true, phase-locking would be impossible and mutual amplification could not be sustained.

- For (b), you should find that mutual amplification is impossible; if the upper wave amplifies the lower, then the lower wave diminishes the upper, and vice versa. This is because the velocity profile (b) lacks an inflection point (section 3.11.1, Appendix 3.15.1).

- For (c), mutual amplification is possible, but phase-locking is not. The waves always travel in opposite directions. This velocity profile violates the Fjørtoft condition: its inflection point represents a minimum, not a maximum, of the absolute shear (section 3.11.2, Appendix 3.15.2, Figure 3.17b).

If the resonance and phase-locking conditions are satisfied, sketch a line connecting adjacent nodes of the upper and lower waves where the vertical velocity has the same sign, as in Figure 3.23, and note that these lines tilt against the background shear.

So, all four general conditions for instability summarized in section 3.11.5 can be understood intuitively in terms of resonant vorticity waves.

3.13 Quantitative Analysis of Wave Resonance

We'll now take a closer look at the interaction of a pair of vorticity waves propagating on the edges of a shear layer. We begin by deriving the dispersion relation for a single wave (isolated from the influence of the other wave). We'll then look at wave interactions *assuming that this dispersion relation remains valid even when the other wave is present*. This will give us an approximate value for the wavenumber of the fastest-growing mode. Finally, we'll dispense with the assumption about the dispersion relation and see what happens when the waves are allowed to interact.

3.13.1 The Vorticity Wave Dispersion Relation

As a simple model for vorticity wave motion, we solve Rayleigh's equation (3.19) for a velocity profile with a single vorticity interface at $z = z_0$ as shown in Figure 3.6. As in section 3.3, the solution can be written in terms of exponential functions. Requiring that

- $\hat{w}$ decay as $z \to \pm\infty$,
- $\hat{w}$ be continuous across the interface, i.e., $[[\hat{w}]]_{z_0} = 0$,

we obtain

$$\hat{w}(z) = Ae^{-k|z-z_0|}. \tag{3.80}$$

The dispersion relation is found by applying the jump condition (3.30) at the vorticity interface $z = z_0$:

$$\boxed{c = u_0 + \frac{\Delta Q}{2k},} \tag{3.81}$$

where u_0 is the background velocity at the interface and $\Delta Q = Q_2 - Q_1$ is the change in the mean vorticity across the interface. (Exercise: Verify this.) We will call the phase speed relative to the mean velocity, $\Delta Q/(2k)$, the intrinsic phase speed. For the example shown in Figure 3.22, $\Delta Q = -U_z < 0$, so the intrinsic propagation is to the left.

3.13.2 The Isolated Wave Approximation

Now let's consider the full shear layer [(3.23) with waves at both the upper and lower vorticity interfaces (Figure 3.23)]. We assume first that each wave propagates with no influence from the other, so that their phase speeds are given by (3.81). For the upper wave, $\Delta Q = -u_0/h$, so the intrinsic propagation is to the left, opposite to the rightward mean flow. Conversely, the lower wave has $\Delta Q = u_0/h$ and therefore propagates intrinsically to the right against the leftward mean flow. (Remember: the intrinsic vorticity wave propagation is always toward the concave side of the velocity profile.)

The net phase speeds of the upper and lower waves,

$$\frac{c}{u_0} = \pm\left(1 - \frac{1}{2kh}\right), \tag{3.82}$$

are shown by the solid curves on Figure 3.25. (This result was stated without proof in section 3.3.2.) When $kh = 0.5$, the intrinsic phase speed of each wave is exactly opposite to the mean flow, so that its net phase speed is zero.

If the phase relationship is right, the waves reinforce each other as shown in Figure 3.20, and more explicitly in Figure 3.23. Now we'll take it to the next level, refining the theory to account for the fact that the waves do not really propagate independently of each other.

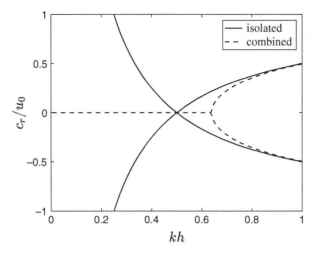

Figure 3.25 Solid: Nondimensional phase speeds of two vorticity waves, isolated from each other, propagating on one or the other edge of a piecewise-linear shear layer, as functions of the nondimensional wavenumber. Dashed: Phase speeds of a pair of counterpropagating stable wave solutions to Rayleigh's equation for a piecewise-linear shear layer, with no assumption of isolation (reproduced from Figure 3.7).

3.13.3 Interaction Effects

With the assumption that the phase speed of each wave is given by (3.81), which means it is unaffected by any other wave (or waves) that may be present, we have identified a single value of kh at which phase-locking occurs, 0.5. But we know that shear instability exists over the whole range $0 < kh < 0.64$ (section 3.3). Moreover, the fastest-growing mode has $kh = 0.4$, not 0.5. Could these discrepancies be explained by the influence of each wave on the phase speed of the other?

In section 3.3, we solved the Rayleigh equation for the full velocity profile, without trying to isolate any part from any other part. The phase speed we obtained is shown by the dashed curves on Figure 3.25 (cf. Figure 3.7a). The phase speeds agree well as $kh \to \infty$, but not so well when kh is finite. Recall that the dispersion relation for c is

$$\frac{c^2}{u_0^2} = \underbrace{\left(1 - \frac{1}{2kh}\right)^2}_{isolated\ waves} - \underbrace{\frac{e^{-4kh}}{4k^2h^2}}_{wave\ interaction} . \tag{3.83}$$

[This is just (3.32) simplified by setting ℓ to zero.] We now see that the first term on the right-hand side is what we would get if the two waves were isolated (cf. 3.82), and we can therefore identify the second term as representing the interaction of the two waves, as foreshadowed in section 3.3. That term is negative and therefore tends to move c toward zero. Moving from large kh (the right-hand side of Figure 3.25) to $kh = 0.64$, the terms in (3.83) come into balance such that the two phase speeds meet at zero. For all smaller values of kh, $c^2 < 0$, meaning that $c_r = 0$ and the waves are phase-locked.

The dependence of the waves' mutual interaction on kh has another important consequence, and that is that the fastest-growing mode is found not at $kh = 0.5$, but rather at $kh = 0.4$ as we discovered in section 3.3. The phase relationship at $kh = 0.4$ is not quite optimal for resonance, but this is compensated for, and then some, by the fact that the waves' amplitude decays more slowly in the vertical at small kh, and hence their ability to amplify each other is greater.

This can all be seen quantitatively by writing the growth rate and the phase speed in terms of the phase relationship between the upper and lower waves. We choose, arbitrarily, to focus on the lower vorticity interface, where the jump condition is

$$(-u_0 - c)[\![\hat{w}_z]\!]_{-h} - \frac{u_0}{h}\hat{w}(-h) = 0, \tag{3.84}$$

giving

$$\frac{c}{u_0} = -1 + \frac{1}{2kh}\left(\frac{B_2}{B_1}e^{-2kh} + 1\right).$$

Let B_1 and B_2 be expressed in polar form $B_1 = A_1 e^{\iota\theta_1}$ and $B_2 = A_2 e^{\iota\theta_2}$. The real constants (A_1, A_2) represent the magnitudes, and (θ_1, θ_2) the phases of the waves at the lower and upper vorticity interfaces. The symmetry of the problem requires that $A_1 = A_2$, hence

$$\frac{c}{u_0} = -1 + \frac{1}{2kh}\left(e^{\iota\Delta\theta - 2kh} + 1\right). \tag{3.85}$$

where we define $\Delta\theta = \theta_2 - \theta_1$. Note that $\Delta\theta = \pi/2 - \phi$, where ϕ is the phase shift defined in Figure 3.20. The real part of (3.85) is

$$\frac{c_r}{u_0} = \underbrace{-1}_{advection} + \overbrace{\frac{1}{2kh}}^{intrinsic} + \underbrace{\cos(\Delta\theta)\frac{e^{-2kh}}{2kh}}_{interaction}. \tag{3.86}$$

The phase speed is composed of three parts: advection by the background profile, the intrinsic propagation speed in isolation, and the change in phase speed due to interaction. When $\Delta\theta = \pi/2$, $\cos\Delta\theta = 0$, and the interaction term is zero. We therefore recover the phase speed of the lower wave in isolation. If that phase speed is zero as in section 3.12.1, then (3.86) gives $kh = 0.5$. But if the waves are allowed to interact, then $\Delta\theta$ is not necessarily $\pi/2$; in fact, $c_r = 0$ requires that

$$\cos(\Delta\theta) = e^{2kh}(2kh - 1), \tag{3.87}$$

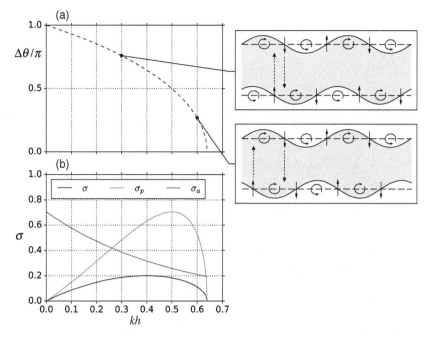

Figure 3.26 (a) Phase difference between the upper and lower vorticity displacements. The two side panels show the phase configuration of the waves for two values of kh on either side of the crossing point of the isolated dispersion relation. Dashed arrows emphasize the contact point of the maximum vertical velocity with the other wave. (b) Total growth rate (black), represented as a product of two factors: the phase difference ($\sigma_p = \sin(\Delta\theta)/\sqrt{2}$, green) and the amplitude decay ($\sigma_a = e^{-2kh}/\sqrt{2}$, blue), with the factor $1/2$ shared, arbitrarily, between them.

which is plotted in Figure 3.26a. Note that $\Delta\theta$ is less than $\pi/2$ (or $\phi < 0$) when $kh > 1/2$ and greater when $kh < 1/2$.

Now take the imaginary part of (3.85) and solve for the scaled growth rate:

$$\frac{h}{u_0}\sigma = \frac{1}{2}\sin(\Delta\theta)e^{-2kh}. \tag{3.88}$$

Evidently the growth rate is the product of two factors. The first, $\sin(\Delta\theta)$, depends on the phase relationship and is a maximum when $\Delta\theta = \pi/2$ (Figure 3.26b). This is the phase relationship described in section 3.12.1. When $\phi = \pi/2 - \Delta\theta = 0$, the waves are naturally phase-locked without any need for interaction, and are in the optimal phase configuration for growth. The second factor, e^{-2kh}, is the reduction in amplitude of the upper wave at the height of the lower wave (since the vertical distance between them is $2h$), and decreases monotonically with kh. The product of the two (with the constant $1/2$) attains its maximum value 0.20 when $kh = 0.4$, as we found in section 3.3. Therefore, the maximum growth rate occurs not in the

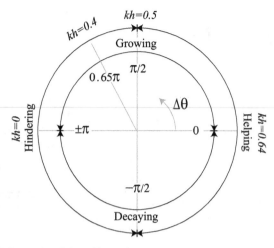

Figure 3.27 Schematic of the effect of wave interactions on the growth and propagation of vorticity waves. Depending on the phase difference, $\Delta\theta$, the waves may cause mutual growth or decay, or they may help or hinder each other's phase propagation. Adapted from Heifetz et al. (2004).

optimal phase relationship $\Delta\theta = \pi/2$, but slightly toward lower kh, where the wave interaction is stronger.

Changes in phase speed and growth rate due to the interaction of two vorticity waves, as quantified in (3.87, 3.88), are diagrammed in Figure 3.27. Depending on the phase difference $\Delta\theta$, we can identify configurations that are optimal for growth ($\Delta\theta = \pi/2$), for decay ($\Delta\theta = -\pi/2$, remembering that every growing mode is accompanied by a decaying mode), and for helping and hindering the intrinsic propagation ($\Delta\theta = 0, \pm\pi$, respectively). In the case of the piecewise shear layer, the bounds of the unstable wavenumber band $kh = 0.64$ and $kh = 0$ coincide with the points of strongest helping/hindering of the phase propagation, $\Delta\theta = 0$ and $\pm\pi$, respectively. A pair of waves with either of these phase relationships will be neutrally stable, but will shift toward either the growth or the decay regime. A pair of waves with $kh = 0.4$ and $\Delta\theta = 0.65\pi$ is optimally configured for growth. Without interaction, the intrinsic propagation term in (3.86) would overcompensate for the advection term, but with $\Delta\theta$ in the "hindering" regime the interaction term brings c_r to zero.

3.14 Summary

Necessary conditions for shear instability in an inviscid, homogeneous fluid:

(i) The mean velocity profile must have an inflection point.
(ii) The inflection point must represent a maximum (not a minimum) of the absolute shear.

Table 3.1 *Spatial and temporal scales of the fastest-growing mode for two example profiles.*

Name	Formula	$\dfrac{\lambda}{2h}$	$\dfrac{\sigma}{u_0/h}$
hyperbolic tangent shear layer	$U = u_0 \tanh(z/h)$	7	0.19
Bickley jet (sinuous mode)	$U = u_0 \operatorname{sech}^2(z/h)$	3.5	0.16

(iii) The mode must have a critical level.

(iv) The phase lines of w' must tilt against the mean shear.

Characteristics of the fastest-growing mode:

(i) The wave vector is directed parallel to the background flow, i.e., there is no variation in the cross-stream direction.

(ii) If h and u_0 are defined as length and velocity scales characteristic of the mean flow $U(z)$, then the wavelength is proportional to h and the growth rate is proportional to u_0/h. Proportionality constants are given in Table 3.1.

Finally, the resonant interaction of vorticity waves provides an interpretation of the mechanism for shear instability, the Rayleigh and Fjørtoft theorems, and the critical level and phase tilt criteria. In the simplest case, the shear layer, instability is driven by the interaction of two vorticity waves. The fastest-growing mode is that for which the waves are phase-locked and mutual amplification is optimized. In more complex parallel shear flows, instability can be understood in terms of the resonance of multiple vorticity waves.

3.15 Appendix: Classical Proof of the Rayleigh and Fjørtoft Theorems

The shear production theorem proven in section 3.11 is original to this text. It includes the classical theorems of Rayleigh and Fjørtoft, as well as some additional detail about *where* shear instability is expected to occur. In this appendix we give the original proofs of the Rayleigh and Fjørtoft theorems.

We begin by writing the Rayleigh equation in the form (3.19):

$$\hat{w}_{zz} = \frac{U_{zz}}{U-c}\hat{w} + \tilde{k}^2\hat{w}. \tag{3.89}$$

Now multiply by $\hat{w}^*$, the complex conjugate of $\hat{w}$, and integrate:

$$\underbrace{\int \hat{w}^*\hat{w}_{zz}\,dz}_{(1)} = \underbrace{\int \hat{w}^*\frac{U_{zz}}{U-c}\hat{w}\,dz}_{(2)} + \underbrace{\int \hat{w}^*\tilde{k}^2\hat{w}\,dz}_{(3)}. \tag{3.90}$$

The integral is understood to cover the entire vertical domain. We next simplify the terms. Term (3) is the easiest:

$$(3) = -\tilde{k}^2 \int |\hat{w}|^2 \, dz.$$

Term (1) can be integrated by parts:

$$(1) = \hat{w}^* \hat{w}_z \bigg| - \int \hat{w}_z^* \hat{w}_z \, dz$$

$$= 0 - \int |\hat{w}_z|^2 \, dz.$$

The first term vanishes because $\hat{w} = 0$ at the boundaries (either impermeable boundaries or boundaries at infinity). Note that the two terms we've worked on are purely real. Now recall that c is a *complex* phase speed:

$$c = \iota \sigma / k = c_r + \iota c_i;$$

hence, term (2) is complex, and we will split it into real and imaginary parts.

$$(2) = \int |\hat{w}|^2 \frac{U_{zz}}{U-c} \frac{U-c^*}{U-c^*} \, dz = \int \frac{|\hat{w}|^2}{|U-c|^2} U_{zz}(U - c_r + \iota c_i) \, dz.$$

Reassembling (3.90) and rearranging, we have

$$\int \frac{|\hat{w}|^2}{|U-c|^2} U_{zz}(U - c_r + \iota c_i) \, dz = -\int |\hat{w}_z|^2 \, dz - \tilde{k}^2 \int |\hat{w}|^2 \, dz. \quad (3.91)$$

3.15.1 Rayleigh's Inflection Point Theorem

Consider the imaginary part of (3.91):

$$c_i \int \frac{|\hat{w}|^2}{|U-c|^2} U_{zz} \, dz = 0. \quad (3.92)$$

For a growing (or decaying) mode, $c_i \neq 0$, and therefore the integral must be zero. Except in the trivial case $\hat{w} = 0$, the integrand must take both positive and negative values in different ranges of z. This requires that U_{zz} change sign at least once in the range of integration, i.e., there must be an inflection point.

3.15.2 Fjørtoft's Theorem

We turn now to the real part of (3.91):

$$\int \frac{|\hat{w}|^2}{|U-c|^2} U_{zz}(U - c_r) \, dz = -\int |\hat{w}|^2 \, dz - \tilde{k}^2 \int |\hat{w}|^2 \, dz. \quad (3.93)$$

If $c_i \neq 0$, then the integral in (3.92) vanishes. We can therefore write:

$$(U_0 - c_r) \int \frac{|\hat{w}|^2}{|U - c|^2} U_{zz} \, dz = 0,$$

where U_0 is an arbitrary, uniform velocity. Subtracting this from (3.93) gives:

$$\int \frac{|\hat{w}|^2}{|U - c|^2} U_{zz}(U - U_0) \, dz = - \int |\hat{w}|^2 \, dz - \tilde{k}^2 \int |\hat{w}|^2 \, dz.$$

Since the right-hand side is negative definite, the integral on the left must be negative, and therefore the integrand must be negative for some z. In other words, the following condition must hold somewhere in the flow:

$$U_{zz}(U - U_0) < 0. \tag{3.94}$$

This is Fjørtoft's theorem.

The geometric meaning of Fjørtoft's theorem is somewhat mysterious due to the arbitrary constant U_0. One way to think of it is that, for some z, U and its second-derivative must have opposite signs. This means that $U(z)$ is wavelike rather than exponential, i.e., it curves back toward zero like a sine or a cosine function. The presence of U_0 guarantees that this is true in any reference frame, as is necessary for any physical law.

A useful choice for U_0 is the velocity at the inflection point, as illustrated in Figure 3.17. We can conclude, for example, that the profile shown in Figure 3.17a may be unstable while that shown in Figure 3.17b is stable.

The form of Fjørtoft's theorem proven in section 3.11.2 may be recovered from (3.94). First, observe that dU_z^2/dz must change sign at an inflection point. Now let $z = z_I$ be an inflection point, and expand the surrounding velocity profile in a Taylor series:

$$U(z) = U(z_I) + U_z(z - z_I) + \cdots,$$

where U_z is evaluated at $z = z_I$. Substituting this, the Fjørtoft criterion (3.94) becomes

$$U_{zz}U_z(z - z_I) < 0,$$

or

$$(z - z_I)\frac{1}{2}\frac{dU_z^2}{dz} < 0. \tag{3.95}$$

Since dU_z^2/dz must change sign at z_I, (3.95) requires that dU_z^2/dz be negative just above z_I and positive just below z_I rather than the reverse, i.e., the inflection point must be a maximum (not a minimum) of U_z^2.

3.16 Further Reading

Holmboe (1962), Baines and Mitsudera (1994), and Carpenter et al. (2013) give detailed discussions of shear instability via wave interactions. Heifetz et al. (1999) calculates the fastest-growing mode of a shear layer explicitly using the wave-interaction mechanism.

The original proofs of the Rayleigh and Fjørtoft theorems were published in Rayleigh (1880) and Fjortoft (1950), respectively.

4

Parallel Shear Flow: the Effects of Stratification

The Earth's oceans and atmosphere tend to be inhomogeneous, with density differences that can strongly affect the motion. The primary reason for the inhomogeneity is heat radiation. At the equator, both ocean and atmosphere gain heat from the sun. Near the poles, the ocean loses heat to the atmosphere, while the atmosphere radiates heat into space at all latitudes. Ocean density is also governed by salinity. All of these processes create inhomogeneities which, under the action of gravity, tend to rearrange themselves into horizontal layers. We therefore say that these fluids are "density-stratified," or just "stratified." Based on our experience

Figure 4.1 Sunrise over Jonesport, Maine, showing Kelvin-Helmholtz billow clouds. Photo courtesy of Gene Hart.

Figure 4.2 Kelvin-Helmholtz billows in a stratified laboratory flow (Thorpe, 1971). The tank is filled with a layer of pure water overlying a layer of denser salt water. When the tank is tilted, the dense lower layer flows downward, forcing the upper layer upward, resulting in an accelerating, stratified shear flow. Eventually, the shear becomes strong enough to overcome the stable stratification, and instability appears.

with motionless equilibria (Chapter 2) we call stratification "statically stable" when buoyant fluid overlies dense fluid. But if a shear flow is present, stability is less easy to predict.

When layers move relative to one another, there is the likelihood of shear instability, as we have seen (Chapter 3). But if the fluid is stably stratified, a growing perturbation (e.g., Figures 4.1 and 4.2) must expend some of its energy doing work against gravity. This tends to reduce the growth rate, and instability may be damped completely by sufficiently strong stratification. In some cases, though, statically stable stratification can interact with shear to create new mechanisms of instability that would not exist in a homogeneous shear flow.

In Figure 4.3, a shear layer is set up by the passage of an internal gravity wave. The wave rides on the slight difference in buoyancy between the surface water and the water below 20 m depth. Internal waves are big and slow: these ones have a period of 10 minutes and a wavelength of several hundred meters. Wave motion causes shear across the buoyancy interface. The shear is strongest on the upstream face of the wave, producing Kelvin-Helmholtz instability. The instability is powered by the shear and resisted by the buoyancy difference. Note, though, that the wave, and therefore the shear, would not exist without the buoyancy difference.

In Figure 4.4, observations at 550 m depth off the Canary Islands show a cold bottom layer where the speed of the tidal current drops to zero. The shear between the bottom layer and the warmer overlying ocean overcomes the buoyancy difference, resulting in Kelvin-Helmholtz instability. The same instability occurs in air, as made visible by the fog layer in Figure 4.5.

In this chapter we will explore stratification effects analytically, taking advantage of the relative simplicity of the inviscid equations. Numerical solution methods will be taken up in Chapter 6, after we have incorporated viscosity and diffusion.

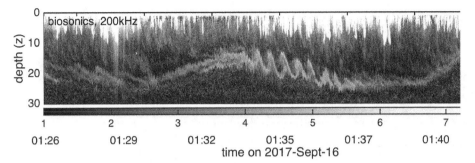

Figure 4.3 Echosounder image showing Kelvin-Helmholtz billows growing on an interfacial gravity wave off the California coast. The wave was visualized by a shipboard echosounder. Similar in principle to a medical ultrasound, it registers sound waves reflected from biota and from tiny changes in water density due to centimeter-scale turbulence. Image is courtesy of J. MacKinnon, J. Colosi, and A. Suanda.

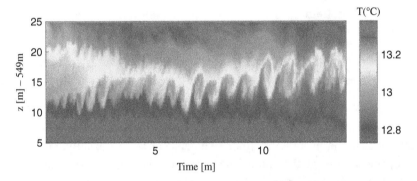

Figure 4.4 Temperature variations in a downslope tidal flow. This time series was constructed using data from multiple temperature sensors spaced vertically on a chain above the sea floor at 550 m depth. Typical wavelengths are inferred to be 75 m. Graphic courtesy H. van Haren (after Van Haren and Gostiaux, 2009).

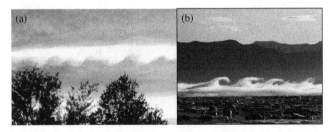

Figure 4.5 (a) Kelvin-Helmholtz instability in a stratified shear flow revealed by a cloud layer over Boston, MA. Photo courtesy of Alexis Kaminski. (b) Instability in a cold fog layer near Nares Strait in northern Canada. Photo courtesy Scott McAuliffe.

4.1 The Richardson Number

A sheared flow has kinetic energy that can power the growth of an instability, as we saw in section 3.10, while statically stable stratification represents an energy *sink*, as a disturbance must do work against gravity in order to grow. On this basis, we may propose a provisional "rule of thumb" for stratified shear flows:

- $U_z \neq 0$ will tend to destabilize the flow, while
- $B_z > 0$ will tend to stabilize the flow.

We will find that, like all rules of thumb, this one is valid often enough to be useful, but can also be dead wrong.

To quantify the relationship between shear and stratification, we define the gradient Richardson number:

$$Ri = \frac{B_z}{U_z^2}. \tag{4.1}$$

[A more general definition, $Ri = B_z/(U_z^2 + V_z^2)$, allows for flow in any horizontal direction.] We can imagine two limiting cases:

- $Ri \gg 1$: stratification dominates, shear is weak, and we don't expect instability.
- $Ri \ll 1$: shear dominates, stratification is weak, and instability is therefore likely.

What do we expect at moderate values of Ri? Is there a critical value of Ri that separates stable and unstable regimes? The answer is yes, usually, but the critical value depends on the details of the flow geometry. We'll get into this later; for now, be content to know that the critical value is typically of order unity.

To illustrate, consider the hyperbolic tangent model for the stably stratified shear layer:

$$U = u_0 \tanh \frac{z}{h} ; \quad B = b_0 \tanh \frac{z}{h}. \tag{4.2}$$

Differentiating, we have

$$B_z = \frac{b_0}{h} \operatorname{sech}^2 \frac{z}{h} ; \quad U_z^2 = \frac{u_0^2}{h^2} \operatorname{sech}^4 \frac{z}{h}.$$

$$\Rightarrow \quad Ri = \frac{b_0 h}{u_0^2} \cosh^2 \frac{z}{h}. \tag{4.3}$$

The profile is shown in Figure 4.6. It is often convenient to define a bulk Richardson number, Ri_b , whose value characterizes the shear flow as a whole. A natural choice is the coefficient $b_0 h/u_0^2$, which in this case is also the minimum value of $Ri(z)$. We will find in section 4.4 that this flow is unstable if and only if $Ri_b < 1/4$.

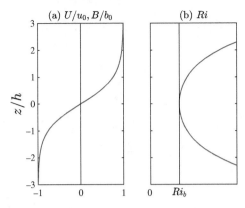

Figure 4.6 (a) Velocity/buoyancy profile for the hyperbolic tangent stratified shear layer (4.2). (b) Gradient Richardson number profile. The vertical line indicates the bulk Richardson number.

4.2 Equilibria and Perturbations

To derive the perturbation theory for stratified, parallel shear flows, we start with the Boussinesq equations for an inviscid, nondiffusive, inhomogeneous fluid. We ignore planetary rotation. The divergence equation is, as usual,

$$\vec{\nabla} \cdot \vec{u} = 0.$$

The momentum equation (1.19), neglecting the Coriolis acceleration and viscosity but restoring buoyancy, is

$$\frac{D\vec{u}}{Dt} = -\vec{\nabla}\pi + b\hat{e}^{(z)}, \tag{4.4}$$

and the buoyancy equation (1.25) is

$$\frac{Db}{Dt} = 0. \tag{4.5}$$

We assume the perturbation solution

$$\vec{u} = U(z)\hat{e}^{(x)} + \epsilon\vec{u}',$$
$$b = B(z) + \epsilon b',$$
$$\pi = \Pi + \epsilon\pi'. \tag{4.6}$$

No assumption is made regarding the background pressure Π. As always, the perturbation velocity has zero divergence: $\vec{\nabla} \cdot \vec{u}' = 0$.

Substituting (4.6) into the momentum equation (4.4) gives

$$\left[\frac{\partial}{\partial t} + U\frac{\partial}{\partial x} + \varepsilon\vec{u}' \cdot \vec{\nabla}\right]\left[U(z,t)\hat{e}^{(x)} + \varepsilon\vec{u}'\right] = -\vec{\nabla}\left[\Pi + \varepsilon\pi'\right] + \left[B(z,t) + \varepsilon b'\right]\hat{e}^{(z)}$$

$$(4.7)$$

With no perturbation ($\varepsilon = 0$), this gives

$$\vec{\nabla}\Pi = B\hat{e}^{(z)},$$

i.e., the background pressure varies only in the vertical, where it maintains hydrostatic balance with the background buoyancy.

The $O(\varepsilon)$ terms in (4.7) give

$$\left(\frac{\partial}{\partial t} + U\frac{\partial}{\partial x}\right)\vec{u}' + U_z w'\hat{e}^{(x)} = -\vec{\nabla}\pi' + b'\hat{e}^{(z)}. \tag{4.8}$$

This is the same as the homogeneous case (3.7) except for the second term on the right-hand side, which describes vertical accelerations due to the perturbation buoyancy.

Substitution of (4.6) into (4.5) gives

$$\left[\frac{\partial}{\partial t} + U\frac{\partial}{\partial x} + \varepsilon\vec{u}' \cdot \vec{\nabla}\right]\left[B(z) + \varepsilon b'\right] = 0. \tag{4.9}$$

For $\varepsilon = 0$, this gives $0 = 0$, so there is no restriction on the background buoyancy profile. The $O(\varepsilon)$ part of (4.9) is:

$$\left(\frac{\partial}{\partial t} + U\frac{\partial}{\partial x}\right)b' + B_z w' = 0. \tag{4.10}$$

It is worthwhile to compare (4.10) with (2.12), the equation for buoyancy perturbations from hydrostatic equilibrium in a motionless, stratified fluid. The final term on the left-hand side describes the advection of the background buoyancy gradient by the vertical velocity perturbation, just as we saw in (2.12). The second term on the left-hand side is new; it describes the advection of buoyancy perturbations by the background flow (which was zero in the motionless case).

4.2.1 Eliminating the Pressure

We eliminate the pressure, as we have done before, by combining the divergence of the momentum equation (4.8) with the Laplacian of its vertical component. The divergence gives a Poisson equation for the pressure:[1]

$$\nabla^2\pi' = -2U_z\frac{\partial w'}{\partial x} + \frac{\partial b'}{\partial z}. \tag{4.11}$$

---

[1] Compare this with equations (2.16) and (3.11).

The vertical component of (4.8) is

$$\left(\frac{\partial}{\partial t} + U\frac{\partial}{\partial x}\right)w' = -\frac{\partial \pi'}{\partial z} + b'. \tag{4.12}$$

Finally, we take the Laplacian of (4.12) and substitute the vertical derivative of (4.11) to obtain:

$$\left(\frac{\partial}{\partial t} + U\frac{\partial}{\partial x}\right)\nabla^2 w' - U_{zz}\frac{\partial w'}{\partial x} = \nabla_H^2 b'. \tag{4.13}$$

In (4.13) and (4.10), we have two equations for the two unknowns w' and b'. We substitute the normal mode forms $w' = \{\hat{w}(z)e^{\sigma t}e^{\iota(kx+\ell y)}\}_r$ and $b' = \{\hat{b}(z)e^{\sigma t}e^{\iota(kx+\ell y)}\}_r$ to obtain a pair of ordinary differential equations:

$$(\sigma + \iota kU)\nabla^2\hat{w} - \iota kU_{zz}\hat{w} = -\tilde{k}^2\hat{b} \tag{4.14}$$
$$(\sigma + \iota kU)\hat{b} + B_z\hat{w} = 0, \tag{4.15}$$

where $\nabla^2 = d^2/dz^2 - \tilde{k}^2$.

4.3 Oblique Modes

Here we will look at obliquity effects using two kinds of Squire transformations. First we extend the approach used previously in section 3.7. We then introduce a new approach based on transforming the velocity profile.

4.3.1 Transforming the Buoyancy

Consider an oblique mode that obeys the equations derived above for stratified shear flow:

$$(\sigma + \iota kU)\nabla^2\hat{w} - \iota kU_{zz}\hat{w} = -\tilde{k}^2\hat{b} \tag{3D1}$$

$$(\sigma + \iota kU)\hat{b} + B_z\hat{w} = 0, \tag{3D2}$$

where

$$\nabla^2 = \frac{d^2}{dz^2} - \tilde{k}^2; \quad \tilde{k} = \sqrt{k^2 + \ell^2}.$$

Suppose also that we have a solution algorithm

$$\sigma = \mathcal{F}(z, U, B_z; k, \ell).$$

The corresponding 2D mode with wave vector $(\tilde{k}, 0)$ obeys:

$$(\sigma + \iota\tilde{k}U)\nabla^2\hat{w} - \iota\tilde{k}U_{zz}\hat{w} = -\tilde{k}^2\hat{b} \tag{2D1}$$

$$(\sigma + \imath \tilde{k} U)\hat{b} + B_z \hat{w} = 0, \tag{2D2}$$

and therefore has the solution algorithm $\sigma_{2D} = \mathcal{F}(z, U, B_z; \ \tilde{k}, 0)$.

Is there a transformation that makes these $(2D)$ equations isomorphic with the $(3D)$? We begin by defining the Squire transformations

$$\sigma = \cos\varphi \ \tilde{\sigma}; \quad \hat{b} = \cos\varphi \ \tilde{\hat{b}}.$$

Now substitute these into $(3D1)$ and divide out the common factor $\cos\varphi$. The result is isomorphic to $(2D1)$

$$(\tilde{\sigma} + \imath \tilde{k} U)\nabla^2 \hat{w} - \imath \tilde{k} U_{zz} \hat{w} = -\tilde{k}^2 \tilde{\hat{b}}. \tag{$\widetilde{3D1}$}$$

Turning to equations $(3D2)$ and $(2D2)$, we now define the additional transformation

$$B_z = \cos^2\varphi \ \tilde{B}_z.$$

Substituting into $(3D2)$ and dividing out $\cos^2\varphi$, we obtain:

$$(\tilde{\sigma} + \imath \tilde{k} U)\tilde{\hat{b}} + \tilde{B}_z \hat{w} = 0. \tag{$\widetilde{3D2}$}$$

With these transformations, $(\widetilde{3D1})$ and $(\widetilde{3D2})$ are isomorphic to $(2D1)$ and $(2D2)$, respectively, and can therefore be solved using the same solution algorithm: $\tilde{\sigma} = \mathcal{F}(z, U, \tilde{B}_z; \ \tilde{k}, 0)$, or

$$\sigma_{3D} = \cos\varphi \times \mathcal{F}\left(z, U, \frac{B_z}{\cos^2\varphi}; \ \tilde{k}, 0\right).$$

The growth rate of the 3D mode is $\cos\varphi$ times that of a corresponding 2D mode that exists *in a fluid with stronger stratification*. In most circumstances this means that the oblique mode will have a slower growth rate, but if stratification should somehow increase the growth rate, and do so rapidly enough to compensate for the obliquity factor $\cos\varphi$, then the oblique mode may grow faster.

Since the angle of obliquity enters only through the function $\cos\varphi$, its sign is irrelevant. Oblique modes therefore come in pairs, identical in every respect except the sign of φ. Such a pair, growing at the same rate, forms a criss-crossing pattern of crests and troughs.

4.3.2 Transforming the Background Velocity

A normal mode perturbation in a shear flow is affected only by the component of the background flow that is parallel to its own wave vector. To see this, inspect (4.14) and (4.15) and note that, wherever U appears, it is multiplied by k. If we

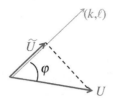

Figure 4.7 Definition sketch for $\tilde{U}$, the component of the background current U in the direction of the wave vector (k, ℓ).

define $\tilde{U}$ such that $kU = \tilde{k}\tilde{U}$ and substitute, we get a pair of equations that is isomorphic to the 2D case with wave vector $(\tilde{k}, 0)$:

$$(\sigma + \imath\tilde{k}\tilde{U})\nabla^2\hat{w} - \imath\tilde{k}\tilde{U}_{zz}\hat{w} = -\tilde{k}^2\hat{b}$$
$$(\sigma + \imath\tilde{k}\tilde{U})\hat{b} + B_z\hat{w} = 0.$$

In fact, the profile $\tilde{U}$ that we just defined is the component of U parallel to the wave vector:

$$\boxed{\tilde{U} = \frac{k}{\tilde{k}}U = U\cos\varphi,}$$

as illustrated in Figure 4.7. So if the growth rate is given by a solution algorithm

$$\sigma = \mathcal{F}(z, U, B_z; k, l),$$

then it is also true that

$$\sigma = \mathcal{F}(z, \tilde{U}, B_z; \tilde{k}, 0).$$

4.4 The Taylor-Goldstein Equation

Based on the results of the previous two sections, we restrict our attention to 2D modes. Replacing $\tilde{k}$ with k, (4.14) and (4.15) become

$$(\sigma + \imath kU)\left(\frac{d^2}{dz^2} - k^2\right)\hat{w} - \imath kU_{zz}\hat{w} = -k^2\hat{b} \qquad (4.16)$$

$$(\sigma + \imath kU)\hat{b} + B_z\hat{w} = 0. \qquad (4.17)$$

If we should need to apply these results to a 3D mode, we simply replace U by $\tilde{U}$ as defined in (4.3.2).

We can derive a single equation for $\hat{w}$ by solving (4.17) for $\hat{b}$ and substituting into (4.16), giving

$$(\sigma + \imath kU)\left(\frac{d^2}{dz^2} - k^2\right)\hat{w} - \imath kU_{zz}\hat{w} = k^2\frac{B_z\hat{w}}{(\sigma + \imath kU)}.$$

Finally, we substitute $\sigma = -\imath kc$ and rearrange to obtain the Taylor-Goldstein (TG) equation:

$$\hat{w}_{zz} + \left\{ \frac{B_z}{(U-c)^2} - \frac{U_{zz}}{U-c} - k^2 \right\} \hat{w} = 0. \tag{4.18}$$

Note that Rayleigh's equation (3.19) can be recovered by setting $B_z = 0$.

Exercise: If you're good with hyperbolic functions (or want to be), try this. Consider the hyperbolic tangent profiles (4.2). Nondimensionalize the problem using the shear scaling, with velocity scale u_0 and length scale h, so that

$$U_\star = \tanh z_\star ; \qquad B_\star = Ri_b \tanh z_\star. \tag{4.19}$$

and $k^\star = kh$. Now suppose that, when Ri_b is not too large, there is a stationary instability like the one we found in the unstratified case (section 3.9.1). Assume also that, if Ri_b is made sufficiently large, the instability is quenched so that $c_\star = 0$. How large must Ri_b be? Try this solution:

$$\hat{w}_\star = (\operatorname{sech} z_\star)^{k_\star} |\tanh z_\star|^{(1-k_\star)}.$$

After a half hour or so of differentiating, you should find that this solution works provided that

$$Ri_b = k_\star(1 - k_\star).$$

This tells us that the stability boundary is an inverted parabola on the $k_\star - Ri_b$ plane with peak at $k_\star = 1/2$, $Ri_b = 1/4$ (illustrated later on figure 6.3). In other words, the critical Richardson number for this flow is 1/4.

Also, as this critical Richardson number is approached, the wavenumber of the fastest-growing mode approaches 1/2, not very different from the value 0.44 found in the unstratified case. More precisely, the ratio of wavelength $2\pi/k$ to shear layer thickness $2h$ approaches 2π, whereas in the homogeneous case the value is 7. In the context of a "rule of thumb," these two values are effectively equal.

4.5 Application to Internal Wave Phenomena

Solutions of (4.18) with real c represent waves: internal gravity waves (section 2.2.1), vorticity waves (section 3.12.2), or some combination of the two. If we set $U = 0$ and $B_z = $ constant in (4.18), we recover the dispersion relation for internal gravity waves in uniform stratification (section 2.2.1).

The limit $k \to 0$ is called the hydrostatic limit. Near that limit, perturbations involve very weak vertical accelerations and are therefore nearly in hydrostatic balance (like the background flow). If we take this limit and also assume $U = 0$, the TG equation becomes

$$\hat{w}_{zz} + \frac{B_z}{c^2}\,\hat{w} = 0.$$

This is the equation for baroclinic normal modes, whose description may be found in any geophysical fluid dynamics text. The hydrostatic limit is a useful description not only for small-amplitude waves but also for nonlinear phenomena such as solitary waves, bores, hydraulic jumps, and gravity currents. For example, in the weakly nonlinear theory of solitary waves in a stratified shear flow, the dependence on x and t is described by the Korteweg-De Vries equation, while the vertical structure is a solution of the TG equation (4.18) in the hydrostatic limit (Lee and Beardsley, 1974).

Although we will not venture far into the realm of waves here, it is important to note that *the numerical methods that we are developing (sections 3.5, 6.2) apply just as well to waves as they do to instabilities.* Those methods are often used to determine gravity wave and baroclinic mode characteristics in realistic situations where the stratification is not uniform and the background current is nonzero.

4.6 Analytical Examples of Instability in Stratified Shear Flows

Like the Rayleigh equation (3.16–3.19), the TG equation (4.18) is easy to solve when the background profiles are sufficiently simple. Here we describe a few examples that show how shear instability is affected by stratification.

4.6.1 Kelvin-Helmholtz and Rayleigh-Taylor Instabilities at an Interface

Imagine an infinitely thin interface at which the velocity and the buoyancy change:

$$U = \frac{\Delta u}{2}\begin{cases} 1, & z > 0 \\ -1, & z < 0 \end{cases} \qquad B = \frac{\Delta b}{2}\begin{cases} 1, & z > 0 \\ -1, & z < 0 \end{cases} \tag{4.20}$$

As expressed in (4.18), the TG equation involves the second-derivative U_{zz} and therefore cannot handle this discontinuity in U. To get around this problem we rephrase the TG equation in terms of the vertical displacement function η', defined by

$$w' = \left(\frac{\partial}{\partial t} + U\frac{\partial}{\partial x}\right)\eta'.$$

In normal mode form this is

$$\hat{w} = (\sigma + \imath k U)\hat{\eta} = \imath k(U - c)\hat{\eta}.$$

With this change of variables (4.18) becomes

$$\left[(U-c)^2\hat{\eta}_z\right]_z + \left[B_z - k^2(U-c)^2\right]\hat{\eta} = 0. \tag{4.21}$$

(You will derive this in homework problem 9.)

The solution is simple because U and B are constant except at the interface. Requiring that $\hat{\eta}$ be continuous and bounded for all z and assuming $k > 0$,

$$\hat{\eta} = Ae^{-k|z|}. \tag{4.22}$$

The discontinuity in $B(z)$ imposes another condition on the solution. Because of that discontinuity, the derivative B_z has the form of a Dirac delta function (section 2.2.4):

$$B_z = \Delta b\delta(z). \tag{4.23}$$

Using property 3 of the delta function (listed in Figure 2.5), you can check that (4.23) integrates to give $B(z)$ as defined in (4.20).

We now apply the integral operation $\lim_{\epsilon\to 0}\int_{-\epsilon}^{\epsilon} dz$ to (4.21). The first term gives

$$\lim_{\epsilon\to 0}\int_{-\epsilon}^{\epsilon}\left[(U-c)^2\hat{\eta}_z\right]_z dz \;=\; \left[\!\left[(U-c)^2\hat{\eta}_z\right]\!\right]_0.$$

Next,

$$\lim_{\epsilon\to 0}\int_{-\epsilon}^{\epsilon} B_z\hat{\eta}\, dz \;=\; \lim_{\epsilon\to 0}\int_{-\epsilon}^{\epsilon} \Delta b\delta(z)\,\hat{\eta}\, dz \;=\; \Delta b\hat{\eta}(0),$$

where property 5 of the delta function has been used. The final term is

$$\lim_{\epsilon\to 0}\int_{-\epsilon}^{\epsilon}\left[k^2(U-c)^2\hat{\eta}\right]dz \;=\; 0.$$

This integral vanishes because the integrand is finite, so when we take the limit $\epsilon\to 0$ the result is zero. Summing these three integrated terms, we have a combined jump condition for a buoyancy change and a velocity change:

$$\boxed{\left[\!\left[(U-c)^2\hat{\eta}_z\right]\!\right]_0 + \Delta b\hat{\eta}(0) = 0.} \tag{4.24}$$

Exercise: Starting from (4.24) recover (2.41), the jump condition for a buoyancy interface in a motionless fluid, by making the appropriate substitutions.

Substituting from (4.22), (4.24) becomes

$$-k\left[\left(\frac{\Delta u}{2}-c\right)^2 + \left(-\frac{\Delta u}{2}-c\right)^2\right] + \Delta b = 0. \tag{4.25}$$

which we can solve for c to get

$$c = \pm\sqrt{\frac{\Delta b}{k} - \frac{\Delta u^2}{4}}.$$

or

$$\sigma = \pm\sqrt{\frac{\Delta u^2}{4}k^2 - \Delta bk}.$$

Evidently Δu has a destabilizing effect regardless of its sign, whereas positive Δb acts to reduce the growth rate. (Remember we have assumed that $k > 0$.)

In terms of the nondimensional variables

$$c^\star = \frac{c}{\Delta u}\ ; \quad k^\star = k\frac{\Delta u^2}{\Delta b}\ ; \quad \sigma^\star = \frac{\Delta u}{\Delta b}\sigma$$

the dispersion relations become

$$c^\star = \pm\sqrt{\frac{1}{k^\star} - \frac{1}{4}}\ ; \quad \sigma^\star = \mp\sqrt{\frac{k^{\star 2}}{4} - k^\star}$$

If $0 \le k^\star \le 4$, $c^\star$ is real and the solution describes two waves moving oppositely (Figure 4.8; blue curves). Otherwise, if $k^\star > 4$ or $k^\star < 0$ (which corresponds to $\Delta b < 0$), we have a growing and a decaying mode (red curves).

- As long as $\Delta u \ne 0$, the flow is unstable, i.e., there is always a range of k in which one solution has $\sigma_r > 0$. These provide the simplest example of the Kelvin-Helmholtz instability: a shear instability partially damped by stable stratification.
- The instability exhibits ultraviolet catastrophe: the shortest waves (largest $k^\star$) have arbitrarily large growth rate.
- Modes with $0 \le k^\star \le 4$ represent interfacial waves.

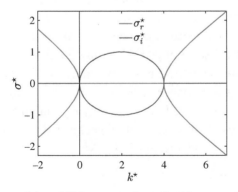

Figure 4.8 Waves and instabilities on a sharp interface at which velocity and buoyancy change discontinuously (4.20). The scaled wavenumber and growth rate are defined as $k^\star = ku_0^2/b_0$ and $\sigma^\star = \sigma u_0/b_0$, respectively.

- If the buoyancy change is *unstable* ($\Delta b < 0$), then the scaled wavenumber is negative (though the dimensional wavenumber is not). The interface is convectively unstable, and all disturbances are amplified. This is a generalization of the Rayleigh-Taylor instability that we explored in section 2.2.4.

4.6.2 The Stratified Shear Layer

The previous example can be made more realistic by considering a shear layer with finite thickness $2h$, as in section 3.3, but retaining the two-layer stratification profile:

$$U(z) = u_0 \begin{cases} 1, & z \geq h \\ z/h, & -h < z < h \\ -1, & z \leq -h \end{cases} \quad \text{and} \quad B(z) = b_0 \begin{cases} 1, & z > 0 \\ 0, & z < 0. \end{cases} \quad (4.26)$$

The Taylor-Goldstein equation (4.18) simplifies considerably, permitting an analytical solution. This is due to the delta function behavior of the vorticity and buoyancy gradient profiles, viz.,

$$U_{zz}(z) = \Delta Q_1 \delta(z - h) + \Delta Q_2 \delta(z + h) \quad \text{and} \quad B_z(z) = b_0 \delta(z), \quad (4.27)$$

where $\Delta Q_1 = -u_0/h$ and $\Delta Q_2 = u_0/h$. In the regions between the interfaces of the flow, where the delta functions in (4.27) are centered, (4.18) reduces to

$$\hat{w}_{zz} - k^2 \hat{w} = 0. \quad (4.28)$$

Requiring that $\hat{w}$ be (first) continuous across each interface and (second) bounded as $z \to \pm\infty$, we can write the solution as

$$\hat{w}(z) = A_0 e^{-k|z|} + A_1 e^{-k|z-h|} + A_2 e^{-k|z+h|}, \quad (4.29)$$

where the A_i are constants to be determined and k is assumed to be positive. Note that this solution is similar to the piecewise shear layer of section 3.3, in that each interface has its own "influence" function associated with it, which decays exponentially over a vertical scale of k^{-1}. The only difference is that we now have a buoyancy interface (with coefficient A_0) in addition to the two vorticity interfaces with coefficients A_1 (upper) and A_2 (lower).

Values of the coefficients are found by applying a jump condition at each interface. We'll do this first for a *general* interface, located at z_j and allowing for *both* a vorticity jump ΔQ_j and a buoyancy jump Δb_j.

General Jump Condition
Multiplying the Taylor-Goldstein equation (4.18) by $(U - c)^2$, and integrating across a small region encompassing an interface at z_j gives

$$\lim_{\epsilon \to 0} \int_{z_j-\epsilon}^{z_j+\epsilon} \left[(U-c)^2 \hat{w}_{zz} - U_{zz}(U-c)\hat{w} + B_z \hat{w} - k^2(U-c)^2 \hat{w} \right] dz = 0. \quad (4.30)$$

The first term is integrated by parts to give $[\![(U-c)^2 \hat{w}_z]\!]_j$. Because $(U-c)^2$ is continuous, this is equivalent to $(U-c)^2 [\![\hat{w}_z]\!]_j$. The second and third terms of (4.30) can be evaluated directly using the properties of delta functions (Figure 2.5), and the last term vanishes as $\epsilon \to 0$. We now have the general jump condition at an interface with a buoyancy jump Δb_j and/or a vorticity jump ΔQ_j:

$$\boxed{(U_j-c)^2 [\![\hat{w}_z]\!]_j - \Delta Q_j(U_j-c)\hat{w}_j + \Delta b_j \hat{w}_j = 0} \quad (4.31)$$

where the j subscripts indicate that the function is evaluated at z_j, e.g., $U_j = U(z_j)$. Exercise: Check that, for the special case of a vorticity jump in a homogeneous environment, $\Delta b_j = 0$, (4.31) reproduces the previous result (3.30). Verify also that, for a buoyancy jump only, (4.31) is equivalent to (4.24) with U continuous and also to (2.41).

We now apply (4.31) at the three interfaces in turn.

- For the interface at $z = h$, $\hat{w} = A_0 e^{-|kh|} + A_1 + A_2 e^{-2|kh|}$, $[\![\hat{w}_z]\!] = -2kA_1$, $U = u_0$, the vorticity jump is $\Delta Q_1 = -u_0/h$, and $\Delta b = 0$. Substituting these expressions into (4.31) gives

$$\left(u_0 - c \right) A_1 + \frac{\Delta Q_1}{2k} \left(A_0 e^{-kh} + A_1 + A_2 e^{-2kh} \right) = 0. \quad (4.32)$$

- Similarly, applying (4.31) at $z = -h$ gives

$$\left(-u_0 - c \right) A_2 + \frac{\Delta Q_2}{2k} \left(A_0 e^{-kh} + A_1 e^{-2kh} + A_2 \right) = 0. \quad (4.33)$$

- At the buoyancy interface $z = 0$, we have $\Delta Q = 0$, $[\![\hat{w}_z]\!] = -2kA_0$, and $U = 0$, so the jump condition results in

$$A_0 c^2 - \frac{b_0}{2k}(A_0 + A_1 e^{-kh} + A_2 e^{-kh}) = 0. \quad (4.34)$$

Combining (4.32, 4.33, 4.34) results in a solution for A_0, A_1, and A_2 and a dispersion relation that is quadratic in c^2:

$$c^4 + B_2 c^2 + B_0 = 0, \quad (4.35)$$

with coefficients

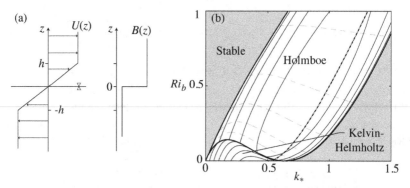

Figure 4.9 (a) Profiles of the piecewise stratified shear layer. The pair of triangles indicates the buoyancy interface. (b) Stability diagram of the piecewise stratified shear layer. Solid contours are of growth rate (every $0.03u_0/h$), and dashed gray contours are of phase speed (every $0.1u_0$), with gray representing stable regions of propagating waves. The resonance approximation from (4.39) is shown as a dash-dot line. Adapted from Carpenter et al. (2013).

$$B_2 = -[(u_0 + c_v)^2 + c_g^2 - c_v^2 e^{-4kh}]$$
$$B_0 = c_g^2 \{(u_0 + c_v)^2 - e^{-2kh}[2(u_0 + c_v)c_v - c_v^2 e^{-2kh}]\}.$$

Here we have defined $c_v = \Delta Q_1/2k = -\Delta Q_2/2k$ and $c_g^2 = b_0/2k$, the intrinsic phase speeds of isolated vorticity and interfacial gravity waves as given in (2.44) and (3.81), respectively.

With the shear scaling $c_\star = c/u_0$, $k_\star = kh$ (cf. section 3.6), the dispersion relation has the form $c_\star(k_\star, Ri_b)$, where $Ri_b = b_0 h/u_0^2$ is a bulk Richardson number. The results of the stability analysis can therefore be plotted on the Ri_b-$k_\star$ plane, as shown in Figure 4.9. This diagram includes contours of growth rate $\sigma_r^\star$ (solid curves) and phase speed $c_r^\star$ (dashed), for the fastest-growing mode at each location on the plane.

For Ri_b less than about 0.07, there is a range of wavenumbers for which the growth rate is real and positive. At $Ri_b = 0$, this unstable regime extends from $k_\star = 0$ to $k_\star = 0.64$. This is the stratified extension of the shear instability described in section 3.3. As in that case, the instability has zero phase speed (i.e., moves with the mean flow speed) in this region. The instability is referred to as Kelvin-Helmholtz, as in section (4.6.1). As Ri_b is increased, the growth rate of the instability is reduced and the band of wavenumbers it occupies shrinks to zero. This is just as we would expect given that a growing mode must do work against gravity.

When Ri_b exceeds 0.07, the Kelvin-Helmholtz instability disappears, but the flow is *not* stable. Instead, a fundamentally different instability is found, the Holmboe instability. This instability is oscillatory; the fastest-growing mode is

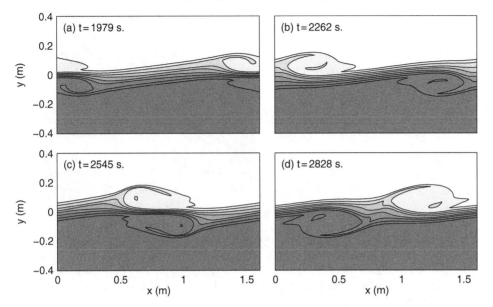

Figure 4.10 Buoyancy contours in a finite-amplitude Holmboe wave. Boundary conditions are horizontally periodic, so that each vortex is one member of an infinite sequence. Parameter values are $Ri_0 = 0.45$; $Re = 300$; $Pr = 9$; $k^\star = 0.35$; $\ell^\star = 0$ (from Smyth and Winters, 2003).

actually two modes with the same growth rate and equal but opposite phase speeds, one stronger in the upper half of the shear layer, the other in the lower half. This behavior is illustrated in Figure 4.10, which shows snapshots from a simulation of the nonlinear wave that grows from Holmboe instability. Each mode grows to form a finite-amplitude vortex. In Figure 4.10(a), the vortices are at the phase of maximal separation. The vortices approach (b), pass each other (c), and move apart (d).

Counterintuitively, the Holmboe instability exists *only* in the presence of statically stable stratification, and its growth rate increases with increasing Ri_b to a maximum at $Ri_b = 0.4$ (Figure 4.9b). To better understand the origin of the Holmboe instability, we look next at a simpler version of the piecewise stratified shear layer first proposed by Baines and Mitsudera (1994).

4.6.3 Holmboe Instability in a Semi-Infinite Shear Layer

Consider the same profiles as in Figure 4.9(a), except now we will remove the lower kink in $U(z)$, so that the shear extends infinitely toward negative z, as shown in Figure 4.11(a). The dispersion relation for this new set of profiles can be easily recovered by setting $A_2 = 0$ in (4.32, 4.34) and discarding (4.33). The result is a cubic equation for c,

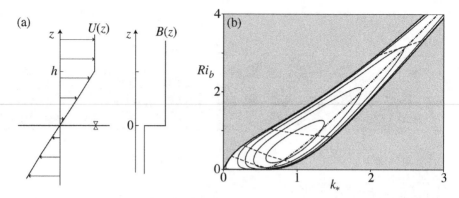

Figure 4.11 (a) Profiles illustrating the simplified setup for Holmboe's instability. (b) Stability characteristics of the Holmboe instability. Solid contours are of growth rate (every $0.03u_0/h$) with gray representing stable regions of propagating waves. All unstable modes are propagating with $c_r \neq 0$ and dashed contours spaced at $0.2u_0$. The resonance approximation from (4.39) is shown as a dash-dot line. Adapted from Baines and Mitsudera (1994) and Carpenter et al. (2013).

$$c^3 - (u_0 + c_v)c^2 - c_g^2 c - c_g^2[u_0 + c_v(1 - e^{-2kh})] = 0, \qquad (4.36)$$

with c_v and c_g defined as above. Again we nondimensionalize and plot contours of growth rate, $\sigma_\star = k_\star c_i^\star$, on the $k_\star$-Ri_b plane in Figure 4.11(b).

The resulting stability diagram does not include a Kelvin-Helmholtz instability region, but displays the Holmboe instability with little alteration. Evidently the lower vorticity interface is crucial for the Kelvin-Helmholtz instability, but unnecessary for the Holmboe instability. This result can be understood when considering the Kelvin-Helmholtz instability to be an extension of the inflectional instability of the homogeneous piecewise shear layer, which relies on the interaction of vorticity waves at the upper and lower vorticity interfaces (section 3.12.3). By removing the inflection point (i.e., without the lower vorticity jump there is no change in the sign of U_{zz}) we have effectively eliminated this instability.

With the addition of a statically stable interface, we have generated a new instability. This is an example of how statically stable stratification can *destabilize* an otherwise stable flow, and is therefore a counterexample to our provisional "rule of thumb" suggested in section 4.1.

The origin of the Holmboe instability can be understood by considering the stable wave modes that are present in this system, and how they may interact to produce instability (as discussed in section 3.12 for homogeneous shear flows). The phase speeds of these wave modes are shown in Figure 4.12 for the example $Ri_b = 2$. Three modes can be identified: the vorticity wave and two internal gravity waves that propagate on the buoyancy interface in opposite directions. The phase

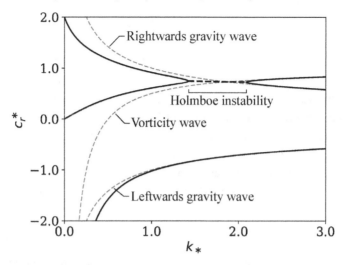

Figure 4.12 Dispersion relation of the Holmboe instability at $Ri_b = 2$. The black curves show c_r^* of the stable waves that exist in the Holmboe dispersion relation (4.36), and the black dashed line the region of the Holmboe instability. Gray dashed lines are the phase speeds of the isolated waves as labeled.

speeds of each of these modes in isolation (i.e., neglecting any interaction) follows from the jump condition (4.31):

$$\text{Vorticity wave:} \quad c = u_0 + \frac{\Delta Q_1}{2k} \quad \Rightarrow \quad c_\star = 1 - \frac{1}{2k_\star} \quad (4.37)$$

and

$$\text{Gravity waves:} \quad c = \pm\left(\frac{\Delta b}{2k}\right)^{1/2} \quad \Rightarrow \quad c_\star = \pm\left(\frac{Ri_b}{2k_\star}\right)^{1/2} \quad (4.38)$$

(cf. 2.44 and 3.81). These phase speeds are plotted in Figure 4.12 as dashed lines. They generally correspond closely to the modes seen in the full Holmboe dispersion relation. The correspondence is inexact, however, due to the interaction between the waves, which is stronger at lower $k_\star$ because the eigenfunctions extend further in the vertical (4.29).

In a mechanism identical to the resonance of two vorticity waves in the piecewise shear layer, the Holmboe instability arises from a resonance of the vorticity wave at $z = h$ and one of the internal gravity waves at $z = 0$. The interaction is centered around the crossing of the waves in the dispersion relation where there is a natural phase-locking, and can clearly be seen in Figure 4.12. This also suggests that we can approximate the location of the band of instability in the Ri_b-$k_\star$ plane by simply equating the speed of the rightward propagating internal gravity wave and the vorticity wave, giving

$$Ri_b = 2k_\star\left(1 - \frac{1}{2k_\star}\right)^2.$$ (4.39)

This curve is plotted in Figure 4.11 and is clearly centered on the band of instability. Referred to as the resonance approximation, this technique is generally useful in identifying the wave resonances responsible for instability, especially in profiles with many possible interactions (Caulfield, 1994).

Finally, this case demonstrates that Rayleigh's inflection point theorem, and others from homogeneous flows, are no longer strictly accurate when stratification is present. In fact, we have seen that no inflection point is needed; stratification can in some sense "complete the inflection point" by providing the necessary wave resonance.

4.6.4 Multi-Layered Shear Flow: the Taylor-Caulfield Instability

Our final example involves no vorticity waves at all, but instead two interfacial gravity waves in a non-inflectional shear flow that resonate just as do the vorticity waves in section 3.12. The role of the shear is to facilitate phase-locking so that the resonance is sustained.

The background state has a three-layered stratification profile (Figure 4.13) with equal buoyancy jumps b_0 across interfaces located at $z = \pm h$. The background velocity is $u_0 z/h$, so the shear is uniform. The dispersion relation is found by applying the jump condition (4.31) at the interfaces:

$$c^4 - 2(u_0^2 + c_g^2)c^2 + (u_0^2 - c_g^2)^2 - e^{-4kh}c_g^4 = 0,$$ (4.40)

with $c_g = \sqrt{b_0/2k}$ as in the previous examples.

The stability diagram (Figure 4.13b) shows a band of instability with growth rates increasing with Ri_b up to $Ri_b = 1$. The instability, called the Taylor-Caulfield instability, is stationary with $c_r = 0$.

Here we have another example of statically stable stratification destabilizing a flow that is stable in the homogeneous case $Ri_b = 0$. The origin of the instability is in the resonance of the leftward propagating gravity wave at the upper interface with the rightward propagating gravity wave at the lower interface. The shear makes it possible for these waves to phase-lock. The phase-locking condition, corresponding to the crossing of the phase-speed curves in the dispersion diagram, can be found by setting the speeds equal to each other, giving[2] $Ri_b = 2k_\star$. This is again found to closely follow the band of instability in Figure 4.13.

---

[2] Try to derive this yourself. Be careful to include the speed of the mean flow at each buoyancy interface.

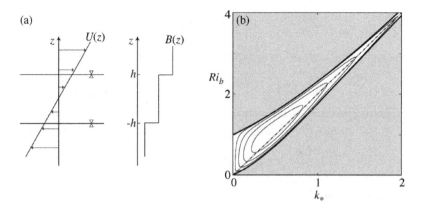

Figure 4.13 (a) Profiles illustrating the simplified setup for the Taylor-Caulfield instability of a sheared multilayered stratification. (b) Stability characteristics of the Taylor-Caulfield instability. Solid contours are of growth rate (every $0.02u_0/h$) with gray representing stable regions of propagating waves. All unstable modes are stationary with $c_r = 0$. Adapted from Carpenter et al. (2013).

4.7 The Miles-Howard Theorem

As we have discussed, the Richardson number $Ri = B_z/U_z^2$ quantifies the competing effects of stratification and shear. If $Ri \gg 1$, stratification dominates and the flow is stable. Conversely, if $Ri \ll 1$, instability is likely. The boundary between stable and unstable flows must lie at some intermediate value of Ri, which we'll call Ri_c. The Miles-Howard theorem tells us that the minimum of $Ri(z)$ must be less than $Ri_c = 1/4$ for instability to be possible. To be precise:

Miles-Howard theorem: *A necessary condition for instability in an inviscid, nondiffusive, stratified, parallel shear flow is that the minimum value of $Ri(z)$ be less than 1/4.*

To prove this, we transform the TG equation via the following change of variables:

$$\hat{w} = (U - c)^{1/2}\phi.$$

The algebra is left as an exercise; the result is:

$$\left[(U - c)\phi_z\right]_z + P(z)\phi = 0, \tag{4.41}$$

where

$$P = \frac{B_z - \frac{1}{4}U_z^2}{U - c} - \frac{1}{2}U_{zz} - k^2(U - c).$$

We now multiply (4.41) by ϕ^* and integrate over the vertical domain:

$$\int_{z_B}^{z_T} \phi^* \left[(U - c)\phi_z \right]_z dz + \int_{z_B}^{z_T} \phi^* P(z)\phi dz = 0,$$

where z_B and z_T may be finite or infinite. Now integrate the first term by parts:

$$\phi^*(U - c)\phi_z \Big|_{z_B}^{z_T} - \int_{z_B}^{z_T} \phi_z^*(U - c)\phi_z dz + \int_{z_B}^{z_T} \phi^* P(z)\phi dz = 0.$$

Because $\hat{w}$ vanishes at the boundaries, the first term drops out and we have

$$-\int_{z_B}^{z_T} (U - c)|\phi_z|^2 dz + \int_{z_B}^{z_T} P(z)|\phi|^2 dz = 0.$$

The imaginary part is

$$c_i \int_{z_B}^{z_T} \left[|\phi_z|^2 + \frac{B_z - \frac{1}{4}U_z^2}{|U - c|^2} |\phi|^2 + k^2|\phi|^2 \right] dz = 0.$$

The first and third terms in the integrand are positive definite. Now suppose that the second term is also positive definite. In that case, the integral is positive, and the equation can only be satisfied if $c_i = 0$. Conversely, the only way c_i can be nonzero is if the second term is negative somewhere, i.e.,

$$B_z - \frac{1}{4}U_z^2 < 0, \quad \text{or} \quad Ri < 1/4,$$

for some z.

It is important to understand just what the Miles-Howard theorem does, and does not, say about the stability of particular profiles $U(z)$ and $B(z)$. Because it is a *necessary* condition for instability, it tells us only that instability is *possible* when $Ri < 1/4$ for some z, not that it actually occurs. In logical terms, what's proven is that *if $c_i \neq 0$ then $Ri < 1/4$ for some z*, not the converse. In practice, if we find that $Ri(z) < 1/4$ at some location and want to know whether instability is present, we must solve the Taylor-Goldstein equation explicitly. On the other hand, the theorem *does* definitively identify states where instability is *not* possible. If $Ri(z) > 1/4$ everywhere, no further analysis is needed; the flow is stable.

We have already seen the example of the continuously stratified shear layer (4.19), where instability requires that the minimum Ri be less than 0.25, consistent with the Miles-Howard theorem. As a second example, suppose the stratification is again $B^\star = Ri_b \tanh z^\star$, but the velocity profile is the Bickley jet $U^\star = \text{sech}^2 z^\star$ as described in section 3.9.2. In this case, instability occurs if and only if the minimum Richardson number is less than 0.231 (Drazin and Howard, 1966; Hazel, 1972), again consistent with the theorem.

The examples shown in sections 4.6.2–4.6.4 are a different matter. At first glance, these appear to contradict the Miles-Howard theorem by exhibiting instability when $Ri_b > 0.25$. The contradiction is resolved by noting that, for these profiles, Ri_b is not the *minimum* value of $Ri(z)$. In fact, $Ri = 0$ for all z except at buoyancy interfaces, regardless of Ri_b. The theorem is therefore irrelevant for these simple models.

In a compressible fluid, the Miles-Howard theorem remains valid provided that B_z is calculated as in (2.64) and (2.65) as shown by Chimonas (1970).

4.8 Howard's Semicircle Theorem

The Miles-Howard theorem described in the previous section provides a condition that the *mean flow* must satisfy if instability is to grow. Here we describe a condition that the *mode* must satisfy.

Howard's semicircle theorem: *In an inviscid, nondiffusive, stably stratified, parallel shear flow, let the background velocity $U(z)$ be bounded by U_{min} and U_{max}. Any unstable normal mode must have complex phase speed c located within the semicircle centered at $c_r = (U_{max} + U_{min})/2$, $c_i = 0$ having radius $(U_{max} - U_{min})/2$, as shown on Figure 4.14.*

A corollary is that the real part c_r must lie within the range of the mean flow. In other words, every unstable mode must have a critical level. This result is a generalization of the critical level theorem (section 3.11.3).

The proof starts off similar to that for the Miles-Howard theorem (in fact it appeared in the same paper), but is more involved. We write the TG equation in terms of the vertical displacement, reproducing (4.21):

$$[(U - c)^2\hat{\eta}_z]_z + [B_z - k^2(U - c)^2]\hat{\eta} = 0 \tag{4.42}$$

(cf. exercise 9).

As before, we multiply by $\hat{\eta}^*$ and integrate over the vertical domain. Integrating by parts, using the boundary condition $\hat{\eta} \to 0$, and rearranging we obtain

$$\int_{z_B}^{z_T} B_z|\hat{\eta}|^2 dz = \int_{z_B}^{z_T} (U - c)^2 \big[|\hat{\eta}_z|^2 + k^2(U - c)^2|\hat{\eta}|^2\big] dz. \tag{4.43}$$

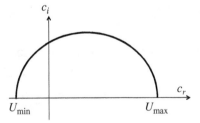

Figure 4.14 Howard's semicircle on the complex c-plane, bounded by the extremal values of the mean velocity. The phase speed of an unstable mode must lie within the semicircle (Howard, 1961).

The imaginary part is

$$0 = -2c_i \int_{z_B}^{z_T} (U - c_r)\left[|\hat{\eta}_z|^2 + k^2|\hat{\eta}|^2\right]dz. \tag{4.44}$$

Because the quantity in square brackets is positive definite, the integral can be zero only if $U - c_r$ changes sign. Two implications of this are worth noting:

- The discussion so far is identical to our earlier proof that an unstable mode must have a critical level in a *homogeneous* shear flow (section 3.11.3). We now see that a critical level is also necessary in *stratified* flow. The stratification term on the left-hand side of (4.43) becomes irrelevant when we take the imaginary part.
- In the special case $U = 0$, this result shows that c_r must be zero. This is relevant to the convective case $B_z < 0$, which we examined in section 2.2.3. We see once again that all unstable modes are stationary.

To extend the proof, we now consider the real part of (4.43):

$$\int_{z_B}^{z_T} B_z|\hat{\eta}|^2 dz = \int_{z_B}^{z_T} (U^2 - 2c_r U + c_r^2 - c_i^2)\left[|\hat{\eta}_z|^2 + k^2|\hat{\eta}|^2\right]dz. \tag{4.45}$$

The second term on the right-hand side can be written as

$$-2c_r \int_{z_B}^{z_T} U\left[|\hat{\eta}_z|^2 + k^2|\hat{\eta}|^2\right]dz$$

which, by (4.44), is equal to

$$-2c_r \int_{z_B}^{z_T} c_r\left[|\hat{\eta}_z|^2 + k^2|\hat{\eta}|^2\right]dz.$$

Therefore, in the second term on the right-hand side of (4.45), we can change the U to c_r, resulting in

$$\int_{z_B}^{z_T} B_z|\hat{\eta}|^2 dz = \int_{z_B}^{z_T} (U^2 - |c|^2)\left[|\hat{\eta}_z|^2 + k^2|\hat{\eta}|^2\right]dz. \tag{4.46}$$

If $B_z > 0$, we have

$$\int_{z_B}^{z_T} (U^2 - |c|^2)\left[|\hat{\eta}_z|^2 + k^2|\hat{\eta}|^2\right]dz > 0. \tag{4.47}$$

Now comes the cute part. Define U_{max} and U_{min} as the maximum and minimum values of $U(z)$. Note that $U_{max} - U \geq 0$ and $U_{min} - U \leq 0$, and therefore

$$(U_{max} - U)(U_{min} - U) \leq 0.$$

As a result,

$$\int_{z_B}^{z_T} (U_{max} - U)(U_{min} - U)\left[|\hat{\eta}_z|^2 + k^2|\hat{\eta}|^2\right]dz \leq 0,$$

or

$$\int_{z_B}^{z_T} \left[U_{max}U_{min} - U(U_{max} + U_{min}) + U^2\right]\left[|\hat{\eta}_z|^2 + k^2|\hat{\eta}|^2\right]dz \leq 0. \qquad (4.48)$$

Note that the integrand is the product of the two factors in square brackets. We will now convert the first of these factors to a constant, which we can then remove from the integral. We do this in two steps. First, as we noted above, (4.44) allows us to replace U with c_r in the second term. Now consider the third term, which contains U^2. By (4.47),

$$\int_{z_B}^{z_T} |c|^2\left[|\hat{\eta}_z|^2 + k^2|\hat{\eta}|^2\right]dz < \int_{z_B}^{z_T} U^2\left[|\hat{\eta}_z|^2 + k^2|\hat{\eta}|^2\right]dz.$$

So if we replace U^2 by $|c|^2$ in (4.48), the inequality is still true:

$$\int_{z_B}^{z_T} \left[U_{max}U_{min} - c_r(U_{max} + U_{min}) + |c|^2\right]\left[|\hat{\eta}_z|^2 + k^2|\hat{\eta}|^2\right]dz \leq 0. \qquad (4.49)$$

Given that the first factor in the integrand is a constant and the second is positive definite, the inequality can be true only if the first factor is negative:

$$U_{max}U_{min} - c_r(U_{max} + U_{min}) + |c|^2 \leq 0.$$

After some juggling, this becomes

$$\left(c_r - \frac{U_{max} + U_{min}}{2}\right)^2 + c_i^2 \leq \left(\frac{U_{max} - U_{min}}{2}\right)^2. \qquad (4.50)$$

This inequality describes the interior of a circle on the complex c-plane whose radius is $(U_{max} - U_{min})/2$ and whose center is on the real axis at $(U_{max} + U_{min})/2$. For any unstable mode, $c_i > 0$ and therefore c must lie in the upper half of the circle; i.e., in Howard's semicircle, as shown in Figure 4.14.

Note that, like the Miles-Howard theorem, the semicircle theorem is a statement that *if* there is an unstable mode, *then* certain conditions are true, not the other way around. A mode can lie within the semicircle and still not be unstable.

4.9 Energetics

To analyze the perturbation kinetic energy, we repeat the calculation of section 3.10, beginning with the momentum equation (4.8) instead of (3.7). The only difference is the buoyancy term $b'\hat{e}^{(z)}$. Therefore, converting to normal mode form gives the same results for the continuity and horizontal momentum equations but the vertical momentum equation now contains the normal mode buoyancy perturbation $\hat{b}$:

$$(\sigma + \iota kU)\,\hat{w} = -\hat{\pi}_z + \hat{b}. \qquad (4.51)$$

As in section 3.10, we multiply the momentum equations by the conjugates of the velocity eigenfunctions, take the real part, and divide by 2. The result is (3.64) with an added term:

$$2\sigma_r K = SP - \frac{d}{dz} EF + BF,$$ (4.52)

where

$$BF = \overline{w'b'} = \frac{1}{2}(\hat{w}^*\hat{b})_r$$

is the buoyancy flux, also called the buoyancy production. As before, the overbar represents a horizontal average over an integer number of wavelengths.

When buoyant fluid rises and dense fluid sinks, $BF > 0$. Therefore, BF is the second term we have found (after SP) that is capable of creating kinetic energy.[3] This term is the source of kinetic energy for convective instability. In contrast $BF < 0$ means that the instability has to do work against gravity in order to grow, i.e., it must lift dense fluid and depress buoyant fluid. Like the shear production (and unlike the energy flux), the buoyancy production can integrate to a nonzero value, and can therefore increase or decrease the net kinetic energy of a disturbance. It is often useful to classify instabilities according to whether their main energy source is SP or BF (Table 4.1).

The buoyancy flux also shows up in the buoyancy variance budget. To derive this budget, we start with (4.15), the buoyancy perturbation equation in normal mode form:

$$(\sigma + \imath kU)\hat{b} + B_z\hat{w} = 0.$$

By analogy with the development of the kinetic energy equation, we multiply through by $\hat{b}^*$, take the real part, and divide by 2. The result is

$$\sigma_r \frac{|\hat{b}|^2}{2} = -\frac{B_z}{2}(\hat{b}^*\hat{w})_r = -B_z BF.$$

The only source of buoyancy variance is a production term that quantifies the interaction of B_z and BF.

A mode can grow only if B_z and BF have opposite sign. In the case of statically stable stratification, $B_z > 0$ and therefore $BF < 0$. Referring back to the kinetic energy budget (4.52) we see that, when $B_z > 0$, the buoyancy flux can *only* act to reduce growth, just as we surmised in section 4.1.

Now here is a paradox. In some cases (e.g., Holmboe instability), growth is possible *only* in the presence of stable stratification, even though the resulting

---

[3] More precisely, it converts potential energy into kinetic energy.

Table 4.1 *Categorizing instabilities by their energy source.*

	$BF < 0$	$BF > 0$
$SP < 0$	stable	convective instability opposed by shear
$SP > 0$	shear instability opposed by buoyancy	sheared convection

buoyancy flux can *only* act to reduce growth. The resolution of this paradox is that, even though some kinetic energy is diverted through BF, the stable stratification causes the velocity field to arrange itself such that the *shear production SP* is positive. Specifically, the resonance between the phase-locked shear and gravity waves causes the phase lines of w' to tilt against the shear as in section 3.11.4. Thus, *while shear is the energy source for the instability, it is buoyancy that allows that energy source to be tapped.*

4.10 Summary

For a parallel shear flow $U(z)$ in an inviscid, nondiffusive fluid with density stratification described by $B(z)$, the following are true:

- The gradient Richardson number $Ri = B_z/U_z^2$ compares the effects of stratification and shear.
- The Taylor-Goldstein equation (4.18) describes a wide array of instabilities and wavelike phenomena.
- A normal mode disturbance is affected only by that component of the background flow that is parallel to its own wave vector.
- Instability is possible only if $Ri(z) < 1/4$ for some z (the Miles-Howard theorem).
- The complex phase speed of a growing mode must lie within the semicircle shown in Figure 4.14 (Howard's semicircle theorem).
- The rule of thumb for a shear layer, i.e., that the wavelength is 7 times the layer thickness, remains valid in the stratified case.

4.11 Further Reading

Miles (1961) is an immense work that covered considerable new ground before arriving at the proof of what we now call the Miles-Howard theorem (section 4.7). That paper was sent to L. Howard for peer review, whereupon Howard discovered the much simpler proof reproduced here (Howard, 1961). For good measure, Howard also proved the semicircle theorem (section 4.8).

See Smyth and Moum (2012) for more information on Kelvin-Helmholtz instability. A review of wave resonance in stratified shear flows is provided in

Carpenter et al. (2013), as well as nice descriptions in Baines and Mitsudera (1994) and Caulfield (1994). It is also worth consulting Holmboe (1962), where it all began.

4.12 Appendix: Veering Flows

In many important cases, the mean flow varies primarily with height, but is not parallel. In an Ekman spiral, for example, both speed and direction change with height. Equatorial mean currents are mostly zonal, but these are exceptional. In most parts of the ocean and atmosphere, the mean current veers with depth.

Happily, the theory that we have already developed for parallel flows is easily extended to veering flows using our results from section 4.3 above: an instability is affected by only the component of the mean flow parallel to its own wave vector, and that component is a parallel shear flow.

Suppose that the mean flow of interest varies in z but has components in both horizontal directions:

$$\vec{u} = U(z)\hat{e}^{(x)} + V(z)\hat{e}^{(y)}.$$

For a given wave vector (k, ℓ), we define the parallel component of the mean flow:

$$\boxed{\tilde{U}(z) = \frac{kU(z) + \ell V(z)}{\tilde{k}}.}$$

Everything we have learned in this chapter about parallel flows is also true for veering flows if we substitute $\tilde{U}$ for U in (4.14) and (4.15).

4.13 Appendix: Spatial Growth

Throughout this book we separate the linearized equations of motion using normal modes in which the dependences on the horizontal coordinates and time take forms like

$$e^{\imath kx + \sigma t}, \quad e^{\imath(kx - \omega t)}, \quad \text{or } e^{\imath k(x - ct)}. \tag{4.53}$$

(We restrict this discussion to 2D modes.) In all cases, we assume that the wavenumber k is purely real, so that the solution is horizontally periodic. Time dependence is then quantified by the complex eigenvalue σ, or ω, or c.

The assumption of horizontal periodicity is not always realistic. Consider, for example, two streams, moving at different speeds, that meet at some point in space (e.g., Figure 4.15). Shear instability grows not in time but rather in the downstream direction, beginning at the point of confluence. In this case, a more realistic assumption is that ω is real (or, equivalently, σ is imaginary) and the streamwise

Figure 4.15 Confluence of the silty Missouri and clearer Mississippi rivers near St. Louis, MO, showing spatially growing instability. Image courtesy United States Geological Survey (USGS).

wavenumber k is allowed to be complex. The imaginary part of k then quantifies the rate of downstream growth.

As a simple model, consider the single interface that we explored in section 4.6.1. Assuming that the streams are side-by-side as in Figure 4.15, we can simplify further by neglecting the buoyancy force, so that the dispersion relation (4.25) becomes

$$\Big(\frac{u_0}{2} - c\Big)^2 + \Big(-\frac{u_0}{2} - c\Big)^2 = 0.$$

Solving for c, we get a purely imaginary result:

$$c = \pm \iota \frac{u_0}{2}.$$

In this model the streams have mean speeds $\pm u_0/2$. The scene is easier to imagine if we choose coordinates so that one stream is stationary while the other has speed u_0, which simply requires adding $u_0/2$ to all velocities. In this new reference frame,

$$c = \frac{u_0}{2}(1 \pm \iota).$$

So the real part of the phase velocity is, not surprisingly, the average of the speeds of the two streams. Now $c = \omega/k$, so we can easily solve for k:

$$k = \frac{\omega}{u_0}(1 \pm \iota).$$

Referring back to (4.53) and assuming that ω is real, we can see that the solution for k with the minus sign describes exponential growth in x. The wavelength is

$2\pi u_0/\omega$, and the growth rate is such that the amplitude grows by a factor $e^{2\pi} = 535$ over that distance!

Both wavenumber and spatial growth rate are proportional to the frequency ω, which is not specified. In reality, billows form quickly at extremely small scales, then merge repeatedly to form the large billows you see in Figure 4.15. The process is essentially chaotic; all that can be predicted reliably is the phase speed. There is much more that could be said regarding spatially growing instabilities that we will not cover in this book. For more details we refer the interested reader to the overview by Huerre (2000).

Exercise: Repeat this calculation with $b_0 \neq 0$, representing streams of different density separated by a horizontal boundary, as in section 4.6.1. You should find that there is a minimum frequency below which spatial instability does not occur.

5

Parallel Shear Flow: the Effects of Viscosity

We now imagine a parallel shear flow in a fluid that is homogeneous but has nonzero viscosity v (e.g., Figure 5.1). Recall from section 1.5 that v is treated as a constant and can represent either molecular viscosity or an effective viscosity due to turbulence.

In the presence of viscosity, the divergence equation is unchanged from (1.17):

$$\vec{\nabla} \cdot \vec{u} = 0. \tag{5.1}$$

The momentum equation (1.19), neglecting buoyancy but retaining viscosity, is

$$\frac{D\vec{u}}{Dt} = -\vec{\nabla}\pi + v\nabla^2\vec{u}. \tag{5.2}$$

The viscosity term raises the order of the system and therefore requires additional boundary conditions. These will be discussed in detail below (section 5.4). For the moment we'll just observe that a viscous fluid must move with the boundary, so for a stationary boundary $u = v = w = 0$.

5.1 Conditions for Equilibrium

Consider a parallel shear flow

$$\vec{u} = U(z)\hat{e}^{(x)}, \tag{5.3}$$

with pressure $\Pi(x, y, z)$. As in Chapter 3, gravity is irrelevant, so z is not necessarily the vertical coordinate. The parallel flow (5.3) satisfies the divergence equation, and its material derivative is zero. Substituting into (5.2) and separating the components, we have

$$\frac{\partial \Pi}{\partial x} = v\frac{\partial^2 U}{\partial z^2} \tag{5.4}$$

$$\frac{\partial \Pi}{\partial y} = \frac{\partial \Pi}{\partial z} = 0 \tag{5.5}$$

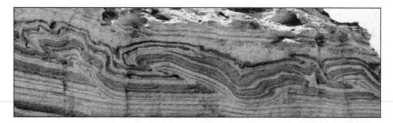

Figure 5.1 Kelvin-Helmholtz instability in paleo-earthquakes revealed in sediment deformations near the Dead Sea. The billow amplitude is ~1 m. These deformations can be modeled by the flow of a viscous fluid. Adapted from Heifetz et al. (2005).

Evidently the pressure gradient is arranged so as to balance the force of viscosity acting on the mean flow. We will call this frictional equilibrium.

According to (5.5), Π can be at most a function of x.[1] Equation (5.4) therefore requires that a function of x equal a function of z, which is possible only if both are constant. A parallel shear flow that is steady in a viscous fluid must therefore obey

$$\frac{d^2 U}{dz^2} = \text{const.} \qquad (5.6)$$

The velocity profile $U(z)$ can take one of three forms: (i) a quadratic function of z if the constant is nonzero, (ii) a linear function if the constant is zero, or (iii) zero, which we examined in Chapter 2. An immediate consequence of (5.6) is that d^2U/dz^2 cannot change sign, i.e., there is no inflection point. In the absence of viscosity, this class of flows would be stable. We consider two examples.

Plane Poiseuille flow: A parallel flow driven by a constant pressure gradient is a quadratic function. With stationary boundaries at $z = 0$ and $z = H$, U has the form

$$U(z) = 4u_0 \frac{z}{H} \left(1 - \frac{z}{H}\right), \qquad (5.7)$$

where u_0 is the maximum velocity, found at $z = H/2$ (Figure 5.2a). The effect of viscosity is balanced by a constant pressure drop in the x direction:

$$\frac{d\Pi}{dx} = -8\nu \frac{u_0}{H^2}.$$

Plane Poiseuille flow has a cylindrical counterpart, the classical engineering problem of flow in a pipe.

Plane Couette flow: If there is no pressure gradient, the velocity profile is a linear function. One way to power a flow in this circumstance is to have one boundary move relative to the other, e.g.,

---

[1] Compare this with the inviscid case, section 3.1.1, where $\nu = 0$ and therefore Π is uniform.

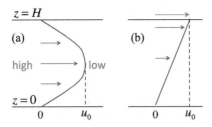

Figure 5.2 Equilibrium states in a viscous fluid. (a) Plane Poiseuille flow maintained by a uniform pressure gradient. (b) Plane Couette flow driven by a moving upper boundary.

$$U(z) = u_0 \frac{z}{H}.$$

In this case the boundary at $z = H$ moves at speed u_0 and the boundary at $z = 0$ is stationary (Figure 5.2b).

5.2 Conditions for Quasi-Equilibrium: the Frozen Flow Approximation

If the mean flow is *not* in equilibrium, we cannot use our methods of stability analysis, because the normal mode solution (3.14) is invalid when the coefficients of the equations are functions of time. So, channeling our inner child, we ask, "But can we do it if the mean flow is *almost* in equilibrium?" The answer is yes, we can, if we're very careful.

Suppose that the mean flow is not exactly in equilibrium; it is evolving slowly, perhaps under the action of viscosity. Suppose also that an instability grows rapidly, so that it reaches large amplitude before the mean flow has time to change very much. We can guess, then, that the evolution of the mean flow will have little effect on the instability. On that basis, we can do the stability analysis with normal modes just as if the mean flow were steady. This is called the frozen flow approximation. But, we must check afterward to make sure that the growth rate really is fast enough to justify the approximation.

Suppose that the mean velocity profile $U(z)$ can be characterized by a length scale h and a velocity scale u_0. Suppose further that it diffuses under the action of viscosity on a time scale T. Now assume that there is no pressure gradient to force the flow; it's just diffusing freely:

$$\frac{\partial U}{\partial t} = \nu \frac{\partial^2 U}{\partial z^2}. \tag{5.8}$$

We can estimate the terms in this equation as follows:

$$\frac{u_0}{T} = \nu \frac{u_0}{h^2},$$

or

$$T = \frac{h^2}{\nu}.$$

Now if

$$\sigma T \gg 1, \tag{5.9}$$

then the instability will grow by many factors of e in the time it takes the mean flow to diffuse. Suppose, for example, that the instability is a shear instability, so its growth rate scales like

$$\sigma = \sigma^\star \frac{u_0}{h}.$$

Then the condition $\sigma T \gg 1$ is equivalent to

$$\sigma^\star \gg \frac{\nu}{u_0 h},$$

or

$$\boxed{\sigma^\star \gg \frac{1}{Re},} \tag{5.10}$$

where

$$\boxed{Re = \frac{u_0 h}{\nu}} \tag{5.11}$$

is the Reynolds number. Any solution we obtain that does not satisfy (5.10) must be interpreted with great caution.

5.3 The Orr-Sommerfeld Equation

We now substitute perturbation forms

$$\vec{u} = U(z)\hat{e}^{(x)} + \epsilon \vec{u}'\,; \quad \pi = \Pi(x) + \epsilon \pi'$$

into the equations of motion (5.1) and (5.2). The procedure is similar to the inviscid case, which we examined in detail in section 3.1.2; we have only to account for the added term representing viscosity in the momentum equation.

Viscosity makes no difference to the divergence condition (3.5):

$$\vec{\nabla} \cdot \vec{u}' = 0. \tag{5.12}$$

Because there is no nonlinearity in the viscous term, the momentum equation is just (3.7) with an extra term:

$$\frac{\partial \vec{u}'}{\partial t} + U(z)\frac{\partial \vec{u}'}{\partial x} + w'\frac{dU}{dz}\hat{e}^{(x)} = -\vec{\nabla}\pi' + \nu\nabla^2\vec{u}'. \tag{5.13}$$

To obtain the Poisson equation for the pressure, we again take the divergence of the momentum equation. Because the divergence of the added viscosity term is zero by (5.12), the result is exactly the same as (3.11):

$$\nabla^2 \pi = -2U_z \frac{\partial w'}{\partial x}.$$

We now substitute the pressure equation into the Laplacian of the vertical component of (5.13). The result is just (3.12) with the added viscosity term:

$$\frac{\partial}{\partial t}\nabla^2 w' + U\frac{\partial}{\partial x}\nabla^2 w' = U_{zz}\frac{\partial w'}{\partial x} + \nu\nabla^4 w'. \tag{5.14}$$

Substituting the normal mode form (3.14) now gives the Orr-Sommerfeld equation:

$$\boxed{\begin{aligned} &\sigma\nabla^2\hat{w} = -\iota k U\nabla^2\hat{w} + \iota k U_{zz}\hat{w} + \nu\nabla^4\hat{w}, \tag{5.15} \\ &\text{where} \\ &\nabla^2 = \frac{d^2}{dz^2} - \tilde{k}^2 ; \quad \tilde{k}^2 = k^2 + \ell^2. \end{aligned}}$$

This may also be written in terms of the phase speed, $c = i\sigma/k$:

$$(U - c)\nabla^2\hat{w} = U_{zz}\hat{w} - \frac{\iota\nu}{k}\nabla^4\hat{w}. \tag{5.16}$$

In the special case $\nu = 0$, this is equivalent to the Rayleigh equation (3.18).

The viscous case differs from the inviscid case in a very important way. Recall that, in the inviscid case, phase velocities occur in complex conjugate pairs so that every decaying mode is accompanied by a growing mode (section 3.4). If all modes are neutral, i.e., $c_i = 0$, or $\sigma_r = 0$, we say that the flow is Lyapunov stable, meaning that the flow remains near the equilibrium state but does not return to it. In the viscous equation (5.16), one term has a complex coefficient, so that c need not occur in complex conjugate pairs. It is possible (common, in fact) for all modes to have $c_i < 0$, or $\sigma_r < 0$. Physically, this means that all perturbations are dissipated by viscosity. Mathematically, we may say that such a flow is asymptotically stable (it eventually returns to equilibrium), or exponentially stable, since departures from equilibrium decay at an exponential rate.

5.4 Boundary Conditions for Viscous Fluid

As usual we assume impermeable boundaries $w' = 0$ (cf. 3.39) above and below the region of interest. But because the Orr-Sommerfeld equation (5.15) is fourth-order (it contains fourth derivatives in the viscous term), two more boundary conditions are needed. These correspond to additional assumptions about the nature of the boundary. There are two plausible choices: solid and frictionless.

5.4.1 The Solid (No-Flow) Boundary

This is the most obvious choice. Physically, we know that velocity goes to zero at a boundary, because fluid molecules become intermingled with the boundary molecules. In the case of a moving boundary, the flow velocity approaches the boundary velocity. Here, we'll assume stationary boundaries, but the generalization to a moving boundary is trivial. So not only is $w' = 0$ at the boundaries, but u' and v' are zero also.

We're not done yet, though. These additional conditions pertain to u' and v', but for (5.15) we need conditions on w'. We arrange this as follows. Because $u' = 0$ *everywhere* on the boundary, and in particular for all values of x, it also follows that $\partial u'/\partial x = 0$ everywhere on the boundary. Likewise, because $v' = 0$ for all y, $\partial v'/\partial y = 0$. Substituting these results into the divergence condition:

$$\frac{\partial u'}{\partial x} + \frac{\partial v'}{\partial y} + \frac{\partial w'}{\partial z} = 0,$$

we find that $\partial w'/\partial z = 0$. When working with normal modes, the added condition at a solid boundary is

$$\boxed{\hat{w}_z = 0,} \tag{5.17}$$

where the subscript indicates the z-derivative.

5.4.2 The Frictionless Boundary

In most flows, the retarding effect of viscosity is restricted to a thin layer adjacent to the boundary within which the velocity goes rapidly to zero. This is called the viscous boundary layer. Outside the viscous boundary layer, the velocity changes much more slowly; the fluid slips past as if the boundary were frictionless.

If the region of interest is much larger than the viscous boundary layer, we can pretend that the outer edge of the layer is actually the boundary, and impose the condition $\partial u'/\partial z = \partial v'/\partial z = 0$ at that location. An equivalent way to state this is that the viscous momentum fluxes $-\nu \partial u'/\partial z$ and $-\nu \partial v'/\partial z$ vanish at the boundary. The boundary may therefore be called "flux-free" (or just "free").

To convert this condition to a condition on $\hat{w}$ for use in the Orr-Sommerfeld equation (5.15), we use the derivative of the divergence condition. Because $\vec{\nabla} \cdot \vec{u}' = 0$ everywhere in the flow, its z-derivative is also zero:

$$\frac{\partial}{\partial z}\left(\frac{\partial u'}{\partial x} + \frac{\partial v'}{\partial y} + \frac{\partial w'}{\partial z}\right) = \frac{\partial^2 u'}{\partial x \partial z} + \frac{\partial^2 v'}{\partial y \partial z} + \frac{\partial^2 w'}{\partial z^2} = 0.$$

Now since $\partial u'/\partial z = 0$ for all x, $\partial^2 u'/\partial x \partial z = 0$, and by the same reasoning $\partial^2 v'/\partial y \partial z = 0$, leaving us with $\partial^2 w'/\partial z^2 = 0$. In normal mode form, the added condition at a free slip (or frictionless) boundary is

$$\boxed{\hat{w}_{zz} = 0.}$$ (5.18)

5.5 Numerical Solution of the Orr-Sommerfeld Equation

Write the Orr-Sommerfeld equation in the form (5.15):

$$\sigma \nabla^2 \hat{w} = -\iota k U \nabla^2 \hat{w} + \iota k U_{zz} \hat{w} + \nu \nabla^4 \hat{w},$$

where

$$\nabla^2 = \frac{d^2}{dz^2} - \tilde{k}^2.$$

Possible choices of boundary conditions include

$$\hat{w} = 0; \quad \hat{w}_z = 0 \quad \text{(solid)} \tag{5.19}$$

$$\hat{w} = 0; \quad \hat{w}_{zz} = 0 \quad \text{(free)}. \tag{5.20}$$

As in the inviscid case, the Laplacian is represented as a matrix A:

$$\nabla^2 \to A = D^{(2)} - \tilde{k}^2 I,$$

where $D^{(2)}$ is the second-derivative matrix incorporating the impermeable boundary condition $\hat{w} = 0$.

Similarly, the right-hand side of the Orr-Sommerfeld equation is expressed using a second matrix, which is just (3.36) with the extra viscous term:

$$B = -\iota k \vec{U} \cdot A + \iota k \vec{U}_{zz} \cdot I + \nu (D^{(4)} - 2\tilde{k}^2 D^{(2)} + \tilde{k}^4 I), \tag{5.21}$$

where $D^{(4)}$ is a fourth-derivative matrix incorporating the boundary conditions (5.19 or 5.20). With free-free boundary conditions, the fourth-derivative matrix turns out to be equivalent to the square of the second-derivative matrix (try it!). If either boundary is solid, though, the fourth-derivative matrix must be designed separately via Taylor series expansions as described in sections 1.4.2 and 1.4.3.

Consider the usual approximation for the second derivative at $i = 1$:

$$f_1'' \simeq \frac{f_0 - 2f_1 + f_2}{\Delta^2}. \tag{5.22}$$

Instead of $f_0 = 0$, we must now assume that $f_0' = 0$. That assumption can be expressed using the one-sided approximation to the derivative developed in project 1:

$$f_0' \simeq \frac{-3f_0 + 4f_1 - f_2}{2\Delta} = 0 \quad \Rightarrow \quad f_0 = \frac{4f_1 - f_2}{3}$$

Substituting this into (5.22), we have

$$f_1'' \simeq \frac{-2f_1 + 2f_2}{3\Delta^2}$$

and the top line of the matrix is therefore $[-2/3 \ \ 2/3 \ \ \ldots]/\Delta^2$. For $i = N$, the result is

$$f_N'' \simeq \frac{2f_{N-1} - 2f_N}{3\Delta^2}.$$

Left-multiplications are performed as in section 3.5.2. The Orr-Sommerfeld equation now becomes a generalized eigenvalue problem:

$$\sigma A \vec{w} = B \vec{w},$$

which can be solved using standard numerical functions.

5.6 Oblique Modes

As we have discussed in sections 3.7 and 4.3, every oblique mode has a corresponding 2D mode. If the oblique mode has wave vector (k, ℓ), then the corresponding 2D mode has wave vector $(\tilde{k}, 0)$ as shown in Figure 3.10. The growth rates are σ and $\tilde{\sigma}$, respectively. We found that, in the absence of viscosity, the growth rates are related by $\sigma = \cos\varphi \, \tilde{\sigma}$, where φ is the angle of obliquity. Here we will see how viscosity affects that relationship.

We begin by writing the Orr-Sommerfeld equation (5.15) more explicitly:

$$(\sigma + \imath k U) \left(\frac{d^2}{dz^2} - \tilde{k}^2 \right) \hat{w} = \imath k U_{zz} \hat{w} + \nu \left(\frac{d^2}{dz^2} - \tilde{k}^2 \right)^2 \hat{w}. \qquad (3D)$$

Suppose we have a solution algorithm for this equation:

$$\sigma = \mathcal{F}(z, U, \nu; \, k, \ell). \qquad (5.23)$$

This equation and solution algorithm pertain to any mode, an oblique mode in particular.

For a corresponding 2D mode, having wave vector $(\tilde{k}, 0)$, $(3D)$ becomes

$$(\sigma + \imath \tilde{k} U) \left(\frac{d^2}{dz^2} - \tilde{k}^2 \right) \hat{w} = \imath \tilde{k} U_{zz} \hat{w} + \nu \left(\frac{d^2}{dz^2} - \tilde{k}^2 \right)^2 \hat{w}, \qquad (2D)$$

with solution algorithm

$$\sigma = \mathcal{F}(z, U, \nu; \, \tilde{k}, 0). \qquad (5.24)$$

Now we start again with $(3D)$ and make the Squire transformations $k = \tilde{k} \cos\varphi$ and $\sigma = \tilde{\sigma} \cos\varphi$ (as we did in the inviscid case; section 3.7), and

$$\boxed{v = \tilde{v}\cos\varphi.}$$

Substituting these into $(3D)$ and dividing out $\cos\varphi$, we have

$$(\tilde{\sigma} + \iota\tilde{k}U)\left(\frac{d^2}{dz^2} - \tilde{k}^2\right)\hat{w} = \iota\tilde{k}U_{zz}\hat{w} + \tilde{v}\left(\frac{d^2}{dz^2} - \tilde{k}^2\right)^2\hat{w}. \qquad (\widetilde{3D})$$

The form $(\widetilde{3D})$ is isomorphic to the special case $(2D)$:

$$(\widetilde{3D}) \leftrightarrow (2D), \text{ under } \tilde{\sigma} \to \sigma \text{ and } \tilde{v} \to v.$$

This means that we can use the same solution algorithm as for the 2D case (5.24):

$$\tilde{\sigma} = \mathcal{F}(z, U, \tilde{v}; \ \tilde{k}, 0),$$

or

$$\sigma = \cos\varphi \times \mathcal{F}\left(z, U, \frac{v}{\cos\varphi}; \ \tilde{k}, 0\right).$$

As in the inviscid case, a general 3D mode with $\varphi \neq 0$ and growth rate σ corresponds to a 2D mode ($\varphi = 0$) whose growth rate is reduced to $\sigma\cos\varphi$. But there is another important difference: the corresponding 2D mode grows on a flow with increased viscosity $\tilde{v} = v/\cos\varphi$.

It is often true that viscosity has the effect of *damping* instability, as we have seen in the convection case (Chapter 2). If that is true, the tendency of 2D modes to grow faster than the corresponding 3D modes is increased by this heightened sensitivity to viscosity. However, it is not impossible that viscosity could act to destabilize a flow. If that were true, and if that viscous destabilization was sufficient to overcome the leading factor $\cos\varphi$, then a 3D mode could grow faster than the corresponding 2D mode.

5.7 Shear Scaling and the Reynolds Number

Here again is the Orr-Sommerfeld equation $(3D)$:

$$(\sigma + \iota k U)\left(\frac{d^2}{dz^2} - \tilde{k}^2\right)\hat{w} = \iota k U_{zz}\hat{w} + v\left(\frac{d^2}{dz^2} - \tilde{k}^2\right)^2\hat{w}, \qquad (3D)$$

with solution algorithm

$$(\sigma, \hat{w}) = \mathcal{F}(z, U, v; k, \ell).$$

(Note that we have included the eigenfunction $\hat{w}$ in the output of $\mathcal{F}$. Previously, this was omitted for simplicity only.)

For shear scaling, we define a length scale h and a velocity scale u_0, just as in section 3.6. For example, the familiar hyperbolic tangent shear layer

$$U = u_0 \tanh \frac{z}{h}$$

can be written as

$$U^* = \tanh z^*,$$

where

$$U^* = \frac{U}{u_0}; \quad z^* = \frac{z}{h},$$

(reproducing 3.47). Other quantities can be scaled as in (3.51), also reproduced here for convenience:

$$\sigma = \sigma^* \frac{u_0}{h}$$

$$\{k, \ell, \tilde{k}\} = \{k^*, \ell^*, \tilde{k}^*\}/h$$

$$\hat{w} = \hat{w}^* u_0$$

$$\frac{d}{dz} = \frac{1}{h} \frac{d}{dz^*} \quad \Rightarrow \nabla^2 = \frac{1}{h^2} \left(\frac{d^2}{dz^{*2}} - \tilde{k}^{*2} \right).$$

With these substitutions, the Orr-Sommerfeld equation ($3D$) can be rewritten as

$$(\sigma^* + \iota k^* U^*) \left(\frac{d^2}{dz^{*2}} - \tilde{k}^{*2} \right) \hat{w}^* = \iota k^* U^*_{z^* z^*} \hat{w}^* + \frac{\nu}{h u_0} \left(\frac{d^2}{dz^{*2}} - \tilde{k}^{*2} \right)^2 \hat{w}^*,$$

$$(3D*)$$

The scaled viscosity appearing in the second term on the right-hand side is the inverse of the Reynolds number (5.11). ($3D*$) is isomorphic to ($3D$) and therefore has the same solution algorithm under the shear scalings listed above:

$$(\sigma^*, \hat{w}^*) = \mathcal{F}(z^*, U^*, 1/Re; k^*, \ell^*). \tag{5.25}$$

5.8 Numerical Examples

5.8.1 Viscous Stabilization of a Shear Layer

Consider a hyperbolic tangent shear layer $U^* = \tanh z^*$ in a viscous fluid (e.g., Figure 5.1). In the limit $Re \to \infty$, viscosity is negligible and the instability is indistinguishable from that found in an inviscid fluid (section 3.9.1). As Re is reduced, the effects of viscosity become evident (Figure 5.3). The growth rate of the fastest-growing mode decreases, and its wavelength increases. (As in the case of convective instability, larger-scale disturbances are better able to resist viscous damping.) The flow is completely stabilized when $Re = 2.7$. That number should be taken with a large grain of salt, however, since the condition $\sigma^* \gg 1/Re$ is not satisfied.

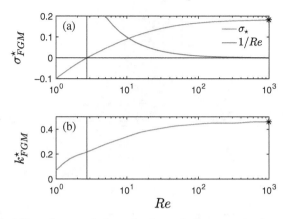

Figure 5.3 Stability characteristics of a hyperbolic tangent shear layer $U^\star = \tanh z^\star$ in a viscous fluid. Frictionless boundaries are placed at $z^\star = \pm 5$. (a) Fastest growth rate of shear layer instability versus Re. The red curve indicates $1/Re$. (b) Wavenumber of fastest-growing mode. Asterisks show the inviscid result (e.g., Figure 3.12).

A useful and valid rule of thumb[2] that we can extract from this analysis is that viscous effects become important when Re drops below ~ 100. Suppose, for example, that the shear layer at the base of the ocean mixed layer is 10 m thick, and the velocity change across it is 0.1 m/s, so that $h = 5$ m and $u_0 = 0.05$ m/s. In that case, the condition $Re < 100$ is equivalent to $\nu > 2.5 \times 10^{-4} \mathrm{m}^2/\mathrm{s}$. That value is in the normal range for turbulent viscosity at the mixed layer base, and we should therefore expect that shear instability in this regime will be affected significantly by ambient turbulence.

5.8.2 Instabilities of Plane Poiseuille Flow

Here we use the matrix-based numerical approach to explore the instabilities of plane Poiseuille flow (5.7; Figure 5.2), which is one of the few shear flows that are truly steady in a viscous fluid. In shear-scaled terms, the mean flow profile is

$$U^\star = 4z^\star(1 - z^\star). \tag{5.26}$$

We begin by setting the spanwise wavenumber $\ell^\star$ to zero and looking at the growth rate as a function of streamwise wavenumber $k^\star$ for a range of Reynolds numbers. A distinct region of instability is found (Figure 5.4), with maximum growth rate at $Re = 10^5$. This is called the Tollmien-Schlichting instability. In this case the fastest-growing mode has $k^\star = 1.55$, or wavelength about 4 times the width of the channel.

---

[2] "It's more of a guideline than a rule." Bill Murray in *Ghostbusters*.

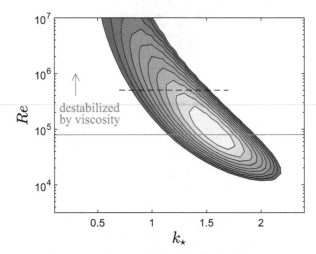

Figure 5.4 Stability characteristics of plane Poiseuille flow (5.26). Colors indicate the growth rate versus wavenumber k^* and Reynolds number Re for 2D modes $\ell^* = 0$. For $Re > 8 \times 10^4$ (above the red line), the growth rate decreases with increasing Re or, in other words, increases with increasing viscosity. The dashed line pertains to Figure 5.6.

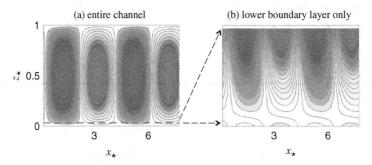

Figure 5.5 Vertical velocity perturbation for the Tollmien-Schlichting instability of plane Poiseuille flow. (a) The entire vertical domain. (b) Blowup of the lower region bounded by the dashed line in (a), showing the up-shear phase tilt.

For $Re > 10^5$, the maximum growth rate increases with decreasing Re, i.e., with increasing viscosity. This is therefore an example of a mode that is destabilized by viscosity. In the absence of viscosity (the limit $Re \to \infty$), the growth rate is zero.

Contours of the cross-stream (or wall-normal) velocity perturbation (Figure 5.5) show that isolines tilt against the background shear near the boundaries. It is this aspect of the perturbation that allows it to access energy from the mean flow despite the absence of an inflection point.

The fact that the Tollmien-Schlichting mode is destabilized by viscosity raises the possibility that the fastest-growing instability may be oblique (on the basis of Squire's theorem). In Figure 5.6, we show a test of this possibility for a value

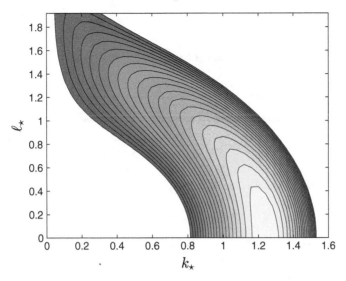

Figure 5.6 Growth rates of oblique modes of plane Poiseuille flow versus scaled wavenumber. $Re = 5 \times 10^5$, indicated in Figure 5.4 by the dashed line. Growth is maximized in the 2D limit despite the destabilizing role of viscosity.

$Re = 5 \times 10^5$, well into the region where the growth rate increases with increasing viscosity (Figure 5.4). The fastest-growing mode is in fact two-dimensional; all oblique modes have lower growth rates.

Test your understanding: Is the criterion $\sigma^\star \gg 1/Re$ relevant for plane Poiseuille flow?

5.9 Perturbation Energetics in Viscous Flow

When exploring the inviscid case, we derived the kinetic energy equation in terms of real perturbation variables u', w', ..., etc., which represent a general disturbance (section 3.10.1). We then applied the result to a normal mode perturbation by integrating quadratic combinations over one wavelength (section 3.10.2). Here we will repeat the derivation with viscous terms included, and we'll allow for fully three-dimensional modes ($\ell \neq 0$). And, just for variety, we'll do the math in normal mode form right from the start.

As in the inviscid case, the perturbation equations describing continuity and momentum are given by (5.12) and (5.13). In normal mode form, these can be written as:

$$\imath k \hat{u} + \imath \ell \hat{v} = -\hat{w}_z, \tag{5.27}$$

$$(\sigma + \imath k U)\hat{u} = -U_z \hat{w} - \imath k \hat{\pi} + \nu \nabla^2 \hat{u} \tag{5.28}$$

$$(\sigma + \imath k U)\hat{v} = -\imath \ell \hat{\pi} + \nu \nabla^2 \hat{v} \tag{5.29}$$

$$(\sigma + \imath k U)\hat{w} = -\hat{\pi}_z + \nu \nabla^2 \hat{w} \tag{5.30}$$

where the subscript z indicates the derivative. We now multiply the momentum equations, (5.28), (5.29), and (5.30), by the complex conjugates of the velocity eigenfunctions, $\hat{u}^*$, $\hat{v}^*$, and $\hat{w}^*$, respectively, and add the results:

$$\overbrace{(\sigma + \iota k U)\left(|\hat{u}|^2 + |\hat{v}|^2 + |\hat{w}|^2\right)}^{(1)} = \overbrace{-U_z\hat{u}^*\hat{w}}^{(2)}\;\overbrace{-\iota k\hat{u}^*\hat{\pi} + \iota\ell\hat{v}^*\hat{\pi} - \hat{w}^*\hat{\pi}_z}^{(3)}$$

$$\overbrace{+\,\nu\left(\hat{u}^*\nabla^2\hat{u} + \hat{v}^*\nabla^2\hat{v} + \hat{w}^*\nabla^2\hat{w}\right)}^{(4)}. \quad (5.31)$$

Next, we take the real part of each term.

(1) The real part of the left-hand side is $4\sigma_r K$, where

$$K = \frac{|\hat{u}|^2 + |\hat{v}|^2 + |\hat{w}|^2}{4} \qquad (5.32)$$

is the perturbation kinetic energy (cf. 3.65).

(2) The real part is twice the shear production (cf. 3.66):

$$SP = -\frac{U_z}{2}\,(\hat{u}^*\hat{w})_r.$$

(3) The first two terms of (3) can be written as

$$(-\iota k\hat{u}^* - \iota\ell\hat{v}^*)\hat{\pi}.$$

The sum in parentheses is equal to $(\iota k\hat{u} + \iota\ell\hat{v})^*$, which is equal to $-\hat{w}_z^*$ by (5.27). Therefore, (3) can be rewritten as

$$-\hat{w}_z^*\hat{\pi} - \hat{w}^*\hat{\pi}_z = -(\hat{w}^*\hat{\pi})_z.$$

After taking the real part, the quantity in parentheses is twice the energy flux (cf. 3.67):

$$EF = \frac{(\hat{w}^*\hat{\pi})_r}{2},$$

so that the real part of (3) is $-2EF_z$.

(4) The fourth term is due to viscosity. It can be split into two parts, which represent distinct physical processes. The three terms in the parentheses of term (4) all have the same form, which can be rewritten as follows:

$$\hat{u}^*\nabla^2\hat{u} = \hat{u}^*\hat{u}_{zz} - \tilde{k}^2\hat{u}^*\hat{u}$$
$$= (\hat{u}^*\hat{u}_z)_z - \hat{u}_z^*\hat{u}_z - \tilde{k}^2\hat{u}^*\hat{u}$$
$$= (\hat{u}^*\hat{u}_z)_z - |\hat{u}_z|^2 - \tilde{k}^2|\hat{u}|^2$$

We now take the real part. The only complex term on the right-hand side is the first one, and its real part is the derivative of

$$(\hat{u}^*\hat{u}_z)_r = \frac{1}{2}(\hat{u}^*\hat{u}_z + \hat{u}\hat{u}_z^*) = \frac{1}{2}(\hat{u}^*\hat{u})_z = \frac{|\hat{u}|_z^2}{2},$$

so we have

$$(\hat{u}^*\nabla^2\hat{u})_r = \frac{|\hat{u}|_{zz}^2}{2} - |\hat{u}_z|^2 - \tilde{k}^2|\hat{u}|^2. \tag{5.33}$$

Adding the corresponding terms involving $\hat{v}$ and $\hat{w}$, we can now assemble term (4):

$$2K_{zz} - |\hat{u}_z|^2 - |\hat{v}_z|^2 - |\hat{w}_z|^2 - 4\tilde{k}^2K.$$

Collecting all four terms of (5.31) and dividing by 2, we have the kinetic energy equation for normal modes in a viscous shear flow:

$$\boxed{2\sigma_r K = SP - EF_z + \nu K_{zz} - \varepsilon.} \tag{5.34}$$

The final term is minus the viscous dissipation rate

$$\boxed{\varepsilon = \frac{\nu}{2}\left(|\hat{u}_z|^2 + |\hat{v}_z|^2 + |\hat{w}_z|^2 + 4\tilde{k}^2K\right).} \tag{5.35}$$

The new terms we have gained via the addition of viscosity (and obliquity) are as follows.

- The perturbation kinetic energy (5.32) now includes the third velocity term, $|\hat{v}|^2$.
- The term νK_{zz} is the convergence of a kinetic energy flux due to viscosity, namely $-\nu K_z$.
- The dissipation term $-\varepsilon$ is negative definite and therefore destroys perturbation kinetic energy (converting it to heat).

None of the viscous processes can create *perturbation kinetic energy.* The viscous flux $-\nu K_z$, like the pressure-driven flux EF, vanishes at the boundaries (check!). So it spreads energy around but does not contribute to the net amount. The dissipation term $-\varepsilon$ is negative definite, and hence can only reduce the kinetic energy. This is a paradox reminiscent of the case of stable stratification (section 4.9). Viscous flows have instabilities that do not exist in the absence of viscosity, e.g., the Tollmien-Schlichting instability, even though viscosity itself acts only as an energy sink for the perturbation. Because viscosity forces the total velocity $\vec{u}$ to go to zero at the boundaries, the perturbation is distorted such that lines of constant w' tilt against the background shear (Figure 5.5) and SP is therefore positive as shown in section 3.11.4. In other words, while viscosity does not create perturbation kinetic energy directly, it configures the perturbation so that it can extract kinetic energy from the mean flow despite the absence of an inflection point.

5.10 Summary

- A parallel shear flow in a viscous fluid can be in equilibrium if either
  - the mean shear is uniform, so that U does not diffuse (as in Couette flow), or
  - viscous smoothing is balanced by a pressure gradient in the x-direction (as in Poiseuille flow).
- The frozen flow approximation permits the use of normal modes provided that the instability grows rapidly compared with the time scale of viscous alteration of the mean profile.
- The Orr-Sommerfeld equation is solved numerically in a manner similar to the Rayleigh equation, but an extra boundary condition (either solid or frictionless) is needed.
- Except in unusual cases, the fastest-growing mode is 2D.
- The effect of viscosity is expressed by the Reynolds number, Re. Viscosity acts to damp shear instability when Re is less than ~ 100.
- Viscosity can *create* shear instability – not directly, but by forcing the disturbance to obey the solid boundary condition and thereby generating positive shear production.

6

Synthesis: Viscous, Diffusive, Inhomogeneous, Parallel Shear Flow

In this chapter we explore equilibria and perturbations in a stratified, parallel shear flow with the effects of viscosity and diffusion included, effectively unifying Chapters 2, 3, 4, and 5. The goal is to develop numerical solution methods that have wide applicability to the study of small-scale processes in the oceans and atmosphere and explore some applications of those methods.

6.1 Expanding the Basic Equations

We start with the Boussinesq equations for a viscous, diffusive, inhomogeneous fluid. The divergence equation is, as usual,

$$\vec{\nabla} \cdot \vec{u} = 0.$$

The momentum equation (1.19), neglecting the Coriolis acceleration but retaining viscosity and restoring buoyancy, is

$$\frac{D\vec{u}}{Dt} = -\vec{\nabla}\pi + b\hat{e}^{(z)} + \nu\nabla^2\vec{u}, \tag{6.1}$$

and the buoyancy equation (1.25) is

$$\frac{Db}{Dt} = \kappa\nabla^2 b. \tag{6.2}$$

We assume the perturbation solution

$$\vec{u} = U(z, t)\hat{e}^{(x)} + \epsilon\vec{u}',$$
$$b = B(z, t) + \epsilon b',$$
$$\pi = \Pi + \epsilon\pi'. \tag{6.3}$$

At this stage, we have not yet assumed that the background state (U, B, Π) is steady. No assumption is made regarding the background pressure Π.

6.1.1 Continuity

As usual we deal with the easiest equation first:

$$\vec{\nabla} \cdot \vec{u}' = 0.$$

6.1.2 Momentum

The momentum equation (6.1), with the perturbation solution (6.3), becomes

$$\left[\frac{\partial}{\partial t} + U\frac{\partial}{\partial x} + \varepsilon\vec{u}' \cdot \vec{\nabla}\right]\left[U(z, t)\hat{e}^{(x)} + \varepsilon\vec{u}'\right]$$

$$= -\vec{\nabla}\left(\Pi + \varepsilon\pi'\right) + \left(B(z, t) + \varepsilon b'\right)\hat{e}^{(z)} + \nu\nabla^2\left[U(z, t)\hat{e}^{(x)} + \varepsilon\vec{u}'\right]. \quad (6.4)$$

With $\varepsilon = 0$, this gives the three component equations:

$$\frac{\partial U}{\partial t} = -\frac{\partial \Pi}{\partial x} + \nu\frac{\partial^2 U}{\partial z^2} \quad (6.5)$$

$$0 = -\frac{\partial \Pi}{\partial y} \quad (6.6)$$

$$0 = -\frac{\partial \Pi}{\partial z} + B. \quad (6.7)$$

The first equation shows that the background flow is governed by the combination of the streamwise pressure gradient and viscosity, as we saw previously in section 5.1. Strict equilibrium ($\partial U/\partial t = 0$) can be satisfied only when $\partial^2 U/\partial z^2 = $ const.[1] As in the homogeneous case, the "frozen flow" approximation holds provided $\sigma^\star \gg 1/Re$ (section 5.2). The second equation shows that the background pressure does not depend on y, while the third describes hydrostatic balance between the vertical pressure gradient and the buoyancy.

At $O(\epsilon)$, (6.4) gives

$$\left(\frac{\partial}{\partial t} + U\frac{\partial}{\partial x}\right)\vec{u}' + U_z w'\hat{e}^{(x)} = -\vec{\nabla}\pi' + b'\hat{e}^{(z)} + \nu\nabla^2\vec{u}'. \quad (6.8)$$

Observe that this perturbation momentum equation combines the terms seen previously in (2.14), (3.7), (4.7), and (5.13).

6.1.3 Buoyancy

Substitution of (6.3) into (6.2) gives

$$\left[\frac{\partial}{\partial t} + U\frac{\partial}{\partial x} + \varepsilon\vec{u}' \cdot \vec{\nabla}\right]\left[B(z, t) + \varepsilon b'\right] = \kappa\nabla^2\left[B(z, t) + \varepsilon b'\right]. \quad (6.9)$$

---

[1] Set $\partial U/\partial t = 0$ in (6.5) then cross-differentiate with (6.7) to eliminate Π. You'll find that $\partial^3 U/\partial z^3 = 0$.

For the unperturbed flow, this is

$$\frac{\partial B}{\partial t} = \kappa \frac{\partial^2 B}{\partial z^2} \tag{6.10}$$

Strict equilibrium requires that both sides vanish, hence

$$\frac{\partial B}{\partial z} = N^2 = \text{const.},$$

as was seen in section 2.7.

Repeating the arguments we used in the previous chapter for inhomogeneous flows, we can show that quasi-equilibrium requires

$$\sigma^\star \gg \frac{1}{Re\ Pr},$$

where $\sigma^\star = \sigma h / u_0$ is the growth rate scaled by the background shear, Re is the Reynolds number, and $Pr = \nu / \kappa$ is the Prandtl number. If $Pr \geq 1$, then the previous condition $\sigma^\star \gg 1/Re$ is sufficient.

The perturbation part of (6.9) is obtained by subtracting (6.10) and omitting terms of order ϵ^2:

$$\left(\frac{\partial}{\partial t} + U\frac{\partial}{\partial x}\right) b' + B_z w' = \kappa \nabla^2 b'. \tag{6.11}$$

Note that (6.11) combines the buoyancy perturbation equations for a motionless, inhomogeneous fluid (2.12), and a stratified, nondiffusive, parallel shear flow (4.10).

6.1.4 *Eliminating the Pressure*

We eliminate the pressure, as we have done before, by combining the divergence of (6.8) with the Laplacian of its vertical component. The divergence gives the pressure equation:[2]

$$\nabla^2 \pi' = -2U_z \frac{\partial w'}{\partial x} + \frac{\partial b'}{\partial z}. \tag{6.12}$$

Taking the Laplacian of the vertical component of (6.8) and substituting the vertical derivative of (6.12), we obtain:

$$\left(\frac{\partial}{\partial t} + U\frac{\partial}{\partial x}\right)\nabla^2 w' - U_{zz}\frac{\partial w'}{\partial x} = \nabla_H^2 b' + \nu\nabla^4 w'. \tag{6.13}$$

In (6.13) and (6.11), we have two equations for the two unknowns w' and b'. It is possible to combine these further into a single equation, but we will not do that

---

[2] Compare this with equations (2.16) and (3.11).

here. Instead, we substitute the normal mode forms $w' = \{\hat{w}(z)e^{\sigma t}e^{\iota(kx+\ell y)}\}_r$ and $b' = \{\hat{b}(z)e^{\sigma t}e^{\iota(kx+\ell y)}\}_r$ to obtain a pair of ordinary differential equations:

$$(\sigma + \iota kU)\nabla^2\hat{w} - \iota kU_{zz}\hat{w} = -\tilde{k}^2\hat{b} + \nu\nabla^4\hat{w} \qquad (6.14)$$

$$(\sigma + \iota kU)\hat{b} + B_z\hat{w} = \kappa\nabla^2\hat{b}, \qquad (6.15)$$

where $\nabla^2 = d^2/dz^2 - \tilde{k}^2$.

6.2 Numerical Solution

Viscous and diffusive effects complicate the stability analysis, and in many problems they are not of central interest. Even so, the added effort is well justified. Computing instabilities of inviscid stratified shear flows requires extremely fine grid resolution, and can therefore be very slow. The inclusion of very weak viscosity, e.g., $Re = 10^6$, stabilizes the numerical algorithm and thereby approximates the inviscid result with much lower computational cost.

The normal mode equations (6.14) and (6.15) can be written in matrix form:

$$\sigma \begin{pmatrix} \nabla^2 & 0 \\ 0 & 1 \end{pmatrix}\begin{pmatrix} \hat{w} \\ \hat{b} \end{pmatrix} = \begin{pmatrix} -\iota kU\nabla^2 + \iota kU_{zz} + \nu\nabla^4 & -\tilde{k}^2 \\ -B_z & -\iota kU + \kappa\nabla^2 \end{pmatrix}\begin{pmatrix} \hat{w} \\ \hat{b} \end{pmatrix}.$$

These are discretized to form a generalized eigenvalue problem with $2N \times 2N$ matrices:

$$\sigma\, A\vec{x} = B\vec{x}.$$

The eigenvector $\vec{x}$ is a concatenation of the discretized forms of $\hat{w}$ and $\hat{b}$:

$$\vec{x} = \begin{pmatrix} \hat{w}_1 \\ \hat{w}_2 \\ \vdots \\ \hat{w}_N \\ \hat{b}_1 \\ \hat{b}_2 \\ \vdots \\ \hat{b}_N \end{pmatrix},$$

and the matrices are

$$A = \begin{pmatrix} \nabla^2 & 0 \\ 0 & I \end{pmatrix}$$

$$B = \begin{pmatrix} -\iota k\vec{U}\cdot\nabla^2 + \iota k\vec{U}_{zz}\cdot I + \nu\nabla^4 & -\tilde{k}^2 I \\ -\vec{B}_z\cdot I & -\iota k\vec{U}\cdot I + \kappa\nabla^2 \end{pmatrix}.$$

Each matrix is composed of four $N \times N$ submatrices, which in turn are combinations of the Laplacian, the squared Laplacian, and input vectors U and B defined in the usual way. Left-multiplications such as $\vec{U} \cdot \nabla^2$ are computed as described in section 3.5.2.

6.2.1 Boundary Conditions

As in the homogeneous case, we assume that $\hat{w} = 0$ at the boundaries. We also have the choice of rigid ($\hat{w}_z = 0$) or frictionless ($\hat{w}_{zz} = 0$) boundaries.

For an inhomogeneous fluid, we also need boundary conditions for the buoyancy, and again there are two choices. First, we can assume that the buoyancy at the boundary is fixed, i.e., the perturbation from equilibrium is zero:

$$\hat{b} = 0.$$

The second choice is that the boundary is insulating, i.e., the diffusive buoyancy flux $-\kappa \partial b / \partial z$ vanishes. In normal mode form, this is expressed as

$$\hat{b}_z = 0.$$

Insulating boundaries, like frictionless boundaries, are often preferable in cases where there is no physical boundary, because they have minimal effect.

For the numerical solution described above, we need a matrix to compute the second-derivative of $\hat{b}$, and that matrix must incorporate the appropriate boundary conditions. In the case of fixed-buoyancy boundaries, the matrix can be exactly the same as we use to compute the second-derivative of $\hat{w}$, because the boundary condition is mathematically identical to $\hat{w} = 0$. But in the case of insulating boundaries, a separate second-derivative matrix is needed. The design of that matrix is the same as that described in section 5.5.

6.2.2 A Note on Applications

The theory encompassed in (6.14) and (6.15) includes every case we have discussed in Chapters 2, 3, 4, and 5 as special cases. For example, we can look at solutions of the Orr-Sommerfeld equation for homogeneous, viscous flows by setting $B_z = 0$. We can solve the Benard problem (section 2.4) by setting $U = 0$ and B_z to a negative constant. We can also combine these problems, e.g., to find the effect of stratification on the Tollmein-Schlichting instability, or the effect of shear on Benard convection (problem 19).

Solutions of the Taylor-Goldstein equation for inviscid stratified shear flows (Chapter 4) are obtained by setting $\nu = \kappa = 0$. These include all of the wavelike phenomena listed in section 4.5: internal gravity waves, vortical waves, baroclinic normal modes as well as the vertical structure functions used in the description of internal solitary waves, bores, and hydraulic jumps. In practice, one sets ν and κ to small but nonzero values. This gives close approximations to the inviscid limit but minimizes distortions due to roundoff error, and is therefore the preferred method for solving the Taylor-Goldstein equation.

The above methods lead to a solution procedure $\mathcal{F}$ for the analysis of viscous, diffusive, stratified, parallel shear flows. It is most commonly a subroutine. Having written and tested the subroutine, we treat it as a "black box," which accepts inputs and gives back outputs but does not need to be modified internally. If used correctly, $\mathcal{F}$ can solve a vast range of problems, based both on idealized models and on observational data. In the following sections 6.3 and 6.4, we will discuss different ways of defining the inputs and interpreting the outputs so that $\mathcal{F}$ will be maximally useful.

6.3 2D and Oblique Modes: Squire Transformations

Here, we consider the effect of stratification, viscosity, and diffusion on the growth rates of oblique modes. Recall that an oblique mode is one whose wave vector (k, ℓ) points at a nonzero angle φ from the x axis, the angle of obliquity. The corresponding 2D mode has a wave vector of the same magnitude, $\tilde{k}$, but parallel to the x axis (Figure 3.10).

6.3.1 Transforming the Buoyancy

Consider an oblique mode that obeys the equations for viscous, diffusive, stratified shear flow:

$$(\sigma + \imath k U)\nabla^2 \hat{w} - \imath k U_{zz}\hat{w} = -\tilde{k}^2 \hat{b} + \nu\nabla^4 \hat{w} \tag{3D1}$$

$$(\sigma + \imath k U)\hat{b} + B_z\hat{w} = \kappa\nabla^2 \hat{b}, \tag{3D2}$$

where

$$\nabla^2 = \frac{d^2}{dz^2} - \tilde{k}^2 ; \quad \tilde{k} = \sqrt{k^2 + \ell^2}.$$

Suppose also that we have a solution algorithm

$$\sigma = \mathcal{F}(z, U, B_z, \nu, \kappa; \ k, \ell). \tag{6.16}$$

The corresponding 2D mode with wave vector $(\tilde{k}, 0)$ obeys:

$$(\sigma + \imath\tilde{k}U)\nabla^2 \hat{w} - \imath\tilde{k}U_{zz}\hat{w} = -\tilde{k}^2 \hat{b} + \nu\nabla^4 \hat{w} \tag{2D1}$$

$$(\sigma + \imath\tilde{k}U)\hat{b} + B_z\hat{w} = \kappa\nabla^2 \hat{b}, \tag{2D2}$$

and therefore has the solution algorithm $\sigma_{2D} = \mathcal{F}(z, U, B_z, \nu, \kappa; \ \tilde{k}, 0)$.

Is there a transformation that makes these $(2D)$ equations isomorphic with the $(3D)$? We begin by defining the Squire transformations

$$\sigma = \cos\varphi\,\tilde{\sigma}; \quad \hat{b} = \cos\varphi\,\tilde{\hat{b}}; \quad v = \cos\varphi\,\tilde{v}.$$

Now substitute these into $(3D1)$ and divide out the common factor $\cos\varphi$. The result is isomorphic to $(2D1)$.

$$(\tilde{\sigma} + \imath\tilde{k}U)\nabla^2\hat{w} - \imath\tilde{k}U_{zz}\hat{w} = -\tilde{k}^2\hat{b} + \tilde{v}\nabla^4\hat{w}. \tag{3$\widetilde{D}$1}$$

Turning to equations $(3D2)$ and $(2D2)$, we now define two more transformations:

$$B_z = \cos^2\varphi\,\tilde{B}_z; \quad \kappa = \cos\varphi\,\tilde{\kappa}.$$

Substituting into $(3D2)$ and dividing out $\cos^2\varphi$, we obtain:

$$(\tilde{\sigma} + \imath\tilde{k}U)\tilde{\hat{b}} + \tilde{B}_z\hat{w} = \tilde{\kappa}\nabla^2\tilde{\hat{b}}. \tag{3$\widetilde{D}$2}$$

With all these transformations, $(3\widetilde{D}1)$ and $(3\widetilde{D}2)$ are isomorphic to $(2D1)$ and $(2D2)$, respectively, and can therefore be solved using the same solution algorithm: $\tilde{\sigma} = \mathcal{F}(z, U, \tilde{B}_z, \tilde{v}, \tilde{\kappa}; \tilde{k}, 0)$, or

$$\sigma_{3D} = \cos\varphi \times \mathcal{F}\left(z, U, \frac{B_z}{\cos^2\varphi}, \frac{v}{\cos\varphi}, \frac{\kappa}{\cos\varphi}; \tilde{k}, 0\right).$$

The growth rate of the 3D mode is $\cos\varphi$ times that of a corresponding 2D mode that exists in a fluid with increased viscosity, diffusivity and stratification. In most circumstances this means that the oblique mode will have a slower growth rate, but if any of those three factors should increase the growth rate, and do so rapidly enough to compensate for the factor $\cos\varphi$, then the oblique mode can grow faster.

6.3.2 Oblique and Veering Flows: Transforming the Velocity

In the inviscid case we found that a normal mode perturbation in a shear flow is affected only by the component of the background flow that is parallel to its own wave vector (section 4.3, Figure 4.7). The same is true in a viscous fluid, as we now confirm. As before, we define the transformed background velocity profile

$$\tilde{U}(z) = \frac{k}{\tilde{k}} U(z) = \cos\varphi\, U(z). \tag{6.17}$$

Now, everywhere in $(3D1, 2)$, replace kU with $\tilde{k}\tilde{U}$:

$$(\sigma + \imath\tilde{k}\tilde{U})\nabla^2\hat{w} - \imath\tilde{k}\tilde{U}_{zz}\hat{w} = -\tilde{k}^2\hat{b} + v\nabla^4\hat{w} \tag{3$\widetilde{D}$1}$$

$$(\sigma + \imath\tilde{k}\tilde{U})\hat{b} + B_z\hat{w} = \kappa\nabla^2\hat{b}. \tag{3$\widetilde{D}$2}$$

This set is isomorphic to $(2D1, 2)$ under the transformation (6.17), and therefore has the same solution algorithm:

$$\sigma_{3D} = \mathcal{F}(z, \tilde{U}, B_z, \nu, \kappa; \tilde{k}, 0). \tag{6.18}$$

For a typical shear instability, this means that the growth rate of the 3D mode will be reduced relative to the corresponding 2D mode. There are cases where this is not true, though. Recall the definitions of the Reynolds and bulk Richardson numbers:

$$Re = \frac{hu_0}{\nu}; \quad Ri_b = \frac{b_0 h}{u_0^2},$$

A decrease in u_0 decreases Re and increases Ri_b, and if either of these changes should increase the growth rate, a 3D mode may be the most unstable (see the example below). Also, some instabilities, such as convection, turn out to be damped by shear, so that reducing u_0 increases the growth rate, with interesting results as you will see in homework project 19.

As in the inviscid case (section 4.12), the solution algorithm (6.18) for parallel flows is easily extended to handle veering flows by replacing (6.17) with the more general definition:

$$\boxed{\tilde{U}(z) = \frac{kU(z) + \ell V(z)}{\tilde{k}}.} \tag{6.19}$$

In practice, one loops over a range of wave vectors (k, ℓ) and repeats the stability analysis for each case using $\tilde{U}(z)$ as given by (6.19).

6.4 Shear and Diffusion Scalings

The normal mode equations for a viscous, diffusive, stratified, parallel shear flow are [reproducing (6.14) and (6.15)]

$$(\sigma + \imath kU)\nabla^2 \hat{w} - \imath kU_{zz}\hat{w} = -\tilde{k}^2\hat{b} + \nu\nabla^4\hat{w} \tag{6.20}$$

$$(\sigma + \imath kU)\hat{b} + B_z\hat{w} = \kappa\nabla^2\hat{b}, \tag{6.21}$$

where $\nabla^2 = d^2/dz^2 - \tilde{k}^2$ as usual. Suppose that we have a solution algorithm

$$[\sigma, \hat{w}, \hat{b}] = \mathcal{F}(z, U, B_z, \nu, \kappa; k, \ell). \tag{6.22}$$

We now introduce a general length scale L and velocity scale V and use them to define nondimensional equivalents for the quantities in (6.20) and (6.22):

$$z^\star = \frac{z}{L}$$

$$U^\star = \frac{U}{V}; \quad \hat{w}^\star = \frac{\hat{w}}{V}$$

$$\sigma^\star = \sigma \frac{L}{V}$$

$$\{k^\star, \ell^\star, \tilde{k}^\star\} = \{k, \ell, \tilde{k}\} L$$

$$\frac{d}{dz} = \frac{1}{L} \frac{d}{dz^\star} \quad \Rightarrow \nabla^2 = \frac{1}{L^2} \left(\frac{d^2}{dz^{\star 2}} - \tilde{k}^{\star 2} \right)$$

$$v^\star = \frac{v}{LV} ; \quad \kappa^\star = \frac{\kappa}{LV}$$

$$B^\star = B \frac{L}{V^2} ; \quad \hat{b}^\star = \hat{b} \frac{L}{V^2}. \tag{6.23}$$

By substitution, one may now write a scaled version of (6.20, 6.21) and show that it is isomorphic to the original version (the student should do this), so that the same solution algorithm holds:

$$[\sigma^\star, \hat{w}^\star, \hat{b}^\star] = \mathcal{F}(z^\star, U^\star, B^\star_{z^\star}, v^\star, \kappa^\star; \ k^\star, \ell^\star). \tag{6.24}$$

Choices for the scales L and V are endless. We will look at two particularly useful choices: shear scaling and diffusion scaling.

While these generalized scalings always "work," i.e., you can input them to the solution procedure and get the correct (scaled) answer as in (6.24), they are not necessarily the most natural scalings for a particular problem. For example, although V^2/L has the right units for a buoyancy scale, it has no logical connection to buoyancy. It may be more natural to define a separate buoyancy scale b_0, so that

$$B = b_0 \beta,$$

where β is a nondimensional function. As an example, perhaps the background buoyancy profile is $B = b_0 \tanh \frac{z}{L}$, in which case $\beta = \tanh z^\star$.

Likewise, we may wish to write the background velocity using a separate velocity scale u_0 and a nondimensional function γ:

$$U = u_0 \gamma,$$

an example being the linear function $U = u_0 z/h$, where $\gamma = z^\star$.

While these choices are easily made, one must then relate them back to their equivalents in the generalized scaling (6.23), for it is these that must be inserted into the solution algorithm (6.24).

6.4.1 Example #1: Shear Scaling

We have used this scaling frequently. The length and velocity scales are:

$$L = h ; \quad V = u_0,$$

and are intended to characterize the vertical variability of the background velocity profile $U(z)$. In this case we find that scaled viscosity and diffusivity defined in (6.23) have recognizable forms:

$$v^\star = \frac{v}{LV} = \frac{v}{hu_0} = \frac{1}{Re}; \quad \kappa^\star = \frac{\kappa}{hu_0} = \frac{v}{hu_0}\frac{\kappa}{v} = \frac{1}{Re\,Pr}$$

Scaling the Buoyancy

Now suppose we choose the alternative form $B = b_0\beta$ for the background buoyancy profile, where b_0 is some buoyancy scale that suits the problem and β is a nondimensional function of $z^\star$. What should we then insert into the buoyancy "slot" in the solution algorithm?

$$B^\star = B\frac{L}{V^2} = b_0\beta\frac{h}{u_0^2} = \frac{b_0 h}{u_0^2}\beta(z^\star) = Ri_b\,\beta(z^\star).$$

The scaled buoyancy gradient is then

$$B^\star_{z^\star} = Ri_b\,\beta_{z^\star}.$$

For example, let $B = b_0\tanh(z/h)$, so that $\beta = \tanh z^\star$. Then the input to the solution procedure would be

$$B^\star_{z^\star} = Ri_b\,\beta_{z^\star} = Ri_b\,\text{sech}^2 z^\star.$$

With the above choices, the format for the solution algorithm is

$$[\sigma^\star, \hat{w}^\star, \hat{b}^\star] = \mathcal{F}(z^\star, U^\star, Ri_b\,\beta_{z^\star}, \frac{1}{Re}, \frac{1}{Re\,Pr}; \kappa^\star, \ell^\star). \qquad (6.25)$$

6.4.2 Example #2: Diffusion Scaling

The shear scaling defined above ($L = h$; $V = u_0$) can be used in any situation as long as the background current is nonzero, i.e., $u_0 \neq 0$. If we need to allow for the possibility that $u_0 = 0$, we may do so by adopting the diffusive scaling. For example, we might wonder what happens to Benard convection when a uniform shear is imposed (Figure 6.1), retaining the option to set the background shear to zero as a special case. We therefore choose the length scale to be the domain height H as in the original Benard problem (section 2.4), and the diffusive velocity scale $V = \kappa/H$.

Using these scales, the nondimensional viscosity and diffusivity defined in (6.23) take very simple forms:

$$v^\star = \frac{v}{VH} = \frac{v}{\kappa} = Pr, \qquad (6.26)$$

$$\kappa^\star = \frac{\kappa}{VH} = 1. \qquad (6.27)$$

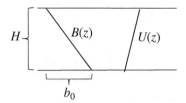

Figure 6.1 Definitions for the sheared Benard problem.

Scaling the Buoyancy

Suppose we choose the scaling $B = b_0 \beta$ for the buoyancy profile (where b_0 is some buoyancy scale that suits the problem, e.g., Figure 6.1, and β is a nondimensional function of z^* as described previously). In the solution algorithm (6.24), the buoyancy gradient input is determined as follows:

$$B^* = B\frac{H}{V^2} = b_0\beta\frac{H}{\kappa^2/H^2} = \frac{b_0 H^3}{\kappa^2}\beta = \frac{b_0 H^3}{\kappa\nu}\frac{\nu}{\kappa}\beta = Ra\,Pr\beta \Rightarrow B^*_{z^*} = Ra\,Pr\,\beta_{z^*}.$$

For the linear profile $B = -b_0 z/H$ (Figure 6.1), $\beta = -z^*$. Then the input to the solution algorithm would be $B^*_{z^*} = Ra\,Pr\,\beta_{z^*} = -Ra\,Pr$.

Scaling the Velocity

Now suppose we want to impose a background velocity

$$U = u_0\gamma,$$

where u_0 is a suitable velocity scale and γ is a nondimensional function. The appropriate input to the solution algorithm would be

$$U^* = \frac{U}{V} = \frac{u_0\gamma}{\kappa/H} = \frac{u_0 H}{\kappa}\gamma = \frac{u_0 H}{\nu}\frac{\nu}{\kappa}\gamma = Re\,Pr\gamma.$$

For example, we may assume a linear background shear as in Figure 6.1: $U = u_0 z/H$. In this case the velocity slot in the solution procedure would be filled by $U^* = Re\,Pr z^*$. The special case of zero background flow corresponds to the limit $Re \to 0$.

In summary, using the diffusive scaling with alternative forms for both buoyancy and velocity, the solution algorithm has the form:

$$[\sigma^*, \hat{w}^*, \hat{b}^*] = \mathcal{F}(z^*, Re\,Pr\,\gamma, -Ra\,Pr\,\beta_{z^*}, Pr, 1;\ k^*, \ell^*). \tag{6.28}$$

Keep in mind that the solution algorithm $\mathcal{F}$ has not changed since it was first defined in (6.16). The procedures outlined in this and the previous section will enable you to use it to solve a wide range of theoretical and observational problems.

6.5 Application: Instabilities of a Stably Stratified Shear Layer

Back in section 4.6, we studied Kelvin-Helmholtz and Holmboe instabilities using a highly simplified analytical model of an inviscid fluid. Here, we study the corresponding phenomena in a more realistic context, including the effects of viscosity and diffusion, using smoothly varying profiles, and solving the equations via the numerical techniques developed in this chapter.

Consider a stratified shear layer in which the velocity and the buoyancy may change over different length scales:

$$U^\star = \tanh z^\star \; ; \quad \beta = \tanh Rz^\star, \tag{6.29}$$

where shear scaling (section 6.4.1) is used and the constant parameter R is the ratio of shear layer thickness (Figure 6.2a) to stratified layer thickness (Figure 6.2b). While the case $R = 1$ is the most common, flows with $R > 1$ are frequently found in seawater, where momentum diffuses faster than buoyancy (Table 1.1). In this section we'll discuss cases with $R = 1$ and $R = 3$.

In the inviscid limit, the Miles-Howard theorem (section 4.7) requires that Ri be less than 1/4 somewhere in the flow. Here we will assume that Re is large so that the frozen flow hypothesis is valid and the Miles-Howard theorem is approximately valid.[3] When $R = 1$, the Richardson number is a minimum at the center of the shear layer (Figure 6.2c, blue). The Miles-Howard criterion therefore requires that $Ri_b < 1/4$. An exact expression is available for the stability boundary on the $k^\star - Ri_b$ plane (section 4.4):

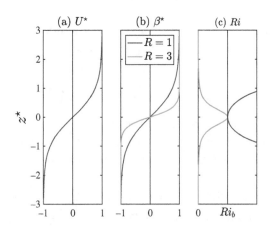

Figure 6.2 (a) Scaled velocity and (b) buoyancy profiles for the Holmboe model (6.29) with $R = 1, 3$. (c) Ri profile for the case $Ri_b = 0.1$, or $Ri_0 = 0.3$. Asterisks indicate shear scaling (section 6.4.1).

---

[3] If $Re \gg 1$, the only effect is a change in the critical Richardson number proportional to $1/Re$ (Thorpe et al., 2013). In the present cases the difference is negligible.

$$Ri_b = k^\star(1 - k^\star). \tag{6.30}$$

The critical state for instability is $k^\star = 1/2$, $Ri_b = 1/4$.

When $R \neq 1$, the value of Ri at the center of the shear layer is different from Ri_b. We call that central value Ri_0; it is equal to $Ri_b R$. In the case $R = 3$, Ri decreases from Ri_0 at the center of the shear layer to zero at infinity (Figure 6.2c, green), so the Miles-Howard criterion is satisfied regardless of how large Ri_0 is. As you might guess, this distinction has a profound effect on the nature of the instability.

6.5.1 Kelvin-Helmholtz Instability

Of all the instabilities of a stratified, parallel shear flow, the Kelvin-Helmholtz instability is probably the most commonly observed in nature. A stationary instability, it grows in place to form a train of co-rotating vortices that visually resemble surface waves breaking on a beach (e.g., Figure 6.4). These finite-amplitude manifestations of the instability are called Kelvin-Helmholtz billows. The underlying vortical structure is revealed more clearly in an echosounder image (Figure 6.5). For other examples, see Figures 4.4, 4.5, and 12.1 and Smyth and Moum (2012), or do a web search on "Kelvin-Helmholtz clouds."

Figure 6.3 shows the growth rate $\sigma^\star(k^\star, Ri_b)$ for the case $R = 1$, computed numerically using the methods of section 6.2. Viscosity is weak ($Re = 200$), and impermeable, frictionless boundaries are placed at $z^\star = \pm 5$. In the homogeneous limit $Ri_b = 0$, this is just the hyperbolic tangent shear layer discussed in section 3.9.1. The growth rate reaches a maximum value $\sigma^\star = 0.19$ at $k^\star = 0.44$, consistent with the previous results.

As Ri_b is increased, the growth rate decreases, indicating the damping effect of the stable buoyancy stratification. The wavenumber of the fastest-growing mode increases slightly. The stability boundary is similar to the inviscid result (6.30), but several differences are evident. For $k^\star > 0.5$, the stability boundary is slightly lower. This shows that, in the presence of viscosity, lower Ri_b is required for growth. Modes with smaller wavenumbers, $k^\star < 0.5$, are less affected by viscosity, but they are more susceptible to the influence of boundaries. The boundaries tend to enhance growth in this regime.

6.5.2 Holmboe Instability

We turn now to the case $R = 3$, i.e., where the stratification covers only the inner 1/3 of the shear layer (Figure 6.2, green curves). In the homogeneous limit $Ri_0 \to 0$, the value of R is irrelevant and we once again have a hyperbolic

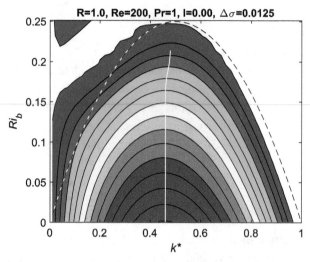

Figure 6.3 Growth rate versus scaled wavenumber and central/bulk Richardson number for the hyperbolic tangent profiles (6.29) with $R = 1$; $Re = 200$; $Pr = 1$. Only two-dimensional modes ($\ell^\star = 0$) are shown. The contour interval is 0.0125, starting at zero. Instability is stationary. The solid white curve indicates the fastest-growing mode at each Ri_b. Dashed curve shows the exact stability boundary for the limit of zero viscosity and infinite domain height. Frictionless, constant-buoyancy boundaries were placed at $z^\star = \pm 10$; grid increment was $\Delta^\star = 0.1$.

Figure 6.4 Kelvin-Helmholtz billow clouds, courtesy Brooks Martner, NOAA.

tangent shear layer with growth rate maximized at $k^\star = 0.44$ (Figure 6.6). If the stratification is sufficiently weak, the shear drives Kelvin-Helmholtz instability not much different from that found in the $R = 1$ case, although the wavenumber decreases markedly with increasing Ri_0. When the stratification is strong, Kelvin-Helmholtz instability is supplanted by Holmboe instability. The boundary

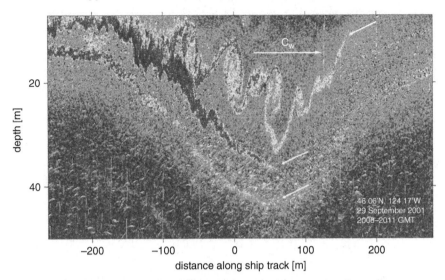

Figure 6.5 Echosounder image of a nonlinear internal wave propagating over the continental shelf toward the Oregon coast. C_w indicates the direction of wave propagation. Upper currents are in the direction of propagation, setting up a shear within the wave. Note the difference in horizontal and vertical scales; at any x, the flow is approximately a parallel shear flow. Short arrows identify three layers where disturbances grow as the wave passes. The uppermost layer forms a clearly resolved train of Kelvin-Helmholtz billows. The lower two layers are too thin for the growing disturbances to be well resolved. Reproduced from Moum et al. (2003).

between the two is approximately (but not exactly) $Ri_0 = 1/4$, and varies strongly with R.

Note the similarity between the stability diagrams in Figures 6.6 and 4.11(b). Both exhibit transitions between Kelvin-Helmholtz and Holmboe instabilities for increasing Richardson numbers. This suggests that the essential mechanisms of the instabilities are captured in the simple piecewise profiles. In our previous discussion of the Holmboe instability (section 4.6.3) we found that it is driven by a resonance of vorticity waves and gravity waves. The threshold stratification corresponds to Ri_0 somewhat greater than 1/4 (Figure 6.6). Above this threshold, the growth rate increases with increasing Ri_0 until a maximum is reached (at $Ri \approx 1$ for the case shown in Figure 6.6). Between these values, it is possible that the fastest growing mode is oblique (see section 6.3).

In the example shown in Figure 6.7, the fastest growing Holmboe instability is in fact oblique, with wave vector directed about 45 degrees from the streamwise direction. By symmetry, there is an accompanying mode with the opposite angle of obliquity. The result, viewed from above or below, is a diamond pattern of standing oscillations growing exponentially in time.

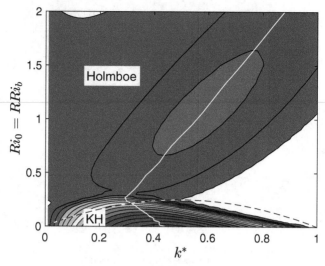

Figure 6.6 Growth rate versus scaled wavenumber and central Richardson number for the Holmboe profiles (6.29) with $R = 3$; $Re = 200$; $Pr = 9$. The contour interval is 0.0125, starting at zero. Instability is stationary in the lower lobe (Kelvin-Helmholtz instability); oscillatory in the higher (Holmboe instability). The dashed curve shows the inviscid stability boundary for $R = 1$, for comparison. Frictionless, constant-buoyancy boundaries were placed at $z^\star = \pm 10$; grid increment was $\Delta^\star = 0.1$. The solid white curve denotes the fastest-growing mode. The black dashed curve is the stability boundary for the case $R = 1$.

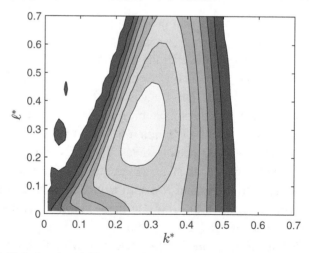

Figure 6.7 Holmboe instability growth rate versus scaled wavenumbers $k^\star$ and $\ell^\star$ for $Re = 200$ and $Ri_0 = 0.3$. The contour interval is 0.002, starting at zero. The fastest-growing mode is oblique.

6.6 Application: Analysis of Observational Data

Figure 6.8 shows profiles of currents and stratification observed just west of the Straits of Gibraltar (Nash et al., 2012). Current profiles (Figure 6.8a) include the zonal and meridional components, and extend to 450 m depth (just above the bottom). Both components show a distinct bottom current, directed southwest into the Atlantic. This is the Mediterranean Outflow, a gravity current driven by the high salinity (and thus high density) of the Mediterranean water. Above the outflow is a weak, nearly zonal return flow of fresher Atlantic water.

The buoyancy profile (Figure 6.8b) shows that the water column is stratified in the upper 150 m and near the bottom, but is much more homogeneous in the upper flank of the outflow and for about 100 m above that, indicating that the flow is, or has recently been, strongly turbulent.

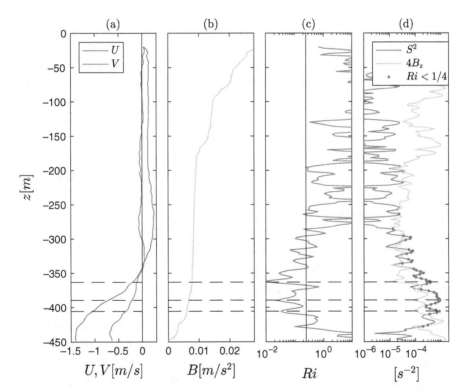

Figure 6.8 Profiles from the Mediterranean Outflow (Nash et al., 2012). (a) Zonal (blue) and meridional (red) current velocity. (b) buoyancy relative to the bottom. (c) Gradient Richardson number. (d) Squared shear magnitude (blue), and 4 times squared buoyancy frequency (yellow). Red dots show depths where $Ri < 1/4$. Horizontal lines show critical levels for the three fastest-growing instabilities. (Data courtesy of Jon Nash, Oregon State University.)

The gradient Richardson number Ri (Figure 6.8c) is high in the upper 300 m and near the bottom, but is low in the Outflow. At several points, $Ri < 1/4$, suggesting the possibility of instability. This possibility is further emphasized by the squared shear (Figure 6.8d, blue), which is high in regions of low Ri (red dots in Figure 6.8d). Shear maxima often coincide with minima of the buoyancy gradient (yellow). Based on these profiles, one could hypothesize that shear instability is common on the upper flank of the Outflow. That hypothesis can be tested via numerical solution of the perturbation equations.

To prepare for this analysis, the profiles were interpolated onto a regular vertical grid with spacing $\Delta = 2$ m and lowpass filtered with cutoff of 12 m. The squared shear, averaged over the water column, is greatest in the direction $\phi = 25°$ from current, roughly the direction of the outflow. Stability analysis was carried out using the current projected onto this direction: $\tilde{U} = U \cos \phi + V \sin \phi$, equivalent to looking for instabilities with wave vectors pointed in this direction (cf. section 4.12).[4]

A grid of wavelengths $\lambda = 2\pi/\tilde{k}$ was chosen, ranging from 32 m to 10 km. The reason for this choice is the rule of thumb for a shear layer instability: the wavelength is about 7 times the thickness of the shear layer (e.g., Table 3.1 or summary of Chapter 4). After applying the 12 m lowpass filter, the thinnest shear layer resolvable has thickness $h \sim 6$ m, so we can guess that it would produce an instability with $\lambda \sim 40$ m. The longest wavelength, 10 km, is intended to approximate a hydrostatic gravity wave whose vertical structure covers the whole water column. That wavelength was arrived at by trial and error. Modes at the short end of this range are poorly resolved and are considered highly approximate. Resolution is discussed in greater detail in Chapter 13.

To represent smoothing by ambient, small-scale turbulence, viscosity and diffusivity were set to $\nu = \kappa = 1 \times 10^{-4} \mathrm{m}^2/\mathrm{s}$. Frictionless, insulating boundary conditions were imposed at the surface and at the bottom. For each λ, the three fastest-growing eigenvalues were saved.

When several modes are retained, the dependence of the resulting growth rates on the wavelength can be difficult to make sense of, especially if the obliquity angle ϕ is also varied. The solution is to examine the depth of the critical level for each mode, as these tend to align themselves into a much more coherent pattern. The critical level depths are approximately independent of wavelength, at least over the range of wavelengths where the growth rate is large. Three depths are shown by horizontal lines on Figure 6.8. Upon close inspection, one can see that each depth coincides closely with a local maximum of the shear magnitude (cf. Shear Production Theorem, section 3.11.2). Therefore, each of the peaks in the

---

[4] This is a bit of corner-cutting. Ideally, one would scan over a range of directions to account for the veering of the flow with depth.

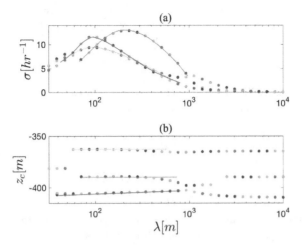

Figure 6.9 (a) Growth rates, and (b) critical level depths, versus wavelength for the three fastest-growing modes. Points are color coded for manual matching of σ's with z_c's.

growth rate shown in Figure 6.9a represents instability growing on a particular shear layer. We refer to such a grouping as a mode family and are most interested in the fastest-growing mode of each family.

For this simple case the growth rates and critical levels were matched by hand using the color-coding of the data points, and thereby organized into mode families as sketched on Figure 6.9. For more extensive datasets, one can automate the process by constructing a histogram of z_c. Mode families are then identified by peaks in the histogram.

Each of the three mode families seen on Figure 6.9a has its own fastest-growing mode and its own critical level. Of these, the most unstable (identified by the blue curve) has z_c at about 390 m depth (middle dashed line on Figure 6.8). Its growth rate is 14 hr^{-1}, so its amplitude grows by a factor $e = 2.718$ in about four minutes. The growth rate peaks at a wavelength a little over 200 m. According to our factor-of-7 rule this suggests a shear layer thickness around 30 m. Close inspection of Figure 6.8b bears this out: a shear maximum is visible at about 390 m depth, and 30 m would be a reasonable estimate of its thickness.

The second fastest mode (highlighted in red on Figure 6.9a) is considerably shorter, with wavelength 100 m, and has critical level at 410 m. The third mode (yellow) is focused at 360 m depth. It also has wavelength 100 m, but a much slower growth rate. Close inspection of Figure 6.8a shows corresponding shear maxima at 410 and 360 m, and in each case the peak shear is roughly consistent with the growth rate (reduced at 410 m, reduced further at 360 m). Both shear maxima are thinner than the one at 390 m, explaining the shorter wavelengths.

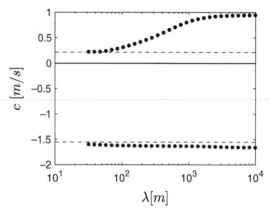

Figure 6.10 Phase velocities of the fastest upstream and downstream neutrally propagating internal waves as a function of wavelength.

None of these modes is associated with the upper flank as a whole[5]. That shear layer is about 100 m thick and would therefore support instabilities of about 700 m wavelength. In reality, these are damped by the close proximity of the bottom. (To learn about boundary effects, try exercise 12.) Instead, instabilities grow on smaller, more intense shear layers that develop more or less at random within the larger shear flow.

One possibility is that each of these instabilities creates turbulence, which then mixes out the shear layer that spawned it, returning Ri to values above $1/4$ until gravity accelerates the flow once again. This process, called cyclic instability, is discussed further in section 12.3. Another possibility is that the "shear layer" is actually a disruption of the measurements by an inquisitive sea creature. (Despite concerns about the impending extinction of ocean life, such encounters are frustratingly common, e.g., Pujiana et al., 2015).

Besides instabilities, one is often interested in neutrally stable wave modes supported by observed profiles. These are obtained by simply rearranging the computed eigenmodes to pick out those for which σ is nearly imaginary, e.g., $|\sigma_r/\sigma_i| < 10^{-3}$. We then compute the phase speed $c = -\iota\sigma/k$. Because the boundaries are fixed, the extremely fast surface and barotropic wave modes are excluded; what this calculation gives us are the baroclinic modes (first, second, etc.) as modified by whatever background current may be present.

For the Mediterranean Outflow profiles, the wave modes with fastest propagation upstream and downstream are shown in Figure 6.10. As is typical of internal gravity waves, the phase speed increases with increasing wavelength, asymptoting to a limiting value when the wavelength is several kilometers or more. Note that

---

[5] Meaning the main shear layer extending between about 330 m and 430 m depth.

both modes have phase speeds outside the range of the background velocity, i.e., they do not have critical levels. This suggests that the Mediterranean outflow is hydraulically subcritical, i.e., information can propagate both upstream and downstream, hence we would not expect to find an internal hydraulic jump, at least at this time and location.

6.7 Summary

Combining results from Chapters 2 through 5, we have developed a versatile set of techniques for the analysis of instabilities and waves in stratified shear flows. Viscosity and diffusion may originate with molecular effects or with turbulence, and may be included either for physical realism or as a numerical strategy to reduce resolution requirements. Besides idealized models like the Kelvin-Helmholtz and Holmboe shear instabilities, we have seen how the techniques may be applied to observational data.

6.8 Further Reading

See Smyth and Moum (2012) for an overview of Kelvin-Helmholtz instability in the ocean. More information on Holmboe instability is in Carpenter et al. (2010). Lab experiments confirming the linear theory of Kelvin-Helmholtz and Holmboe instabilities may be found in Thorpe (1973) and Tedford et al. (2009b). Further examples of normal mode analysis applied to observational data are Putrevu and Svendsen (1992), Einaudi and Finnigan (1993), Sun et al. (1998), Tedford et al. (2009a), Moum et al. (2011), Smyth et al. (2011), and Smyth et al. (2013).

7

Nonparallel Flow: Instabilities of a Cylindrical Vortex

The cylindrical (or columnar) vortex is the simplest example of a non-parallel shear flow, and is a useful model for tornados and other geophysical vortices. Here we'll examine two classes of vortex instabilities: (1) barotropic instabilities are closely analogous to the instabilities of a parallel shear flow, while (2) axisymmetric instabilities resemble convection, but with the centrifugal force playing the role of gravity.

Consider a cylindrical coordinate system with radial, azimuthal, and axial coordinates r, θ, and z (Figure 7.1) and corresponding velocities $u = dr/dt$, $v = r\,d\theta/dt$, and $w = dz/dt$. In geophysical applications, the axial direction usually coincides with the vertical.

Assuming inviscid, homogeneous flow, the equations are

$$\frac{1}{r}\frac{\partial}{\partial r}ru + \frac{1}{r}\frac{\partial v}{\partial \theta} + \frac{\partial w}{\partial z} = 0. \tag{7.1}$$

$$\frac{Du}{Dt} = \frac{v^2}{r} - \frac{\partial \pi}{\partial r} \tag{7.2}$$

$$\frac{Dv}{Dt} = -\frac{uv}{r} - \frac{1}{r}\frac{\partial \pi}{\partial \theta} \tag{7.3}$$

$$\frac{Dw}{Dt} = \frac{\partial \pi}{\partial z}, \tag{7.4}$$

where the material derivative is

$$\frac{D}{Dt} = \frac{\partial}{\partial t} + u\frac{\partial}{\partial r} + \frac{v}{r}\frac{\partial}{\partial \theta} + w\frac{\partial}{\partial z}, \tag{7.5}$$

[e.g., Smyth (2017), appendix I; Kundu et al. (2016), appendix B.6].

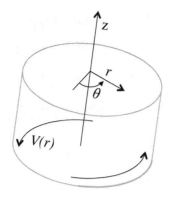

Figure 7.1 Axisymmetric (circular) vortex with cylindrical coordinates.

7.1 Cyclostrophic Equilibrium

We now seek an equilibrium state in which the flow is purely azimuthal: $u = 0$, $w = 0$, $v = V(r)$. This flow geometry is nondivergent, i.e., (7.1) is satisfied automatically. The momentum equations (7.2–7.4) become

$$\frac{V^2}{r} = \frac{\partial \Pi}{\partial r}$$

$$\frac{\partial \Pi}{\partial \theta} = 0$$

$$\frac{\partial \Pi}{\partial z} = 0.$$

The pressure field can vary only in r, and is related to the azimuthal velocity by

$$\frac{d}{dr}\Pi(r) = \frac{V(r)^2}{r}. \tag{7.6}$$

This balance between the pressure gradient and the centrifugal force is called cyclostrophic equilibrium.

It is also useful to define the angular velocity:

$$\Omega(r) = \frac{V}{r},$$

the axial vorticity

$$Q(r) = \frac{1}{r}\frac{d}{dr}(rV),$$

and the streamfunction $\Psi(r)$ such that

$$V(r) = -\frac{d\Psi}{dr}.$$

7.2 The Perturbation Equations

Now imagine a small perturbation to cyclostrophic equilibrium:

$$u = \epsilon u' ; \quad v = V(r) + \epsilon v' ; \quad w = \epsilon w' ; \quad \pi = \Pi(r) + \epsilon \pi'.$$

Substituting into (7.1–7.4) and linearizing, we obtain at $O(\epsilon)$:

$$\frac{1}{r}\frac{\partial}{\partial r}ru' + \frac{1}{r}\frac{\partial v'}{\partial \theta} + \frac{\partial w'}{\partial z} = 0. \tag{7.7}$$

$$\left[\frac{\partial}{\partial t} + \Omega\frac{\partial}{\partial \theta}\right]u' = 2\Omega v' - \frac{\partial \pi'}{\partial r} \tag{7.8}$$

$$\left[\frac{\partial}{\partial t} + \Omega\frac{\partial}{\partial \theta}\right]v' = -Qu' - \frac{1}{r}\frac{\partial \pi'}{\partial \theta} \tag{7.9}$$

$$\left[\frac{\partial}{\partial t} + \Omega\frac{\partial}{\partial \theta}\right]w' = -\frac{\partial \pi'}{\partial z}. \tag{7.10}$$

Exercise: Fill in the algebra.

Since the coefficients of the linearized equations depend on r, we seek a normal mode solution with the r-dependence undetermined:

$$u' = \hat{u}(r)e^{\sigma t}e^{\iota(\ell\theta + mz)},$$

where ℓ is an integer and only the real part is physically relevant. Substituting into the linearized equations (7.7–7.10) gives

$$\frac{1}{r}\frac{d}{dr}(r\hat{u}) + \frac{\iota\ell}{r}\hat{v} + \iota m\hat{w} = 0. \tag{7.11}$$

$$(\sigma + \iota\ell\Omega)\,\hat{u} = 2\Omega\hat{v} - \frac{d\hat{\pi}}{dr} \tag{7.12}$$

$$(\sigma + \iota\ell\Omega)\,\hat{v} = -Q\hat{u} - \frac{\iota\ell}{r}\hat{\pi} \tag{7.13}$$

$$(\sigma + \iota\ell\Omega)\,\hat{w} = -\iota m\hat{\pi}. \tag{7.14}$$

Two classes of perturbation are important and relatively easy to deal with.

- A barotropic perturbation has $m = 0$, i.e., no dependence on z. The first-mode barotropic instability has $\ell = 1$ (Figure 7.2a), and shifts the entire vortex horizontally. Higher-order barotropic modes ($\ell = 2, 3, \ldots$, Figure 7.2b,c, Figure 7.3) leave the vortex in place but distort its circular shape in increasingly ornate ways.
- The second important class of disturbances is the axisymmetric modes, also called centrifugal instability. These have $\ell = 0$, and hence no dependence on θ, but $m \neq 0$ (Figure 7.4a).

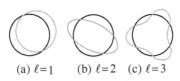

(a) $\ell=1$ (b) $\ell=2$ (c) $\ell=3$

Figure 7.2 Barotropic perturbations of a circular vortex, seen in plan view. In all cases $m = 0$.

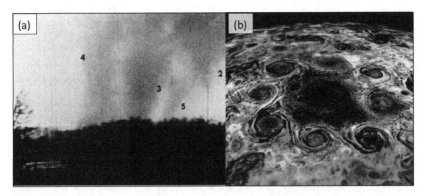

Figure 7.3 (a) Secondary vortices in a tornado suggestive of barotropic instability. Photo by W. Hubbard, WISH Indianapolis, from Snow (1978). (b) Instability with $\ell = 8$ surrounding Jupiter's north polar vortex (courtesy NASA). Structures appear to be barotropic, extending as deep as 3000 km (Adriani et al., 2018; Kaspi et al., 2018).

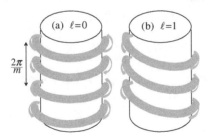

Figure 7.4 (a) Axisymmetric perturbation of a circular vortex. The mode takes the form of counter-rotating secondary vortices. (b) General, normal mode perturbation. In this case $\ell = 1$.

In each of these special cases there is a (relatively) easy way to collapse (7.11–7.14) into a single equation.

7.3 Barotropic Modes ($m = 0$)

For barotropic modes (Figures 7.2 and 7.3), the trick is to recognize that the perturbation flow is two-dimensional and nondivergent, and can therefore be represented by a streamfunction. We define $\hat{\psi}$ such that

$$\hat{u} = \frac{\iota\ell}{r}\hat{\psi} ; \quad \hat{v} = -\frac{d\hat{\psi}}{dr}.$$

Note that, with $m = 0$, (7.11) is satisfied exactly, and (7.14) gives $\hat{w} = 0$. Remaining are two equations for the two unknowns $\hat{\psi}$ and $\hat{\pi}$. These combine to form

$$(\sigma + \iota\ell\Omega)\left[\frac{d}{dr}\left(r\frac{d\hat{\psi}}{dr}\right) - \frac{\ell^2}{r}\hat{\psi}\right] = \iota\ell\frac{dQ}{dr}\hat{\psi}. \tag{7.15}$$

Exercise: Show this.

Exercise: Compare (7.15) with the Rayleigh equation (3.18). Identify and interpret the differences in each term.

7.3.1 Boundary Conditions for Barotropic Modes

- An impermeable boundary can be placed at any radius, say $r = r_1$. Impermeability requires that the radial velocity be zero at that boundary, i.e., $\hat{u} = 0$, and assuming $\ell \neq 0$,

$$\hat{\psi}(r_1) = 0.$$

- If the inner boundary is to be placed at $r = 0$, then we need an approximate solution for (7.15) that becomes exact as $r \to 0$. Suppose that $\hat{\psi}$ is proportional to r^α. Substituting into (7.15) and multiplying through by $r^{1-\alpha}$, we get

$$(\sigma + \iota\ell\Omega)(\alpha^2 - \ell^2) = \iota\ell r\frac{dQ}{dr}\hat{\psi}. \tag{7.16}$$

The background profiles $\Omega(r)$ and $Q(r)$ are not yet specified, but as long as dQ/dr is finite, the right-hand side goes to zero as $r \to 0$, and therefore as long as $\sigma - \iota\ell\Omega \neq 0$, $\alpha^2 - \ell^2 = 0$. We choose the solution that is bounded as $r \to 0$ and end up with

$$\hat{\psi} \propto r^\ell, \quad \text{or } \hat{\psi}(0) = 0.$$

In numerical calculations it is not a problem to have the inner boundary at $r = 0$, even though r appears in the denominator of (7.15). This is because $r_0 = 0$ is a ghost point, so nothing is ever actually evaluated there.

- If the outer boundary is at infinity, we can again assume that $\hat{\psi} \propto r^\alpha$, resulting again in (7.16). If we now assume that dQ/dr decays to zero faster than $1/r$ as $r \to \infty$, then the right-hand side goes to zero, and if $\sigma - \iota\ell\Omega \neq 0$ we again have $\alpha = \pm\ell$. The bounded solution is now $\hat{\psi} \propto r^{-\ell}$, and the boundary condition becomes

$$\lim_{r\to\infty} \hat{\psi}(r) = 0. \tag{7.17}$$

- In numerical calculations, we cannot actually place the outer boundary at infinity, so we place it at some large but finite radius (hopefully where dQ/dr has decreased almost to zero) and apply the asymptotic condition

$$\frac{d\hat{\psi}}{dr} = -\ell\hat{\psi}.$$

The perturbation equation (7.15) can then be reduced to a generalized eigenvalue problem using derivative matrices as in the case of parallel flows.

The matrix solution of (7.15) is analogous to the case of parallel shear flow. We first replace the derivatives with derivative matrices incorporating the appropriate boundary conditions. We then arrange the equation as an eigenvalue equation and find the eigenvalues and eigenvectors numerically.

Admonition: It may be tempting to define a first-derivative matrix $\mathbf{D}^{(1)}$, then use it twice to form the second-derivative. Don't do this – it effectively replaces the grid spacing Δ by 2Δ, degrading the accuracy of the results. In (7.15), the first term in the brackets should be computed in the expanded form

$$\mathbf{D}^{(1)} + r \cdot \mathbf{D}^{(2)},$$

rather than the simpler but less accurate

$$\mathbf{D}^{(1)} r \cdot \mathbf{D}^{(1)}.$$

7.3.2 Stability Theorem for Barotropic Modes

We rewrite (7.15) as

$$\frac{d}{dr}\left(r\frac{d\hat{\psi}}{dr}\right) - \frac{\ell^2}{r}\hat{\psi} = \frac{\iota\ell\hat{\psi}}{\sigma + \iota\ell\Omega}\frac{dQ}{dr}, \tag{7.18}$$

then apply the operator $\int_{r_1}^{r_2} \hat{\psi}^* dr$. The radii r_1 and r_2 are the boundaries of the domain. The inner boundary radius may be $r_1 = 0$, and the outer may be $r_2 = \infty$.

The first term on the left gives

$$\int_{r_1}^{r_2} \hat{\psi}^* \frac{d}{dr}\left(r\frac{d\hat{\psi}}{dr}\right) dr = \hat{\psi}^* r \frac{d\hat{\psi}}{dr}\bigg|_{r_1}^{r_2} - \int_{r_1}^{r_2} r\left|\frac{d\hat{\psi}}{dr}\right|^2 dr.$$

Using the boundary conditions derived in the previous subsection, the first term vanishes, leaving

$$\int_{r_1}^{r_2} \hat{\psi}^* \frac{d}{dr}\left(r\frac{d\hat{\psi}}{dr}\right) dr = -\int_{r_1}^{r_2} r\left|\frac{d\hat{\psi}}{dr}\right|^2 dr.$$

The second term on the left of (7.18) is just

$$\int_{r_1}^{r_2} \frac{\ell^2}{r} |\hat{\psi}|^2 dr.$$

Note that both of the above integrals are real. Applying the integral operator to the right-hand side and taking the imaginary part, we have

$$0 = \Im \int_{r_1}^{r_2} \frac{\imath \ell \, |\hat{\psi}|^2}{\sigma + \imath \ell \Omega} \frac{dQ}{dr} dr.$$

Multiplying and dividing the integrand by the complex conjugate $\sigma^* - \imath \ell \Omega$ to isolate the imaginary part gives

$$0 = \ell \sigma_r \int_{r_1}^{r_2} \frac{|\hat{\psi}|^2}{|\sigma + \imath \ell \Omega|^2} \frac{dQ}{dr} dr.$$

For a growing (or decaying) mode, $\sigma_r \neq 0$, and therefore the integral must vanish, i.e., dQ/dr must change sign at least once in $r_1 < r < r_2$.

> **Theorem** *Given an inviscid, homogeneous, circular vortex, a necessary condition for barotropic instability is that the vorticity gradient dQ/dr change sign somewhere in the domain $r_1 < r < r_2$.*

Note the similarity between this and the inflection point theorem for parallel shear flows (section 3.11.1 or 3.15). As we will see later in this chapter, barotropic vortex instabilities and parallel shear flow instabilities have many properties in common. For these instabilities, it is not entirely wrong to think of the vortex as a parallel shear flow bent to form a circle. However, the curvature also introduces an important new effect: the centrifugal force. This effect is understood most simply in the context of axisymmetric modes, which we investigate next.

Exercise: Derive a Fjørtoft-type condition for barotropic vortex instabilities.

7.4 Axisymmetric Modes ($\ell = 0$)

In the axisymmetric case $\ell = 0$, (7.11–7.14) can be combined into a single equation for the radial velocity perturbation $\hat{u}$:

$$\sigma^2 \left\{ \frac{d}{dr} \left[\frac{1}{r} \frac{d}{dr} (r\hat{u}) \right] - m^2 \hat{u} \right\} = \Phi m^2 \hat{u}, \tag{7.19}$$

where

$$\Phi(r) = 2\Omega Q$$

is called the Rayleigh discriminant.

7.4.1 Boundary Conditions for Axisymmetric Modes

- If an impermeable boundary is placed at some r_1, then the radial velocity must vanish there, i.e., the boundary condition is just $\hat{u}(r_1) = 0$.
- Now suppose there is no inner boundary, so we need a virtual boundary condition at $r = 0$. Assume that, for r near zero, $\hat{u}$ is proportional to r^α. Substituting into (7.19) and multiplying through by $r^{2-\alpha}$, we obtain

$$\sigma^2\{\alpha^2 - 1 - m^2 r^2\} = m^2 r^2 \Phi.$$

Assuming that Φ remains finite, the right-hand side must vanish as $r \to 0$. Therefore, for nonzero σ, the quantity in braces must vanish as $r \to 0$, hence $\alpha = \pm 1$. To keep the solution bounded, we choose $\alpha = 1$, i.e., $\hat{u} \propto r$. The boundary condition at $r = 0$ is therefore

$$\hat{u}(0) = 0.$$

- If there is no outer boundary, we employ an asymptotic boundary condition. We will assume that the vortex is *isolated*, meaning that if you go far enough away, the vortex motion vanishes. More specifically, $\Phi \to 0$ as $r \to \infty$. In that case, for sufficiently large r, (7.19) becomes

$$\frac{d}{dr}\left[\frac{1}{r}\frac{d}{dr}(r\hat{u})\right] - m^2\hat{u} = 0$$

This is the *modified Bessel equation* (Spiegel, 1968), and its bounded solution is the first-order modified Bessel function:

$$\hat{u} = K_1(mr).$$

As $r \to \infty$, K_1 can be approximated using Stirling's formula

$$K_1(mr) \approx \frac{e^{-mr}}{\sqrt{2\pi mr}}; \quad \text{for } mr \gg 1.$$

Therefore, $\hat{u} \to 0$ in the limit $r \to \infty$.

- An asymptotic condition is also available for use in numerical calculations where the domain must be finite. Taking the logarithmic derivative of the Stirling approximation to K_1,

$$\frac{1}{\hat{u}}\frac{d\hat{u}}{dr} = \frac{d}{dr}(\ln \hat{u}) = \frac{d}{dr}\left[-mr - \frac{1}{2}\ln(2\pi mr)\right] = -m - \frac{1}{2r}.$$

So if the computation domain ends at $r = R$, the asymptotic boundary condition is

$$\frac{d\hat{u}}{dr} = -\left(m + \frac{1}{2R}\right)\hat{u}.$$

7.4.2 Stability Theorem for Axisymmetric Modes

We now apply the integral operator $\int_{r_1}^{r_2} r\hat{u}^* dr$ to (7.19). Here, r_1 and r_2 are the boundaries of the domain. The inner boundary radius may be $r_1 = 0$, and the outer may be $r_2 = \infty$. We'll apply this operator individually to the two terms on the left-hand side and the single term on the right. The first term on the left, omitting the factor σ^2 for now, gives

$$\int_{r_1}^{r_2} r\,\hat{u}^* \frac{d}{dr}\left[\frac{1}{r}\frac{d}{dr}(r\hat{u})\right] dr = \hat{u}^* \frac{d}{dr}(r\hat{u})\Big|_{r_1}^{r_2} - \int_{r_1}^{r_2} \frac{d}{dr}(r\hat{u}^*)\frac{1}{r}\frac{d}{dr}(r\hat{u})dr$$

$$= -\int_{r_1}^{r_2} \frac{1}{r}\left|\frac{d}{dr}(r\hat{u})\right|^2 dr.$$

Here, the boundary conditions derived in the previous subsection have been used. The second term (setting aside the factor $-\sigma^2 m^2$) is

$$\int_{r_1}^{r_2} r\,\hat{u}^*\hat{u}\,dr = \int_{r_1}^{r_2} r\,|\hat{u}|^2 dr.$$

Finally, the right-hand side (omitting m^2) is

$$\int_{r_1}^{r_2} r\,\Phi|\hat{u}|^2 dr.$$

Combining these results and restoring the various constants, we have

$$\sigma^2\left\{\int_{r_1}^{r_2} \frac{1}{r}\left|\frac{d}{dr}(r\hat{u})\right|^2 dr + m^2\int_{r_1}^{r_2} r\,|\hat{u}|^2 dr\right\} = -m^2\int_{r_1}^{r_2} r\,\Phi|\hat{u}|^2 dr$$

or, with some rearranging

$$\sigma^2\int_{r_1}^{r_2} \frac{1}{r}\left|\frac{d}{dr}(r\hat{u})\right|^2 dr = -m^2\int_{r_1}^{r_2} r\,|\hat{u}|^2(\sigma^2 + \Phi)\,dr. \qquad (7.20)$$

For $\sigma^2 > 0$ the integral on the right must be negative, and therefore $\sigma^2 + \Phi$ must be negative for some r. Therefore the minimum value of $\Phi(r)$ must be less than $-\sigma^2$, or

$$\sigma < \sqrt{-\min_z(\Phi)}. \qquad (7.21)$$

Instability is possible provided that $\min_z(\Phi) < 0$. This class of unstable modes is called centrifugal instability.

> **Theorem** *Given an inviscid, homogeneous, circular vortex, a necessary condition for centrifugal instability is that the Rayleigh discriminant $\Phi(r) = 2\Omega(r)Q(r)$ be negative for some r. Moreover, (7.21) gives an upper bound on the growth rate.*

Centrifugal instability is closely analogous to convection. To see this, note the similarity between (7.19) and (2.29), the equation for convective instability in an inviscid, nondiffusive fluid with arbitrary stratification $B_z(z)$. The Rayleigh discriminant $\Phi(r)$ is the analog of stratification. In the convective case, if $B_z < 0$, the fluid possesses gravitational potential energy that can be converted to kinetic energy. Here, a variant of potential energy due to the centrifugal force is available for conversion to kinetic energy wherever $\Phi < 0$. Also compare the growth rate bound (7.21) for centrifugal instability with the upper bound on the convective growth rate, (2.34). This analogy is discussed in greater detail later in section 7.8.

7.5 Analytical Example: the Rankine Vortex

The Rankine vortex has uniform vorticity $2\Omega_0$ inside a radius R and zero vorticity outside (Figure 7.5, black curves). It is a useful model for localized vortices such as tornadoes. The vorticity gradient is given by

$$\frac{dQ}{dr} = -2\Omega_0\delta(r - R) \tag{7.22}$$

There is no radius at which the vorticity gradient changes sign, so there is no possibility of barotropic instability. How about centrifugal instability? The azimuthal velocity profile is

$$V(r) = \begin{cases} \Omega_0 r, & \text{for } r \leq R \\ \Omega_0\dfrac{R^2}{r}, & \text{for } r \geq R \end{cases} \quad \Rightarrow \quad \Phi(r) = \begin{cases} 4\Omega_0^2, & \text{for } r \leq R \\ 0, & \text{for } r \geq R \end{cases}. \tag{7.23}$$

With no negative values of Φ, there can be no centrifugal instability.

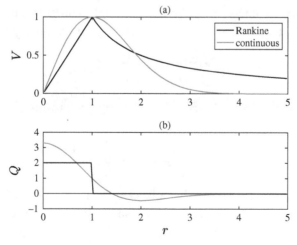

Figure 7.5 Profiles of velocity (a) and vorticity (b) for the Rankine vortex (7.23) with $\Omega_0 = R = 1$ and the continuous vortex (7.29).

Barotropic Waves on a Rankine Vortex

Although the Rankine vortex is stable, its barotropic wave modes are of interest. (Finding the wave modes of the axisymmetric case is left as an exercise.) We will find it convenient to define the complex angular velocity of the perturbation,

$$\omega = \frac{\iota\sigma}{\ell},$$

and rewrite (7.15) as

$$\frac{d}{dr}\left(r\frac{d\hat{\psi}}{dr}\right) - \frac{\ell^2}{r}\hat{\psi} = \frac{dQ/dr}{\Omega - \omega}\hat{\psi}. \tag{7.24}$$

Except at $r = R$, the right-hand side of (7.24) is zero. As was noted in section 7.3.1, solutions then have the form $\hat{\psi}(r) \propto r^{\pm\ell}$. Applying the boundary conditions $\hat{\psi} \to 0$ as $r \to 0$ and ∞ and requiring continuity across $r = R$ leads to

$$\hat{\psi}(r) = A \begin{cases} (r/R)^{\ell} & ,r < R \\ (r/R)^{-\ell} & ,r > R \end{cases} \tag{7.25}$$

with A an arbitrary constant.

The dispersion relation is obtained as in the analysis of both convection at an interface (section 2.2.4) and the instability of a piecewise-linear shear layer (section 3.3). A jump condition is found by integrating (7.24) across the delta function in (7.22):

$$\lim_{\epsilon\to 0}\int_{R-\epsilon}^{R+\epsilon}\frac{d}{dr}\left(r\frac{d\hat{\psi}}{dr}\right)dr - \lim_{\epsilon\to 0}\int_{R-\epsilon}^{R+\epsilon}\frac{\ell^2}{r}\hat{\psi}dr = \lim_{\epsilon\to 0}\int_{R-\epsilon}^{R+\epsilon}\frac{-2\Omega_0\delta(r-R)}{\Omega - \omega}\hat{\psi}dr. \tag{7.26}$$

The first integral is trivial, and the second goes to zero as its range of integration vanishes because its integrand is finite. The right-hand side simplifies by using properties of the delta function (Figure 2.5), leaving us with the jump condition:

$$\left[\!\left[r\frac{d\hat{\psi}}{dr}\right]\!\right]_R = \frac{-2\Omega_0}{\Omega_0 - \omega}\hat{\psi}(R). \tag{7.27}$$

After substituting $[\![r\,d\hat{\psi}/dr]\!]_R = -2A\ell$ and $\hat{\psi}(R) = A$ from (7.25), we arrive at the dispersion relation

$$\omega = \Omega_0 + \frac{\Delta Q_0}{2\ell} = \Omega_0\left(1 - \frac{1}{\ell}\right). \tag{7.28}$$

This describes a vorticity wave being advected around the core of the vortex with angular velocity Ω_0 while propagating with an intrinsic phase speed $-\Delta Q_0/2\ell = \Omega_0/\ell$. Note the similarity to the dispersion relation of the vorticity waves from section 3.13.1. The wave propagates upstream relative to the vortex. The fundamental mode $\ell = 1$, with intrinsic phase speed $-\Omega_0$, is stationary. Modes with higher wavenumbers ($\ell > 1$) are unable to keep up with the advective speed and therefore precess in the same sense as the vortex.

More complex profiles can support multiple wave modes. Like instability in an inviscid shear layer, barotropic instability of a circular vortex can result from the resonant interaction of these waves, as is described below in section 7.7.

7.6 Numerical Example: a Continuous Vortex

We now consider a vortex with a continuous azimuthal velocity profile, nondimensionalized so that both the maximum flow and the radius of maximum flow are unity (Figure 7.5, blue curves):

$$V = r\,e^{-\frac{1}{2}(r^2-1)}\,; \quad Q = (2 - r^2)\,e^{-\frac{1}{2}(r^2-1)}. \tag{7.29}$$

7.6.1 Barotropic Modes

Because the vorticity gradient changes sign at $r = 2$, barotropic instability is possible (section 7.3.2). In fact, the barotropic mode with $\ell = 2$ is unstable as shown in Figure 7.6a. The streamfunction eigenfunction has maximum amplitude just inside $r = 2$, the inflectional radius, and the phase shifts rapidly near this radius. The sign of the phase shift is such that phase lines of the radial velocity tilt against the vorticity. This is the circular analog of the instability of a parallel shear flow (Chapter 3).

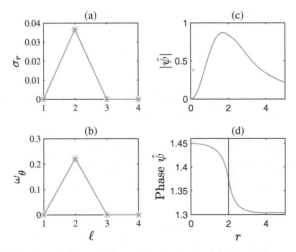

Figure 7.6 Growth rate (a) and frequency (b) versus azimuthal wavenumber for barotropic modes of (7.29). Amplitude (c) and phase (d) profiles for the fastest-growing barotropic mode. Vertical line shows the radius of minimum vorticity (cf. Figure 7.5).

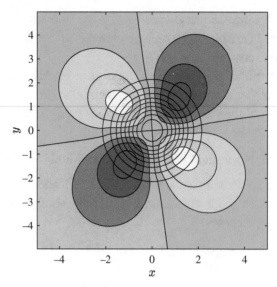

Figure 7.7 Streamfunction for the fastest-growing barotropic mode of (7.29). Circles are streamlines of the background flow.

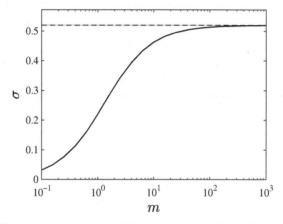

Figure 7.8 Growth rate versus axial wavenumber for axisymmetric modes. Dashed line shows Rayleigh's upper bound $\sqrt{-\Phi_{min}}$.

The velocity perturbation causes the vortex to bulge inward and outward as in Figure 7.2b. The mode is not stationary; it precesses around the vortex with azimuthal velocity about one-fifth that of the maximum flow speed (Figure 7.6b).

7.6.2 Axisymmetric Modes

The Rayleigh discriminant $2\Omega Q$ is negative for $r > \sqrt{2}$ (where $Q < 0$, Figure 7.5b). We therefore suspect axisymmetric instability, and that suspicion is confirmed in the numerical results (Figure 7.8). There is no preferred axial scale:

the growth rate increases monotonically with increasing axial wavenumber. This is broadband instability, as we found previously for convective instability of an inviscid fluid (section 2.2). As $m \to \infty$, the growth rate approaches the maximum value $\sqrt{-\Phi_{min}}$.

The radial velocity is greatest near $r = \sqrt{3}$, where Φ is most negative. As m is increased, the eigenfunction becomes more tightly concentrated near that radius. The result is a stack of counter-rotating vortices surrounding the background vortex, as sketched in Figure 7.4a.

Exercise: Derive a perturbation kinetic energy budget analogous to (3.56) based on (7.7–7.10).

7.7 Wave Interactions in Barotropic Vortices

Recall from Chapter 3 that instabilities of a parallel shear flow may be interpreted as resonant wave interactions. Here we develop an equivalent view for barotropic modes of a circular vortex. Consider a background profile $V(r)$ that has a concentric, piecewise-uniform vorticity distribution. The vorticity gradient is composed of a series of delta functions:

$$\frac{dQ}{dr} = \sum_i \Delta Q_i \delta(r - R_i), \tag{7.30}$$

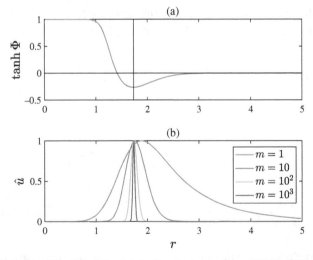

Figure 7.9 (a) Profile of the Rayleigh discriminant $\Phi = 2\Omega Q$ of (7.29), scaled using the tanh function to make the minimum visible. The vertical line indicates $r = \sqrt{3}$, where Φ is a minimum. (b) Eigenfunction of the radial velocity for various m.

where ΔQ_i is the jump in vorticity across each vorticity interface, located at $r = R_i$. The advantage of choosing this type of profile (equivalent to the piecewise-linear representation of a parallel shear flow) is that it replaces the governing equation with the simpler form

$$\frac{d}{dr}\left(r\frac{d\hat{\psi}}{dr}\right) - \frac{\ell^2}{r}\hat{\psi} = 0. \tag{7.31}$$

This form applies between the interfaces, and has a solution that was already found in our look at the Rankine vortex in section 7.5. Writing the solution (7.25) in a slightly more general form, for a single interface at $r = R_i$ we have

$$\hat{\psi}(r) = A_i G(r, R_i) \quad \text{where} \quad G(r, R_i) = \begin{cases} (r/R_i)^\ell & , r < R_i \\ (r/R_i)^{-\ell} & , r > R_i \end{cases}. \tag{7.32}$$

The function $G(r, R_i)$ can be thought of as an "influence function"[1] describing the decay of the interfacial disturbance from its peak at $r = R_i$ (Figure 7.10). Note that, as in the case of the shear layer, the amplitude of G decays more rapidly with increased wavenumber (ℓ, in this case), so that longer waves are "felt" over a greater distance. The solution for N interfaces is

$$\hat{\psi}(r) = \sum_{i=1}^{N} A_i G(r, R_i). \tag{7.33}$$

The remaining step is to connect this solution, which applies *between* the interfaces, with jump conditions that apply *at* the interfaces. The required jump condition is given above in (7.27). Substituting (7.33), we have

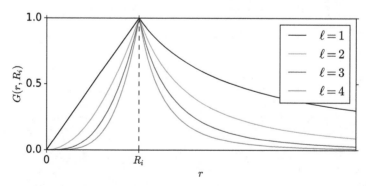

Figure 7.10 Structure of the influence function, $G(r, R_i)$, in a barotropic vortex arising from a vorticity interface located at $r = R_i$ (dashed line).

---

[1] More precisely, this is the Green's function for the linear differential operator in (7.31). The use of Green's functions is a more general approach to solving this type of problem.

$$- 2\ell A_i = \frac{\Delta Q_i}{2\ell(\Omega_i - \omega)} \sum_{j=1}^{N} A_j G(R_i, R_j). \tag{7.34}$$

With a little rearranging, we can write this as

$$\omega A_i = \sum_{j=1}^{N} \left\{ \Omega_i I_{ij} + \frac{\Delta Q_i}{2\ell} G_{ij} \right\} A_j, \tag{7.35}$$

where I is the $N \times N$ identity matrix and $\mathbf{G}$ is the influence matrix, defined by $G_{ij} = G(R_i, R_j)$. The quantity enclosed in braces is an $N \times N$ matrix, each of whose N eigenvalues is the angular velocity ω for one of the N eigenmodes. If ω has a positive imaginary part, the mode is unstable. The corresponding eigenvector contains the coefficients A_i that define the radial dependence of the amplitude.

Example: a General Two-Interface Vortex

We now look at a general example of a barotropic vortex that consists of two vorticity interfaces, located at $r = R_1$ and $r = R_2$ and having magnitudes ΔQ_1 and ΔQ_2. The influence matrix is

$$\begin{bmatrix} 0 & \delta^\ell \\ \delta^\ell & 0 \end{bmatrix}$$

where $\delta = R_1/R_2$.

According to (7.35), the phase velocities are given by the eigenvalue equation

$$\begin{bmatrix} \Omega_1 + \dfrac{\Delta Q_1}{2\ell} & \dfrac{\Delta Q_1}{2\ell}\delta^\ell \\ \dfrac{\Delta Q_2}{2\ell}\delta^\ell & \Omega_2 + \dfrac{\Delta Q_2}{2\ell} \end{bmatrix} \begin{bmatrix} A_1 \\ A_2 \end{bmatrix} = \omega \begin{bmatrix} A_1 \\ A_2 \end{bmatrix}. \tag{7.36}$$

In the special case of a single interface at $r = R_1$, with vorticity change $\Delta Q_1 = -2\Omega_1$, we set $\delta = \Omega_2 = \Delta Q_2 = 0$. Our eigenvalue problem (7.36) then simplifies to

$$\omega_1 = \Omega_1 + \frac{\Delta Q_1}{2\ell} \tag{7.37}$$

which is equivalent to (7.28). Alternatively, the single interface could be located at R_2, in which case

$$\omega_2 = \Omega_2 + \frac{\Delta Q_2}{2\ell}. \tag{7.38}$$

In terms of those frequencies, the eigenvalues of (7.36) are

$$\omega = \frac{\omega_1 + \omega_2}{2} \pm \left[\left(\frac{\omega_1 - \omega_2}{2} \right)^2 + \Delta Q_1 \Delta Q_2 \frac{\delta^{2\ell}}{4\ell^2} \right]^{1/2}. \tag{7.39}$$

In order to have instability the vorticity jumps must have opposite sign: $\Delta Q_1 \Delta Q_2 < 0$. Compare this result with the theorem proven in section 7.3.2.

Note also that we recover the undisturbed phase speeds if the strength of the interaction between the two waves, described by the factor $\delta^{2\ell}$, goes to zero. This is equivalent to increasing the distance between the two interfaces indefinitely so that the velocity perturbations decay to zero and the waves become uncoupled.

7.8 Mechanisms of Centrifugal and Convective Instabilities

As we noted in section (7.4.2), the stability equations (7.19) for centrifugal instability and (2.29) for convection are very similar. In fact, the only differences are due to the cylindrical geometry of the former. Here we will describe the mechanisms of the two instabilities in terms that will highlight the parallels between the two.

In the convectively unstable background state sketched in Figure 7.11, dense fluid overlies light fluid.

(i) If a downward flow w' is initiated at some point (thick blue arrow), the density at that point will increase in time.

(ii) The resulting change in the buoyancy force, F', is directed downward (thin blue arrow), and hence accelerates the downward flow.

(iii) Consistent with mass conservation, this downward motion is accompanied by upward motion at some other location. There, the reverse process happens: the buoyancy force is perturbed so as to accelerate the upward motion (red arrows).

In the case of centrifugal instability, the background azimuthal velocity at some radius decreases with distance from the vortex center. On the right-hand side of the vortex sketched in Figure 7.12, the background flow is directed into the page.

(i) A radially outward flow is initiated at some point (thick red arrow). It carries with it a positive perturbation in azimuthal velocity, and thus an increase in the centrifugal force (thin red arrow).

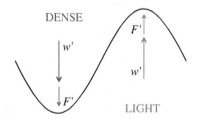

Figure 7.11 Perturbations involved in the convective instability of a statically unstable buoyancy distribution.

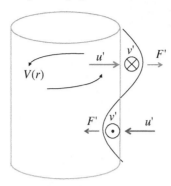

Figure 7.12 Perturbations involved in the centrifugal instability of a circular vortex. The radial motion u advects the background azimuthal velocity to create a perturbation, F, in the centrifugal force.

(ii) The disturbance in the force is directed outward, and hence accelerates the outward flow.

(iii) The outward motion is accompanied by inward motion at some other location. There, the reverse process happens (blue arrows): the centrifugal force is perturbed so as to accelerate the inward motion.

Exercise: Examine the perturbations equations for each of these instabilities and identify the terms that correspond to the three-part processes described above.

7.8.1 Universality of the Fastest-Growing Mode

When inspecting Figure 7.8, you may have noticed that the growth rate actually reaches the upper bound $\sqrt{-\min_r(\Phi)}$ in the limit $m \to \infty$. There is nothing special about the profile (7.29); in fact, it appears that this is a general property of both centrifugal and convective instabilities in an inviscid fluid. Specifically, the upper bound we have derived for the growth rate is actually reached in the limit of large wavenumber ($\tilde{k}$ for convection, m for centrifugal instability), regardless of the details of the profile B_z (or Φ), provided only that it includes at least one negative local minimum as required by the stability theorem. An example for convective instability is found in homework problem 16 (Appendix A).

To see why this may be so, consider the convective example illustrated in Figure (7.13), which shows a negative local minimum of B_z. If we zoom in to a small enough scale, the variability of B_z becomes negligible, and the solution of the perturbation equation (2.29) should be similar to the solution for constant B_z (section 2.2.2). In the limit of large wavenumber, the motions are locally vertical, and the growth rate is equal to the upper bound, $\sqrt{-B_z}$. Therefore, the fastest-growing mode has growth rate $\sqrt{-\min_z(B_z)}$ regardless of the detailed shape of $B(z)$. In project B.7, you will have the opportunity to examine this result rigorously.

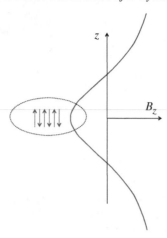

Figure 7.13 Convective instability near a negative local minimum in a generic stratification profile $B_z(z)$.

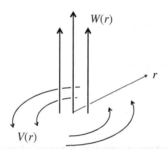

Figure 7.14 Cylindrical vortex with axial flow.

Note that:

- The result holds only in the inviscid limit. In a viscous fluid, motions on suffi-ciently small scales are damped. As a result, there is a preferred wavenumber having growth rate smaller than the inviscid upper bound.
- This class of instabilities (convective, centrifugal, and others that we'll encounter in Chapter 8) bypasses the usual turbulent energy cascade. Rather than begin-ning a sequential process in which motions excite successively smaller motions until viscous dissipation takes over, the instability transfers energy directly to the smallest-scale motions allowed by viscosity.

7.9 Swirling Flows

Vortical flows in nature are frequently accompanied by flow in the axial (typically vertical) direction. Tornadoes and hurricanes, for example, are powered largely by rising air in their centers. Conversely, deep convection in both atmosphere and ocean usually involves some degree of rotation. Figure 7.14 shows a simple model in which nothing varies in the axial or azimuthal directions, but the radially varying azimuthal background flow $V(r)$ is accompanied by an axial component $W(r)$.

To keep the math simple we will assume that the disturbance, like the background flow, is axisymmetric:

$$u' = \epsilon u'(r, z, t), \quad v = V(r) + \epsilon v'(r, z, t), \quad w = W(r) + \epsilon w'(r, z, t),$$
$$\pi = \Pi(r) + \epsilon \pi'(r, z, t). \tag{7.40}$$

Substituting into (7.1–7.5), we find that the cyclostrophic equilibrium condition (7.6) is unchanged. The equations for axisymmetric perturbations can be simplified using the normal mode solution

$$u'(r, z, t) = \hat{u}(r)e^{\iota m(z-ct)},$$

where only the real part is retained and similar forms apply for v', w', and π'.

The linearized continuity equation is

$$\frac{1}{r}\frac{d}{dr}(r\hat{u}) + \iota m \hat{w} = 0. \tag{7.41}$$

We can therefore write the radial and axial perturbations in terms of a streamfunction:

$$\hat{u} = -\iota m \, \hat{\psi}(r); \quad \hat{w} = -\frac{1}{r}\frac{d}{dr}r\hat{\psi}(r).$$

Note that we do not assume that the flow is two-dimensional. The radial and axial velocities can be described by a streamfunction because the azimuthal perturbation, while nonzero, is independent of θ, and therefore does not enter into (7.41).

After the usual manipulations, which the student is encouraged to check, we arrive at

$$\frac{d}{dr}\left[\frac{1}{r}\frac{d}{dr}\left(r\frac{d\hat{\psi}}{dr}\right)\right] + \left[\frac{\Phi}{(W-c)^2} - \frac{Z}{W-c} + m^2\right]\hat{\psi} = 0, \tag{7.42}$$

where

$$Z = -\frac{1}{r}\frac{d}{dr}r\frac{dW}{dr}.$$

Now, here is something amazing. Ready?

Take a close look at (4.18), the Taylor-Goldstein equation for stratified shear flow, and compare it term by term with (7.42). The two are practically isomorphic, the only difference being that the form of the r-derivatives is modified due to the cylindrical geometry. In place of B_z we have the Rayleigh discriminant Φ, showing again that Φ represents a gradient in the centrifugal force having the same effect as the buoyancy gradient. The place of U_{zz} is taken by Z, the radial gradient of the vorticity due to the axial parallel shear flow $W(r)$, and the axial wavenumber m takes the place of k.

Basically everything we learned about inviscid parallel shear flows in Chapters 3 and 4 can be turned on its side, bent around into a circle, and applied to swirling flows.

- If the axial flow contains an inflection point, it can produce shear instability just as $U(z)$ does, the only difference being that the resulting vortices are circular. Smoke rings are an example.
- If $\Phi > 0$, the centrifugal force tends to oppose the instability just as gravity does with Kelvin-Helmholtz instability when $B_z > 0$. In this case there is an analog to the Miles-Howard theorem: $\Phi/(dW/dr)^2$, the analog of the gradient Richardson number, must be $< 1/4$ at some r for instability to be possible (Howard and Gupta, 1962).
- The axial phase velocity c must lie within a semicircle bounded by the maximum and minimum of W (cf. Howard's semicircle theorem).

7.10 Summary

A circular vortex exhibits two relatively simple instability types corresponding to barotropic ($m = 0$) and axisymmetric ($\ell = 0$) disturbances. The instabilities are related to shear and convective instabilities, respectively. Both the mechanisms of the instabilities and the general theorems that govern them follow precisely as in the previous discussions (Chapters 2 and 3).

7.11 Further Reading

See Terwey and Montgomery (2002) for a more detailed analysis of barotropic instabilities of the concentric, piecewise-uniform vorticity distribution. The original theory of swirling flow instabilities is in Howard and Gupta (1962).

8

Instability in a Rotating Environment

8.1 Frontal Zones

Imagine a fluid whose buoyancy varies in the horizontal. You might expect that such a buoyancy distribution could not be sustained; the dense fluid would flow under the buoyant fluid, and the buoyant over the dense, until the buoyancy gradient became purely vertical.[1] On a rotating planet, though, a horizontal buoyancy gradient can be maintained by the Coriolis force. One example is the atmospheric polar front, where cold polar air flows alongside warmer mid-latitude air. Because the fluid is moving, the Coriolis force pulls it to one side (right in the northern hemisphere; left in the southern), and that force can balance the effect of the buoyancy gradient (Figure 8.1). Most major ocean currents have the same property;

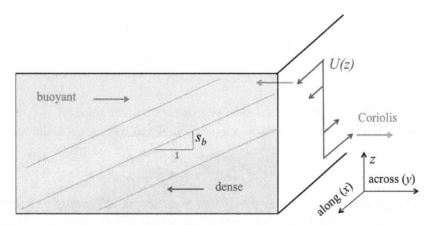

Figure 8.1 A baroclinic frontal zone in thermal wind balance. The buoyancy contrast creates a pressure gradient that is balanced by the Coriolis force. The isopycnal slope is s_b. Coordinates indicate the along-front (x) and across-front (y) directions and the vertical, z. The figure assumes that we're in the northern hemisphere and the Coriolis acceleration is therefore to the right.

[1] As in, for example, Thorpe's tilted tank experiment (Figure 4.2).

for example, the Gulf Stream carries warm equatorial water into the cold North Atlantic.

The resulting equilibrium state is generally unstable: the Coriolis effect can maintain the density distribution in the mean, but any small disturbance will upset the balance locally. Mid-latitude weather systems result from instabilities of the polar front, while the Gulf Stream continually spins off mesoscale eddies (Figure 1.1). Those instabilities are the focus of this chapter.

Figure 8.1 describes a frontal zone. The domain of interest covers only a small part of the front, so that the buoyancy variation can be approximated as linear. With the same justification, we use Cartesian coordinates. The geographical orientation of the x and y directions is arbitrary, but for definiteness we'll imagine that the view is to the west, with south at the left. With apologies to our friends "down under," we assume we're in the northern hemisphere. The force balance is easiest to envision if we imagine a zonal flow changing direction from westward near the ground to eastward aloft. The resulting Coriolis force changes from northward to southward, balancing the buoyancy gradient.

8.2 Geostrophic Equilibrium and the Thermal Wind Balance

Our goal in this chapter is to explore equilibria and perturbations in a frontal zone. For this purpose, we'll neglect viscosity and diffusion, but retain buoyancy and re-introduce the Coriolis acceleration:

$$\vec{\nabla} \cdot \vec{u} = 0 \tag{8.1}$$

$$\frac{D\vec{u}}{Dt} = -\vec{\nabla}\pi + b\hat{e}^{(z)} + \vec{u} \times f\hat{e}^{(z)} \tag{8.2}$$

$$\frac{Db}{Dt} = 0. \tag{8.3}$$

To describe a baroclinic frontal zone like that shown in Figure 8.1, we seek an equilibrium state in which buoyancy is a function of y as well as z, and the velocity is purely zonal:

$$b = B(y, z) ; \quad \vec{u} = U\hat{e}^{(x)}.$$

No assumption is made about the spatial variation of U, but (8.1) requires $\partial U/\partial x = 0$, hence $U = U(y, z)$. The buoyancy equation (8.3) is satisfied automatically because there is no x-dependence for the zonal current to advect. For the same reason, the left-hand side of the momentum equation (8.2) is zero. The three components of (8.2) are then

$$\frac{\partial \Pi}{\partial x} = 0 \tag{8.4}$$

$$\frac{\partial \Pi}{\partial y} = -fU \tag{8.5}$$

$$\frac{\partial \Pi}{\partial z} = B. \tag{8.6}$$

In order, these equations tell us that:

- The pressure, like U, can depend only on y and z.
- The meridional pressure gradient is balanced by the Coriolis force, i.e., the pressure is in geostrophic[2] balance with the current.
- The vertical pressure gradient is in hydrostatic balance with the buoyancy.

Eliminating Π between (8.5) and (8.6), we obtain the thermal wind balance:[3]

$$\boxed{f\frac{\partial U}{\partial z} = -\frac{\partial B}{\partial y}.} \tag{8.7}$$

The horizontal variation of buoyancy is referred to as baroclinicity. In the literature, you will sometimes see $\partial B/\partial y$ abbreviated as M^2 in analogy with $\partial B/\partial z = N^2$. A useful measure of the strength of the baroclinicity is the isopycnal slope:

$$s_b = -\frac{\partial B/\partial y}{\partial B/\partial z},$$

the slope of a surface on which buoyancy is uniform (Figure 8.1). Using (8.7), we can also write

$$s_b = \frac{f\partial U/\partial z}{\partial B/\partial z}.$$

8.3 The Perturbation Equations

We now linearize (8.1–8.3) by applying perturbations $\vec{u}'$, π', and b'. As always, the perturbation is incompressible:

$$\vec{\nabla} \cdot \vec{u}' = 0. \tag{8.8}$$

The linearized momentum equation is

$$\left(\frac{\partial}{\partial t} + U\frac{\partial}{\partial x}\right)\vec{u}' + \underbrace{\frac{\partial U}{\partial y}v'\hat{e}^{(x)}}_{(1)} + \frac{\partial U}{\partial z}w'\hat{e}^{(x)} = -\vec{\nabla}\pi' + b'\hat{e}^{(z)} + \underbrace{\vec{u}' \times f\hat{e}^{(z)}}_{(2)}, \tag{8.9}$$

---

[2] Literally "Earth turning."
[3] The name reflects the meteorological origins of the concept, but it is equally relevant in any rotating, stratified fluid.

while the buoyancy equation becomes

$$\left(\frac{\partial}{\partial t}+U\frac{\partial}{\partial x}\right)b'+\frac{\partial B}{\partial y}v'+\underbrace{\frac{\partial B}{\partial z}w'}_{(3)}=0. \tag{8.10}$$

The braced terms in (8.9) and (8.10) have not appeared in previous models (e.g., 4.8, 4.10) where there was no ambient rotation and the mean state varied only with z. The final term in (8.9), marked (2), is the Coriolis acceleration, and can be expanded as $fv'\hat{e}^{(x)}-fu'\hat{e}^{(y)}$. The remaining new terms, (1) and (3), represent advection of the horizontal gradients of U and B by the cross-front velocity perturbation v'.

8.4 Energetics

As in section 3.10.1, we derive the equation for the perturbation kinetic energy by dotting $\vec{u}'$ onto the perturbation velocity equation which, in this case, is (8.9). The result is

$$\left(\frac{\partial}{\partial t}+U\frac{\partial}{\partial x}\right)\frac{|\vec{u}'|^2}{2}+\frac{\partial U}{\partial y}u'v'+\frac{\partial U}{\partial z}u'w'=-\vec{u}'\cdot\vec{\nabla}\pi'+b'w'+\vec{u}'\cdot(\vec{u}'\times f\hat{e}^{(z)}). \tag{8.11}$$

The final term vanishes, because the cross product is perpendicular to $\vec{u}'$. In physical terms, this reflects the fact that the Coriolis acceleration acts at right angles to the flow (e.g., to the right in the northern hemisphere). It therefore affects the *direction* of the flow but not the *magnitude*. The kinetic energy, being a measure of the magnitude, is unaffected by the Coriolis acceleration.

Rotation therefore has this property in common with stable stratification (section 4.9) and viscosity (section 5.9): it cannot, by itself, transfer energy to the perturbation. It can, however, alter the form of the perturbation such that it gains energy via the shear or buoyancy production mechanisms.

We next use (8.1) to convert $\vec{u}'\cdot\vec{\nabla}\pi'$ to $\vec{\nabla}\cdot(\vec{u}'\pi')$ and bring the second and third terms to the right-hand side:

$$\left(\frac{\partial}{\partial t}+U\frac{\partial}{\partial x}\right)\frac{|\vec{u}'|^2}{2}=-\vec{\nabla}\cdot(\vec{u}'\pi')-\frac{\partial U}{\partial y}u'v'-\frac{\partial U}{\partial z}u'w'+b'w'. \tag{8.12}$$

Applying a horizontal average and rearranging a little, we arrive at

$$\boxed{\frac{\partial}{\partial t}\frac{\overline{|\vec{u}'|^2}}{2}=-\frac{\partial}{\partial z}\overline{w'\pi'}-\frac{\partial U}{\partial z}\overline{u'w'}+\overline{b'w'}-\frac{\partial U}{\partial y}\overline{u'v'}.} \tag{8.13}$$

Most of this should look very familiar. The first and second terms on the right were described in section 3.10.1: the first is the convergence of the vertical energy flux, which vanishes when integrated in the vertical; the second is the shear production.

The third term is the buoyancy flux that we discussed (in normal mode form) in section 4.9: it is positive (i.e., adding to the kinetic energy of the perturbation) if buoyant fluid is rising and dense fluid is falling. In the reverse case, the perturbation must do work against gravity to grow, so this term exerts a damping influence.

The fourth term is new, but it will be easily recognized as a second shear production term. Through it, the instability can exchange energy with the mean flow by advecting the horizontal shear $\partial U / \partial y$.

8.5 The Vertical Vorticity Equation

The planet's rotation can be thought of as a vorticity whose vertical component is f. Measured in an inertial reference frame, the total vorticity would be that of the flow as we measure it plus the extra contribution from the planet, a circumstance that affects the flow profoundly. In this chapter we will pay special attention to factors affecting vorticity. Since we have made the f-plane approximation ($f = $ const.), the vertical component of vorticity is the most important.

The perturbation vertical vorticity is given by

$$q' = \frac{\partial v'}{\partial x} - \frac{\partial u'}{\partial y}. \tag{8.14}$$

We derive an evolution equation for q' from the x and y components of the perturbation momentum equation (8.9), namely:

$$\left(\frac{\partial}{\partial t} + U \frac{\partial}{\partial x} \right) u' + \frac{\partial U}{\partial y} v' + \frac{\partial U}{\partial z} w' = -\frac{\partial \pi'}{\partial x} + f v' \tag{8.15}$$

and

$$\left(\frac{\partial}{\partial t} + U \frac{\partial}{\partial x} \right) v' = -\frac{\partial \pi'}{\partial y} - f u'. \tag{8.16}$$

Subtracting the y derivative of (8.15) from the x derivative of (8.16), we obtain

$$\left(\frac{\partial}{\partial t} + U \frac{\partial}{\partial x} \right) q' = \underbrace{v' \frac{\partial}{\partial y} \left(\frac{\partial U}{\partial y} \right) + w' \frac{\partial}{\partial z} \left(\frac{\partial U}{\partial y} \right)}_{advection} + \underbrace{\frac{\partial U}{\partial z} \frac{\partial w'}{\partial y}}_{tilting} + \underbrace{f_a \frac{\partial w'}{\partial z}}_{stretching}. \tag{8.17}$$

The first two terms on the right-hand side describe advection of the background vorticity $-\partial U / \partial y$ by the velocity perturbations v' and w'. The third represents vortex tilting as shown in Figure 8.2a. The vertical shear of the mean flow carries y-vorticity, which is tilted toward the vertical by differences in vertical motion. The

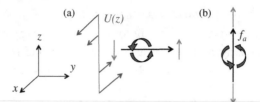

Figure 8.2 Mechanisms governing the perturbation vorticity as in (8.17). Red arrows depict the vertical velocity w'. (a) Tilting of the mean vertical shear by cross-front variations in w'. (b) Stretching of the absolute vertical vorticity by the vertical strain $\hat{w}'_z$.

final term of (8.17), illustrated in Figure 8.2b, represents vortex stretching.[4] The total vertical vorticity (that of the planet plus the mean flow, as would be measured by an observer in outer space) is written as

$$f_a = f - \frac{\partial U}{\partial y}.$$ (8.18)

In the final term of (8.17), that total vorticity is stretched or compressed by the vertical strain $\partial w'/\partial z$.

8.6 Analytical Solution #1: Inertial and Symmetric Instabilities

Here we consider two cases in which nothing varies in the along-front (x) direction. This eliminates advection of the perturbation by the along-front background flow, i.e., $U\partial/\partial x = 0$, an extreme simplification that enables two interesting analytical solutions, both of which are important in the Earth's oceans and atmosphere.

First, we will neglect buoyancy in order to focus on the Coriolis effect. The result is the inertial instability. In the second case we will restore buoyancy effects and thereby examine symmetric instability.

8.6.1 Inertial Instability

For zonal shear flow $u = U\hat{e}^{(x)}$ in a homogeneous fluid in a rotating environment, the equilibrium state is described by

$$\frac{\partial \Pi}{\partial x} = 0, \quad \frac{\partial \Pi}{\partial y} = -fU, \quad \frac{\partial \Pi}{\partial z} = 0,$$ (8.19)

---

[4] In case you are unfamiliar with vortex stretching, the strain $\partial w/\partial z$, when positive, causes fluid parcels to become taller and thinner. Because each parcel has vorticity f_a directed vertically, that vorticity is amplified to conserve angular momentum, like the classical example of a figure skater pulling their arms inward to spin faster. If $\partial w/\partial z < 0$, the opposite happens: the fluid parcel is compressed vertically and the vertical vorticity is reduced.

A meridional pressure gradient maintains geostrophic balance. The pressure does not vary in x or z, and therefore (by differentiating the middle equation with respect to x or z) neither does the velocity, i.e., $U = U(y)$.

The perturbation equations are now

$$\frac{\partial v'}{\partial y} + \frac{\partial w'}{\partial z} = 0 \tag{8.20}$$

$$\frac{\partial u'}{\partial t} = -\frac{\partial U}{\partial y} v' + f v' \tag{8.21}$$

$$\frac{\partial v'}{\partial t} = -\frac{\partial \pi'}{\partial y} - f u' \tag{8.22}$$

$$\frac{\partial w'}{\partial t} = -\frac{\partial \pi'}{\partial z}. \tag{8.23}$$

The only coefficient is $\partial U / \partial y$, which varies only in the y direction; hence, we can use a normal mode solution of the form

$$v' = \hat{v}(y)e^{\sigma t + \imath m z}.$$

Note that $k = 0$ in accordance with our assumption that the perturbations do not vary in x. (Remember also that only the real part is relevant, and that corresponding expressions are used for w', u', b', and π'.) The resulting normal mode equations are

$$\hat{v}_y + \imath m \hat{w} = 0 \tag{8.24}$$

$$\sigma \hat{u} = (f - U_y)\hat{v} \tag{8.25}$$

$$\sigma \hat{v} = -f \hat{u} - \hat{\pi}_y \tag{8.26}$$

$$\sigma \hat{w} = -\imath m \hat{\pi} \tag{8.27}$$

where the subscript y indicates the derivative. These can be reduced by combining (8.24) and (8.27) to get

$$\hat{\pi} = -\frac{\sigma}{m^2} \hat{v}_y.$$

Substituting this into (8.26) along with (8.25) yields the single stability equation

$$\sigma^2 (\hat{v}_{yy} - m^2 \hat{v}) = m^2 f (f - U_y)\hat{v}. \tag{8.28}$$

Notice that this equation is isomorphic with the convection equation (2.29) after the substitutions $\hat{w} \to \hat{v}$, $z \to y$, $\tilde{k} \to m$, and $B_z \to f(f - U_y)$. Making the same substitutions in the convective growth rate bound (2.34), we see that the growth rate is bounded from above by

$$\sigma < \sqrt{-\min_y \{\, f(f - U_y) \,\}}, \tag{8.29}$$

and instability is possible only if $f(f - U_y) < 0$ at some latitude. In other words, the relative vorticity must be *anticyclonic*, meaning in this case that $-U_y$ has sign opposite to f, and it must be greater in magnitude than f. In the northern hemisphere, for example, U must increase (become more eastward) toward the pole.

By analogy with the convective and centrifugal instabilities (section 7.8.1), inertial instability can grow wherever there is a negative local minimum in $f(f - U_y)$. The fastest-growing mode has arbitrarily small vertical wavelength and growth rate that approaches the upper bound (8.29).

Inertial Instability – Physical Mechanism

The mechanism of inertial instability is similar to the mechanisms of the convective and centrifugal instabilities described in section 7.8, but there is a slight complication. In those previous cases, the destabilizing force acted in only one direction: gravity downward, centrifugal force outward. In contrast, the Coriolis force acts in two directions, zonal and meridional, adding a new element to the process.

In the inertially unstable flow geometry considered here, the background zonal velocity at some latitude decreases with distance from the pole (Figure 8.3).

(i) Suppose a northward flow is initiated at some altitude (thick red arrow). It advects with it a negative (westward) perturbation in zonal velocity, u'.

(ii) The westward perturbation sets up a northward Coriolis force (F', thin red arrow), which accelerates the northward flow, completing a positive feedback loop.

(iii) Now here is the complication: the northward flow (red arrow) *also* experiences an eastward Coriolis acceleration (in the northern hemisphere). If that acceleration dominates, then u' is eastward and the feedback is negative. Only if the advective effect dominates, i.e., if $U_y > f$, is the net zonal acceleration westward and the flow unstable.

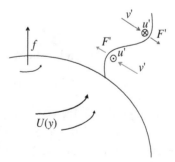

Figure 8.3 Perturbations involved in inertial instability. The meridional motion v advects the background zonal velocity to create a perturbation, F', in the Coriolis force.

The northward motion is accompanied by southward motion at some other altitude. There, the reverse process happens (blue arrows). In the southern hemisphere, the signs are reversed in the above argument, but the criterion $f(f - U_y) < 0$ remains generally valid.

Exercise: Examine the perturbation equations and identify the terms that correspond to the three-part processes described above.

8.6.2 Symmetric Instability

We now reintroduce the frontal stratification (Figure 8.4), so that U must vary in z to maintain thermal wind balance:

$$f \frac{\partial}{\partial z} U(y, z) = -\frac{\partial}{\partial y} B(y, z). \tag{8.30}$$

We retain the assumption that nothing varies in the along-front (x) direction.

For this discussion we make two additional simplifying assumptions:

- To facilitate a normal mode solution, all mean gradients (U_y, U_z, B_y, and B_z) are assumed to be uniform, with $fU_z = -B_y$ in accordance with (8.30).
- The perturbation is *quasi-hydrostatic* in the sense that b' can be approximated by $\partial \pi'/\partial z$.[5] This assumption is easily relaxed; it merely simplifies the algebra.

With these assumptions, the coefficients of the perturbation equations (8.1, 8.9, 8.10) are all constants, and we can seek a solution using normal modes of the form

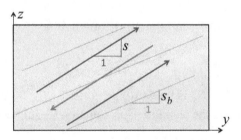

Figure 8.4 Definition sketch for symmetric instability. The thin green lines are isopycnals, with slope $s_b = -B_y/B_z$. Arrows show buoyant (red) and dense (blue) currents whose slope is $s = -\ell/m$. The flow is invariant in the along-front (x) direction.

---

[5] Admonition: This does *not* mean that $\hat{w} = 0$, as a look at the vertical momentum equation,

$$(\sigma + \imath k U)\hat{w} = -\hat{\pi}_z + \hat{b},$$

might suggest (cf. 8.23). Vertical motions, slow as they may be, play a critical role in the other equations by advecting the vertical gradients U_z and B_z. The assumption we make here is only that the left-hand side of (8.23) is a small difference between two large quantities $\hat{b}$ and $\hat{\pi}_z$, so we can approximate the latter as equal.

$$u' = \hat{u} e^{\sigma t} e^{\iota(\ell y + mz)},$$

where the complex amplitude $\hat{u}$ is a constant.

The normal mode equations are now

$$\iota \ell \hat{v} + \iota m \hat{w} = 0 \tag{8.31}$$

$$\sigma \hat{u} + U_y \hat{v} + U_z \hat{w} = f \hat{v} \tag{8.32}$$

$$\sigma \hat{v} = -\iota \ell \hat{\pi} - f \hat{u} \tag{8.33}$$

$$\iota m \hat{\pi} = \hat{b} \tag{8.34}$$

$$\sigma \hat{b} + B_y \hat{v} + B_z \hat{w} = 0. \tag{8.35}$$

To express this system as an eigenvalue equation, we solve (8.31) for $\hat{w}$ and (8.34) for $\hat{\pi}$, then substitute the results into the remaining equations. This leaves us with three homogeneous equations for the three unknowns $\hat{u}$, $\hat{v}$, and $\hat{b}$. The characteristic equation is easily found:

$$\sigma^2 = \frac{\ell^2}{m^2} B_z - 2\frac{\ell}{m} f U_z - f(f - U_y), \tag{8.36}$$

where use has been made of the thermal wind relation $B_y = -f U_z$ (8.7, 8.30).

A few abbreviations will be useful here. First, note that the wave vector components ℓ and m appear only as the ratio ℓ/m, which is minus the ratio $\hat{w}/\hat{v}$ according to (8.31). As shown in Figure 8.4, this is the slope of the planes to which motion is restricted:

$$s = -\frac{\ell}{m}.$$

The slope of the isopycnals (thin lines in Figure 8.4) is

$$s_b = -\frac{B_y}{B_z} = \frac{f U_z}{B_z}.$$

Finally, the quantity $f - U_y$ is the absolute vorticity f_a, as defined in section 8.5. With these abbreviations the characteristic equation (8.36) becomes

$$\sigma^2 = -B_z s^2 + 2 B_z s_b s - f f_a. \tag{8.37}$$

Differentiation with respect to s shows that the growth rate is a maximum when

$$s = s_b,$$

and that

$$\sigma_{max} = |f| \left(\frac{U_z^2}{B_z} - \frac{f_a}{f} \right)^{1/2}. \tag{8.38}$$

Exercise: Derive $s = s_b$ and (8.38) from (8.37).

Exercise: Repeat the whole derivation without using the quasi-hydrostatic approximation. You will find that the latter is equivalent to assuming that the slopes s and s_b are small. This makes sense if you examine (8.23) and recognize that small slope implies small vertical velocity.

Remembering that B_z/U_z^2 is the Richardson number Ri, the condition for σ_{max} to be real can be written as

$$Ri < \frac{f}{f_a}. \tag{8.39}$$

For example, if there is no across-front shear, then $f_a = f$ and the condition for instability is $Ri < 1$.

A few points to consider:

- Unlike the similar condition $Ri < 1/4$ for shear instability, (8.39) is a necessary *and sufficient* condition, i.e., instability is guaranteed if the condition is satisfied.
- Like convection in an inviscid, unbounded fluid, symmetric instability has no preferred length scale (cf. section 2.2.2). The growth rate depends only on the orientation of the wave vector, and symmetric instability is therefore a broadband instability.
- The across-front shear U_y can have either a stabilizing or a destabilizing influence depending on its sign. If U_y has the same sign as f (i.e., $U_y > f > 0$ in the northern hemisphere), the across-front shear reduces the absolute vorticity f_a. From (8.38) we see that this increases the growth rate, while (8.39) shows that the criterion for Ri is relaxed. The reverse is true if U_y has sign opposite to f.
- If we reduce the buoyancy gradients to zero, we recover inertial instability as described in the previous section.

Symmetric Instability – Physical Mechanism

Recall that inertial instability grows when meridional flow advects the velocity gradient U_y, setting up a Coriolis acceleration that reinforces the original meridional flow. The only difference in the symmetric case is that the velocity gradient has a second component, U_z, that can be advected if the perturbation velocity has a vertical component. In fact, the instability can grow even if $U_y = 0$, solely by advecting the vertical shear.

Note that buoyancy has no direct effect on the fastest-growing symmetric instability. When $s > s_b$, as in Figure 8.4, the instability does work against gravity, whereas the opposite is true if $s < s_b$. For the fastest-growing symmetric instability, motion is along isopycnals ($s = s_b$) and the mode therefore exchanges no energy with the gravitational field. The energy source can only be the kinetic energy of the along-front current.

8.7 Analytical Solution #2: Baroclinic Instability

For perturbations that vary in the along-front direction (as distinct from inertial and symmetric instability), analytical solutions are available provided that we make the set of simplifying assumptions that define a quasigeostrophic flow. The essential assumption is that the flow varies slowly relative to the Earth's rotation, i.e., on a time scale much greater than a day. The conditions for quasigeostrophy to hold are described in much more detail elsewhere, e.g., Pedlosky (1987). Here we will give only enough detail to make the approximation plausible. We will find, however, that the characteristics of the predicted instability correspond well with those of (1) mid-latitude storms and (2) oceanic mesoscale eddies.

8.7.1 *The Quasigeostrophic Potential Vorticity Perturbation*

The vertical component of the perturbation vorticity, introduced in section 8.5, is

$$q' = v'_x - u'_y \tag{8.40}$$

(writing partial derivatives as subscripts). Its evolution equation (8.17) is written as

$$q'_t = \underbrace{-Uq'_x}_{advection} + \underbrace{U_z w'_y}_{tilting} + \underbrace{f_a w'_z}_{stretching} . \tag{8.41}$$

We now make two critical assumptions. First, we assume that the vortex tilting effect is negligible in comparison with the stretching effect, i.e., $|U_z w'_y| \ll |f_a w'_z|$ (Figure 8.2). Second, we neglect U_y relative to f so that f_a is replaced by f.

This leaves us with the approximate vertical vorticity equation

$$\boxed{q'_t + Uq'_x = fw'_z.} \tag{8.42}$$

This is a single equation for two unknowns (q' and w'), so we need more information (i.e., assumptions) to make a complete theory. The strategy is to approximate both q' and $\hat{w}'_z$ in terms of the pressure perturbation π'.

The Left-Hand Side of (8.42)

We begin with an assumption concerning the perturbation momentum equations (8.15) and (8.16). In each of those equations, it is often true that the individual terms on the right-hand side, i.e., the pressure gradient and the Coriolis acceleration, are large in magnitude compared with the terms on the left-hand side. In other words, the left-hand side is not zero, but it is a small difference of large numbers. It follows that the pressure gradient and Coriolis terms are nearly equal. If those terms are, in fact, equal, the perturbation is in geostrophic balance, and the horizontal velocity components can be represented entirely in terms of the pressure perturbation:

$$u^{(g)} = -\frac{\hat{\pi}'_y}{f} \; ; \quad v^{(g)} = \frac{\hat{\pi}'_x}{f}. \tag{8.43}$$

In fact, the horizontal velocity perturbation can be thought of as the sum of a *geostrophic part* and an *ageostrophic part*:

$$u' = u^{(g)} + u^{(a)} \; ; \quad v' = v^{(g)} + v^{(a)}.$$

We can do the same with the perturbation vorticity:

$$q' = q^{(g)} + q^{(a)}$$

where the geostrophic part

$$q^{(g)} = v^{(g)}_x - u^{(g)}_y = \frac{1}{f}\nabla^2_H \pi'.$$

Our assumption is that the ageostrophic part is negligible, i.e., $|q^{(a)}| \ll |q^{(g)}|$, so that

$$q' = \frac{1}{f}\nabla^2_H \pi'. \tag{8.44}$$

The Right-Hand Side of (8.42)

Suppose we try to approximate w'_z in the same way, by assuming that it is dominated by its geostrophic part. From continuity we have

$$w'_z = -(u'_x + v'_y) = -(u^{(g)}_x + u^{(a)}_x + v^{(g)}_y + v^{(a)}_y)$$

$$= -\left(-\frac{\pi'_{yx}}{f} + u^{(a)}_x + \frac{\pi'_{xy}}{f} + v^{(a)}_y\right)$$

$$= -(u^{(a)}_x + v^{(a)}_y).$$

So, we cannot assume that the ageostrophic part of w'_z is negligible because it's the only part (the geostrophic part being zero). Instead, we invoke the buoyancy equation (8.10), which we write in the form

$$B_z w' = -B_y v' - b'_t - U b'_x. \tag{8.45}$$

We assume once again that v' is dominated by its geostrophic part $v^{(g)}$, and moreover that the perturbation is in hydrostatic balance $b' = \pi'_z$. Making the appropriate substitutions, (8.45) becomes

$$B_z w' = -\frac{B_y}{f}\pi'_x - \pi'_{zt} - U\pi'_{zx} \tag{8.46}$$

and, after a differentiation,

$$B_z w'_z = -\frac{B_y}{f}\pi'_{xz} - \pi'_{zzt} - U\pi'_{zzx} - U_z\pi'_{zx}.$$

Remembering the thermal wind balance $fU_z = -B_y$, we see that the first and last terms on the right-hand side cancel, and therefore

$$w'_z = -\frac{1}{B_z}(\pi'_{zzt} - U\pi'_{zzx}).$$
(8.47)

With (8.47) and (8.44), we have accomplished our goal of approximating both q' and $\hat{w}'_z$ in terms of π'.

The Potential Vorticity Equation

Inserting (8.47) and (8.44) into (8.42), we have

$$\xi'_t + U\xi'_x = 0,$$
(8.48)

where ξ' is the linearized, quasigeostrophic potential vorticity

$$\boxed{\xi' = \nabla_H^2 \pi' + \frac{f^2}{B_z}\pi'_{zz}.}$$
(8.49)

Equation (8.48) states that, to an observer moving with the mean along-front current U, the potential vorticity of the perturbation remains constant. If the perturbation is in fact growing exponentially, then its potential vorticity can have only one value and that value is zero.

We next reconfigure (8.48, 8.49) so that it can be solved in a finite vertical domain with impermeable upper and lower boundaries. We assume that π' has the normal mode form

$$\pi' = \hat{\pi}(z)e^{\iota k(x-ct)+\iota\ell y},$$

and similarly for ξ'. Equation (8.48) becomes

$$\iota k(U-c)\hat{\xi} = 0$$

or, as anticipated, $\hat{\xi} = 0$ for $c \neq 0$. From (8.49) we now obtain an ordinary differential equation

$$\boxed{\hat{\pi}_{zz} - \mu^2\hat{\pi} = 0,}$$
(8.50)

where

$$\mu = \frac{\tilde{k}}{P}$$
(8.51)

is a scaled vertical wavenumber. The Prandtl ratio P is defined as

$$P = \frac{|f|}{\sqrt{B_z}},$$
(8.52)

and $\tilde{k}^2 = k^2 + \ell^2$ as usual.

The Impermeable Boundary Condition

We must now express the boundary condition $\hat{w} = 0$ in terms of $\hat{\pi}$. To this end we write (8.46) in the normal mode form

$$B_z\hat{w} = -\frac{B_y}{f}\iota k\hat{\pi} - \iota k(U - c)\hat{\pi}_z,$$

or, invoking thermal wind balance $B_y = -fU_z$,

$$B_z\hat{w} = \iota k[U_z\,\hat{\pi} - (U - c)\hat{\pi}_z].$$

The boundary condition $\hat{w} = 0$ is therefore equivalent to

$$\boxed{(U - c)\hat{\pi}_z - U_z\hat{\pi} = 0.} \tag{8.53}$$

8.7.2 Eady Waves

We can solve (8.50) with (8.53) imposed at upper and lower boundaries, or with one boundary at infinity. In the latter case, the solution describes Eady waves, which are analogous to the vorticity waves discussed in section 3.12.2. And, like the vorticity waves, a pair of Eady waves can resonate to drive exponential growth.

Before deriving the Eady wave dispersion relation, we give a qualitative description of the mechanism. In Figure 8.5a, we are looking across a frontal zone from the warm (buoyant) side, shaded in red. Now consider an imaginary horizontal surface, and suppose that the fluid is displaced upward and downward by some means so that the surface varies sinusoidally in the along-front direction. At the top of the domain is an impenetrable horizontal boundary where the amplitude of the displacement must decrease to zero as shown by the thin curves.

Because the vertical motion drops to zero at the boundary, the fluid above each trough of the perturbation experiences an extensional strain, $w'_z > 0$, which stretches, and thereby amplifies, the ambient vorticity f. The result is a counterclockwise increment of vorticity as shown at the top of Figure 8.5b by the curved arrows. In contrast, the strain above each crest is compressive, effectively reducing the ambient vorticity, so that the increment is clockwise. Between these strips of oppositely signed vorticity increments are cross-front currents that carry alternately dense (blue) and buoyant (red) fluid.

In each cross-front current, gravity acts to accelerate the fluid vertically: upward in buoyant currents, downward in dense currents (Figure 8.5c). As a result, the nodes of the sinusoidal disturbance are displaced vertically such that the whole pattern propagates to the left.

This intrinsic leftward propagation is relative to the mean along-front current, which must be rightward to maintain thermal wind balance. At the *lower* boundary,

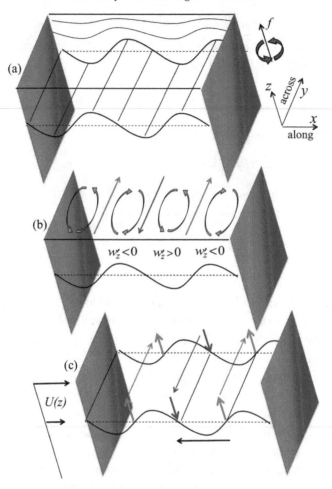

Figure 8.5 Mechanics of Eady wave propagation. (a) Sinusoidal perturbation of a frontal zone in thermal wind balance. Shading on the cross-sections indicates the mean buoyancy distribution: dense (blue) and buoyant (red). The figure assumes $f > 0$, i.e., that we are in the northern hemisphere. (b) Circulations induced by stretching and compression of the ambient vorticity, and the resulting cross-front circulations. (c) Advected buoyancy drives rising and sinking motions, which cause the pattern to propagate to the left, counter to the mean along-front flow U.

the same dynamic supports a right-going wave in a leftward background current. It is therefore possible for both waves to be stationary and, if their relative phases are such that the vertical motion of one wave reinforces the crests and troughs of the other, positive feedback leads to a growing instability, just as we saw with vorticity waves in Chapter 3 (Figure 3.23).

To derive a dispersion relation for the Eady wave shown in Figure 8.5, we solve (8.50), with (8.53) imposed at the upper boundary and the condition that the solution remain bounded as $z \to -\infty$. The general solution of (8.50) is

$$\hat{\pi} = \alpha e^{\mu z} + \beta e^{-\mu z}, \tag{8.54}$$

where α and β are constants. Boundedness as $z \to -\infty$ requires that $\beta = 0$. In this case $\hat{\pi}_z = \mu \hat{\pi}$, and the boundary condition becomes

$$(U_u - c)\mu - U_z = 0,$$

where U_u is the mean along-front velocity at the boundary. Solving for c, we have

$$c = U_u - \frac{U_z}{\mu}. \tag{8.55}$$

For a wave at a lower boundary, we carry out the same steps, this time requiring boundedness as $z \to +\infty$. In this case $\alpha = 0$, $\hat{w}_z = -\mu \hat{w}$, and the boundary condition gives the dispersion relation

$$c = U_l + \frac{U_z}{\mu}, \tag{8.56}$$

where U_l is the mean along-front velocity at the lower boundary.

The upper and lower waves propagate oppositely relative to the mean flow at their respective boundaries. It is therefore possible that their phase speeds could be equal. In that case, the two waves might resonate and drive exponential growth. To see if this possibility is in fact true, we must solve (8.50) with (8.53) imposed at both the upper and lower boundaries.

8.7.3 The Eady Mode of Baroclinic Instability

We consider a finite domain, with coordinates chosen such that boundaries are at $z = \pm H/2$ and the mean along-front velocity is $U = U_z z$. The general solution of (8.54) must satisfy the boundary conditions

$$\left(-\frac{U_z H}{2} - c \right) \hat{\pi}_z + U_z \hat{\pi} = 0 \quad \text{at } z = -\frac{H}{2}$$

and

$$\left(\frac{U_z H}{2} - c \right) \hat{\pi}_z + U_z \hat{\pi} = 0 \quad \text{at } z = \frac{H}{2}.$$

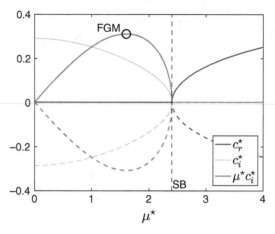

Figure 8.6 Phase speed (real = blue, imaginary = green) and scaled growth rate $\mu^\star c_i^\star$ (red) versus vertical length scale for the Eady model of baroclinic instability, based on (8.57). Scalings are defined by $\mu^\star = \mu H$ and $c^\star = c/U_z H$. Annotations mark the fastest-growing mode (FGM) and the stability boundary (SB).

Substituting (8.54) and solving for c, we obtain

$$c^{\star 2} = \frac{1}{4} - \frac{\coth \mu^\star}{\mu^\star} + \frac{1}{\mu^{\star 2}}, \tag{8.57}$$

where the nondimensional phase speed and vertical scale are

$$c^\star = \frac{c}{U_z H}; \quad \mu^\star = \mu H. \tag{8.58}$$

The solution (8.57) is shown in Figure 8.6. For $\mu^\star > 2.40$, c is real and represents oppositely propagating modes. In the limit $\mu^\star \to \infty$ (in which the domain height is large compared to the vertical scale μ^{-1}), these correspond to the isolated Eady waves described in section 8.7.2.[6]

As $\mu^\star$ decreases from infinity, the two Eady waves become close in phase speed. At $\mu^\star = 2.40$, both phase speeds reach zero, i.e., the waves become phase locked.

---

[6] For $\mu^\star \gg 1$, $\coth \mu^\star$ approaches 1. Therefore,

$$c^{\star 2} \approx \frac{1}{4} - \frac{1}{\mu^\star} + \frac{1}{\mu^{\star 2}} = \left(\frac{1}{2} - \frac{1}{\mu^{\star 2}} \right)^2.$$

In dimensional terms,

$$c \approx \pm \left(\frac{U_z H}{2} - \frac{U_z}{\mu} \right),$$

which is equivalent to the dispersion relations (8.55, 8.56) for the upper and lower Eady waves derived in section 8.7.2.

For $\mu^\star < 2.40$, c is imaginary and we therefore have unstable modes with real growth rate σ. The growth rate is proportional to the product $\mu^\star c_i^\star$:

$$\sigma = k c_i = \frac{k}{\tilde{k}} \, \tilde{k} \, c_i^\star U_z H = \frac{k}{\tilde{k}} \frac{\mu}{P} \, c_i^\star U_z H = \frac{k}{\tilde{k}} \frac{U_z}{P} \, \mu^\star c_i^\star, \tag{8.59}$$

where (8.51) and (8.58) have been used.

Instead of oppositely propagating waves, the solutions represent one growing and one decaying mode (Figure 8.6). The product $\mu^\star c_i^\star$ reaches a maximum of 0.31 at $\mu^\star = 1.61$ (circle on Figure 8.6), then drops to zero at $\mu^\star = 0$. This band of unstable modes $0 < \mu^\star \leq 2.40$ is the Eady mode of baroclinic instability, which we will call the Eady mode for short.

We will be most interested in the fastest-growing Eady mode. The optimal nondimensional length scale $\mu^\star = 1.61$ corresponds to

$$\tilde{k} = \frac{1.61}{H} P. \tag{8.60}$$

The wavelength is

$$\lambda = \frac{2\pi}{\tilde{k}} = \frac{3.9}{P} H.$$

This is sometimes written as

$$\lambda = 3.9 L_d,$$

where

$$L_d = \frac{H}{P}$$

is called the deformation radius.

For the optimal value of $\mu^\star$, the dimensional growth rate is given by (8.59):

$$\sigma = 0.31 \frac{k}{\tilde{k}} P U_z.$$

Like shear instability, the Eady mode grows fastest when $k = \tilde{k}$, or $\ell = 0$, i.e., when the wave vector is aligned with the mean along-front flow. The maximum growth rate is then proportional to the thermal wind shear and also to the Prandtl ratio:

$$\sigma = 0.31 P U_z. \tag{8.61}$$

There are two interesting alternative ways to express this growth rate. First:

$$\sigma = 0.31 \sqrt{B_z} \, s_b,$$

i.e., the growth rate depends on the strength of the stratification and the degree to which the isopycnals are tilted. Second:

$$\sigma = 0.31 \frac{|f|}{\sqrt{Ri}}.$$ (8.62)

8.7.4 Terrestrial Examples

The Prandtl ratio

We now look at some typical dimensional parameter values for baroclinic instability, beginning with the Prandtl ratio $f/\sqrt{B_z}$. A typical mid-latitude value for f is $10^{-4}\mathrm{s}^{-1}$ (section 1.5). In the troposphere, a typical value for B_z is $10^{-4}\mathrm{s}^{-2}$. Coincidentally, this value is also typical of the upper ocean. Therefore, the Prandtl ratio is near 0.01 in both fluids.

The Prandtl ratio exerts a control on the aspect ratio (i.e., the height to length ratio) of the eddy structures that result from baroclinic instability. This is in line with the wave length scaling in (8.60) above, which states that $H/\lambda \propto f/\sqrt{B_z}$. An illustration of this feature can be seen in Figure 8.7, where mesoscale eddies, thought to arise from baroclinic instability, have been observed from a mooring that measures the horizontal velocity of the upper 2000 m of the Arctic Ocean from a position fixed to the sea bed. The sudden bursts of enhanced horizontal velocity indicate the passage of eddies at various depths over a number of years. Due to the strong stratification in the upper 200 m, the Prandtl ratio confines the shallower eddies in this region to small vertical scales. This is not the case in the deeper (> 200 m) ocean, where the much weaker stratification allows the eddies to extend thousands of meters in the vertical. Notice that some eddies also have cores

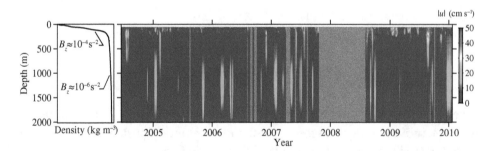

Figure 8.7 Observations of horizontal velocity magnitude from a mooring in the Arctic Ocean together with a representative density profile (left). The different vertical signatures of the eddies illustrate the dependence of the vertical scale height to the Prandtl ratio, $f/\sqrt{B_z}$, which changes by an order of magnitude. The horizontal scale of the eddies varies between 10 and 40 km [adapted from Carpenter and Timmermans (2012), see also Zhao and Timmermans (2015)].

that are centered at the transition between the two different stratification regimes so that they have very asymmetrical velocity signatures – being able to extend to great depths once they are able to "puncture" through to the weakly stratified layer.

The troposphere: If we take H to be the atmospheric scale height, which is near 10 km, then the deformation radius is 1000 km and the wavelength of the fastest-growing Eady mode is about 4000 km. Now suppose that the wind speed changes by 10 m/s over the scale height H, so that $U_z = 10^{-3}\text{s}^{-1}$. The growth rate is then $3 \times 10^{-6}\text{s}^{-1}$, for an e-folding time of 4 days. These estimates are "in the ballpark" for mid-latitude weather systems, which develop over a few days and have longitudinal extents of a few thousand kilometers.

The ocean thermocline: In the ocean, the depth over which baroclinic instability acts is more like 1 km, so the predicted wavelength is 1/10 of the atmospheric value, say 400 km. This is a typical length scale for mesoscale eddies. If the current changes by 0.1 m/s over 1 km depth, then U_z is 10^{-4}s^{-1}, and the predicted e-folding time is about a month, not too different from the time scale for mesoscale eddy growth.

The ocean mixed layer: The upper few tens of meters of the ocean is well mixed by wind and waves. But while buoyancy and other properties vary little in the vertical, they can vary substantially in the horizontal. The result is a class of baroclinic and symmetric instabilities that currently goes by the general name *mixed layer instability* (Boccaletti et al., 2007). Because stratification is weak, the Prandtl ratio is relatively large, $P = 0.1$ perhaps. The depth of the layer is of order 100 m, so the Eady model predicts a wavelength of 4 km, in the *submesoscale* range. If we assume a velocity scale of 0.1 m/s, we get an e-folding time of 9 hours.

8.8 Numerical Solution Method

Numerical solution of (8.8–8.10) is more complicated than in the non-rotating cases we have studied previously for two main reasons:

(1) elimination of the pressure is harder because the pressure equation acquires a Coriolis term, and
(2) the variation of the background state in both y and z complicates the application of normal modes.

In the analytic solutions above we sidestepped those problems by making various assumptions (no variation in x for inertial and symmetric instabilities, the quasi-geostrophic approximation for the Eady mode). Numerical methods allow us to relax those assumptions, at least in part. In what follows we will describe the two problems in turn, and then describe the solution.

8.8.1 The Pressure Equation

Taking the divergence of (8.9) gives the pressure equation

$$\nabla^2 \pi' = -2U_y \frac{\partial v'}{\partial x} - 2U_z \frac{\partial w'}{\partial x} + \frac{\partial b'}{\partial z} + f \left(\frac{\partial v'}{\partial x} - \frac{\partial u'}{\partial y} \right). \tag{8.63}$$

With $f \neq 0$, u' and v' appear on the right-hand side, hence our usual tactic of using the pressure equation to eliminate those variables will have to be reconsidered.

8.8.2 Limitations of the Normal Mode Approach

To solve the perturbation equations (8.1, 8.9, 8.10) we assume solutions of the form

$$u' = \hat{u}(z) e^{\sigma t} e^{\iota(kx+\ell y)}. \tag{8.64}$$

This requires that the coefficients depend only on z.

Now consider the second terms in (8.9) and (8.10). These describe advection of the perturbation by the background flow, and have the form U times $\partial/\partial x$ of something. If the perturbation is independent of x, as in the inertial and symmetric instabilities (section 8.6), those terms vanish regardless of the form of U. For a general perturbation, however, the factor U must depend only on z, i.e.,

$$U_y = 0. \tag{8.65}$$

In addition, thermal wind balance requires that $B_{yy} = fU_{yz} = 0$, hence

$$B = B_y y + \vartheta(z), \tag{8.66}$$

where B_y is a constant and ϑ is an arbitrary function.

8.8.3 Reduction to an Eigenvalue Problem

With the constraints (8.66) and (8.65) satisfied, we substitute (8.64) into (8.8–8.10) to obtain the normal mode perturbation equations

$$\iota k \hat{u} + \iota \ell \hat{v} + \hat{w}_z = 0 \tag{8.67}$$

$$(\sigma + \iota k U)\hat{u} + U_z \hat{w} = -\iota k \hat{\pi} + f \hat{v} \tag{8.68}$$

$$(\sigma + \iota k U)\hat{v} = -\iota \ell \hat{\pi} - f \hat{u} \tag{8.69}$$

$$(\sigma + \iota k U)\hat{w} = -\hat{\pi}_z + \hat{b} \tag{8.70}$$

$$(\sigma + \iota k U)\hat{b} + B_y \hat{v} + \vartheta_z \hat{w} = 0. \tag{8.71}$$

Instead of using $\hat{u}$ and $\hat{v}$, we define new variables

$$\hat{q} = \iota k \hat{v} - \iota \ell \hat{u} ; \quad \hat{\chi} = \iota k \hat{u} + \iota \ell \hat{v}. \tag{8.72}$$

The first of these is just the normal mode form of the perturbation vertical vorticity defined in section 8.5. Its evolution equation, arrived at by cross-differentiating (8.68) and (8.69), is

$$(\sigma + \iota k U)\hat{q} - \iota \ell U_z \hat{w} = f \hat{w}_z. \tag{8.73}$$

This is the normal mode form of (8.17). [The limitation $U_y = 0$ has been imposed in accordance with (8.65), so that y-derivatives of U vanish and $f_a = f$.]

Our next goal is to eliminate $\hat{\pi}$ from the vertical momentum equation (8.70). The troublesome pressure equation (8.63) can be written as

$$\nabla^2 \hat{\pi} = -2\iota k U_z \hat{w} + \hat{b}_z + f \hat{q} \tag{8.74}$$

where, as usual,

$$\nabla^2 = \frac{d^2}{dz^2} - \tilde{k}^2, \quad \text{and } \tilde{k} = \sqrt{k^2 + \ell^2}.$$

In section 3.1.2, we discussed in detail the procedure of eliminating pressure by (1) deriving a Poisson equation like the one above, (2) taking the Laplacian of the vertical velocity equation, and combining the two. Thanks to our use of $\hat{q}$, that procedure will work here. Starting with (8.70) and applying ∇^2, we obtain

$$(\sigma + \iota k U)\nabla^2 \hat{w} + 2\iota k U_z \hat{w}_z = -\nabla^2 \hat{\pi}_z + \nabla^2 \hat{b},$$

remembering that $U_{zz} = 0$. Differentiating (8.74) and substituting leads, after some gratifying cancellations, to

$$(\sigma + \iota k U)\nabla^2 \hat{w} = -\tilde{k}^2 \hat{b} - f \hat{q}_z. \tag{8.75}$$

In (8.73) and (8.75) we have two equations for the three unknown functions $\hat{q}$, $\hat{w}$, and $\hat{b}$. To close the system we will add the buoyancy equation (8.71), but note that (8.71) involves dependence on $\hat{v}$. To remove this dependence, we solve the pair of equations (8.72) for $\hat{v}$:

$$\hat{v} = -\frac{\iota k}{\tilde{k}^2}\hat{q} - \frac{\iota \ell}{\tilde{k}^2}\hat{\chi},$$

and remember that $\hat{\chi} = -\hat{w}_z$ by (8.67). With this substitution, (8.71) becomes

$$(\sigma + \iota k U)\tilde{k}^2 \hat{b} = \iota k B_y \hat{q} - \iota \ell B_y \hat{w}_z - \tilde{k}^2 \vartheta_z \hat{w}. \tag{8.76}$$

Exercise: Compare (8.75, 8.76) with (4.14, 4.15) and also with (6.14, 6.15). Note all differences, and make sure you can explain each in terms of the different assumptions that have been made.

We now have three equations, (8.73), (8.75), and (8.76) for the three unknowns $\hat{q}$, $\hat{w}$, and $\hat{b}$, which we can write as a matrix differential equation:

$$\sigma \begin{pmatrix} 1 & 0 & 0 \\ 0 & \nabla^2 & 0 \\ 0 & 0 & 1 \end{pmatrix} \begin{pmatrix} \hat{q} \\ \hat{w} \\ \hat{b} \end{pmatrix} = \begin{pmatrix} -\imath k U & \imath \ell U_z + f D^{(1w)} & 0 \\ -f D^{(1q)} & -\imath k U \nabla^2 & -\tilde{k}^2 \\ \dfrac{\imath k}{\tilde{k}^2} B_y & -\dfrac{\imath \ell}{\tilde{k}^2} B_y D^{(1w)} - \vartheta_z & -\imath k U \end{pmatrix} \begin{pmatrix} \hat{q} \\ \hat{w} \\ \hat{b} \end{pmatrix},$$

$$(8.77)$$

with

$$\nabla^2 = D^{(2)} - \tilde{k}^2$$

and

$$D^{(1)} = \frac{d}{dz} \; ; \quad D^{(2)} = \frac{d^2}{dz^2}.$$

The first-derivative operators in (8.77) are marked with the superscripts w and q to indicate that their matrix equivalents may be different depending on the choice of boundary conditions (see below). We now replace the derivative operators $D^{(1)}$ and $D^{(2)}$ with derivative matrices as defined previously (sections 1.4.2, 3.5, 5.5, and 6.2), remembering to incorporate the appropriate boundary conditions. The result is a generalized eigenvalue problem with $3N \times 3N$ matrices:

$$\sigma \, \mathsf{A}\vec{x} = \mathsf{B}\vec{x}.$$

The eigenvector $\vec{x}$ is a concatenation of the discretized forms of $\hat{q}$, $\hat{w}$, and $\hat{b}$.

8.8.4 Boundary Conditions

For $\hat{w}$, we use the impermeable boundary $\hat{w} = 0$. Application of this boundary condition should be familiar by now. For buoyancy, (8.77) does not require a boundary condition since $\hat{b}$ is not differentiated.

The new variable is the vertical vorticity q. Differentiating the definition (8.14), we have

$$\frac{\partial q'}{\partial z} = \frac{\partial}{\partial z}\left(\frac{\partial v'}{\partial x} - \frac{\partial u'}{\partial y} \right) = \frac{\partial}{\partial x}\frac{\partial v'}{\partial z} - \frac{\partial}{\partial y}\frac{\partial u'}{\partial z}. \tag{8.78}$$

If we now assume that the boundary is *frictionless*, as described in section 5.4.2, then this reduces to $\partial q'/\partial z = 0$ or, in normal mode form,

$$\hat{q}_z = 0.$$

We now need a first-derivative matrix for $\hat{q}$ that incorporates this boundary condition. The necessary modifications are derived in section 5.5; the top and bottom rows are $[-2/3 \;\; 2/3 \;\; \ldots]/\Delta$ and $[\ldots \;\; 2/3 \;\; -2/3]/\Delta$.

8.8.5 Shear Scaling

Suppose that (8.77) has a solution algorithm:

$$[\sigma, \hat{q}, \hat{w}, \hat{b}] = \mathcal{F}(z, U_z, \vartheta_z, f; k, \ell).$$

Now let us choose the time scale to be $1/U_z$ and the length scale to be the domain height H. The nondimensionalization is straightforward. Scalings of particular interest are

$$f^\star = \frac{f}{U_z},$$

and

$$\vartheta^\star_{z^\star} = \frac{\vartheta_z}{U_z^2} = Ri(z),$$

the gradient Richardson number. Note that Ri can vary with height.

The scaled equations are isomorphic to (8.77), and therefore

$$[\sigma^\star, \hat{q}^\star, \hat{w}^\star, \hat{b}^\star] = \mathcal{F}(z^\star, 1, Ri(z^\star), f^\star; k^\star, \ell^\star).$$

Time scales other than $1/U_z$ are possible, e.g., $1/|f|$. If the stratification ϑ_z has an identifiable "characteristic" value B_z, then $1/\sqrt{B_z}$ is also a viable time scale.

8.9 Instability in the Ageostrophic Regime

Numerical solutions of (8.77) as described above can approximate the analytically derived inertial, symmetric, and baroclinic instabilities (sections 8.6 and 8.7), as well as allowing us to relax some simplifying assumptions for greater realism. In the example shown in Figure 8.8, variables are scaled by the shear and the domain height (see section 8.8.5). We set the Prandtl ratio to the typical terrestrial value $P = 0.01$ and the Richardson number to $Ri = 0.94$. At the largest values of $k^\star$ and $\ell^\star$ (upper right), instabilities are oscillatory, but for smaller wavenumbers (lower left) stationary modes are found.

The symmetric mode is visible at the left, where $k^\star = 0$. Modes with small $\ell^\star$ (and $k^\star = 0$, lower left) are damped in the presence of boundaries, but at large $\ell^\star$ the growth rate is consistent with the analytical solution $P\sqrt{1 - Ri}$. The fastest-growing mode along the bottom edge ($\ell^\star = 0$), indicated by "AB," is a stationary baroclinic instability similar to the Eady mode (section 8.7). For this example Ri has been set close to the value at which the symmetric and baroclinic modes have equal growth rate. For comparison, the symbol "QB" shows the wavenumber $1.61 P$, the predicted value for the baroclinic mode in the quasigeostrophic regime $Ri \gg 1$.

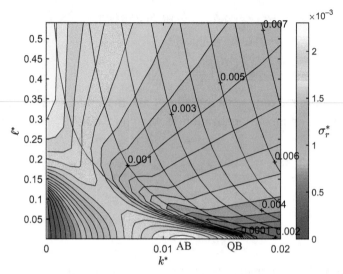

Figure 8.8 Growth rates of instabilities far from the quasigeostrophic regime. All background gradients are uniform. Shear scaling is used, with parameter values $P = 0.01$, $Ri = 0.94$. Filled (plain) contours show the growth rate (frequency). Symbols "AB": ageostrophic baroclinic instability; "QB": quasigeostrophic (Eady) baroclinic instability. Numerical solution employs $N = 100$ grid points.

Exercise: The student is invited to develop the code and explore the instabilities further. Useful papers for reference are Stone (1966) and Stamper and Taylor (2016).

8.10 Summary

- A stratified fluid in a rotating environment is in equilibrium if the pressure is hydrostatic and the thermal wind balance holds: $fU_z = -B_y$.
- The perturbation equations can be solved numerically after the introduction of the vertical vorticity perturbation.
- Inertial instability
  - Arises from a sheared current in a rotating environment in the absence of stratification.
  - Instability requires that the shear produces a relative vorticity that is greater than, and in opposition to the planetary vorticity, i.e., $|U_y| > |f|$ and $-U_y f > 0$.
  - There is no preferred wavelength, and the motions are purely horizontal.
  - The instability is caused by a continual alignment of the perturbation velocity so that it is able to extract the shear of the equilibrium current.
- Symmetric instability
  - Instability requires $Ri < f_a/f$.

- Perturbations do not vary in the along-front (x) direction.
- Motion is along isopycnals in the ($y - z$) plane.
- There is no preferred wavelength.
- The physical mechanism of the instability is the same as inertial instability, except it has an additional influence of the vertical shear in the thermal wind equilibrium.
- Baroclinic instability (the Eady model)
  - Flow is assumed to be quasigeostrophic.
  - Perturbations do not vary in the cross-front (y) direction.
  - Motion is sinusoidal, with phase velocity stationary with respect to the central plane.
  - The wavelength $\lambda = 3.9H/P$, where $P = |f|/\sqrt{B_z}$.
  - The growth rate $\sigma = 0.31PU_z = 0.31|f|/\sqrt{Ri}$.
  - The instability can be understood as a resonance of Eady wave trains focused near the upper and lower boundaries.

8.11 Further Reading

Theoretical details are developed further in Pedlosky (1987), Haine and Marshall (1998), and Thomas et al. (2013). For more details of wave resonance in baroclinic instability see Bretherton (1966) and Heifetz et al. (2004).

9

Convective Instability in Complex Fluids

So far we have assumed that the equation of state is trivial: buoyancy is proportional to a single fluid property (e.g., temperature) that varies only due to advection and diffusion (section 1.5). Real geophysical fluids may be much more complicated than this. For example, buoyancy may be controlled by two or more scalar constituents with different chemical properties, or by phase changes (e.g., freezing). The equation of state may include nonlinearities; for example, a mixture of two constituents may be lighter or denser than either constituent individually. The scalars that control buoyancy may have complex properties such as being alive. All of these properties affect the stability of the fluid. In this chapter we consider a few simple examples in which buoyancy changes lead to convection.

9.1 Conditional Instability in a Moist Atmosphere or a Freezing Ocean

Instability may result when the fluid undergoes a phase change. For example, atmospheric motions may be greatly affected by humidity. Humidity is in some ways an analog of salinity in the ocean but, unlike salt, water in the atmosphere can exist in liquid, solid, and gas phases. Changes between these phases affect the temperature of the air, and therefore its buoyancy.

If the pressure in an air parcel drops, so does its capacity to hold water vapor in the gas state. If the humidity is close to its saturation value, pressure reduction causes water vapor to condense to the liquid state, releasing latent heat into the air (and creating rain). A rising air parcel, if moist enough, will be warmed by condensation and therefore rise even faster. This is called *conditional* instability, as it depends not only on the vertical temperature gradient but also on the humidity. In a falling air parcel, rising pressure increases the air's capacity to hold water vapor, allowing liquid water to evaporate, but any liquid water present is likely to precipitate as rain before this can happen.

222

Figure 9.1 Two convective phenomena visible from Kume Island, near Japan. In the foreground, lava has cooled from above, forming hexagonal and pentagonal cells (cf. section 2.5). In the distance, cumulus clouds are driven by conditional convective instability (section 9.1). Credit: Ippei Naoi/ Moment/ Getty Images

Conditional instability is quantified in terms of a *saturated* (or *moist*) adiabatic lapse rate Γ_s, whereas the quantity defined in section 2.7 as Γ is actually the *dry* adiabatic lapse rate, henceforth called Γ_d. Γ_s varies considerably with temperature and pressure but is always less than Γ_d; a typical value is 6°C/km. Saturated air can therefore be in one of three states depending on the measured lapse rate $\Gamma = -\partial T/\partial z$:

(i) $\Gamma < \Gamma_s$, stable,

(ii) $\Gamma_s < \Gamma < \Gamma_d$, conditionally unstable,

(iii) $\Gamma > \Gamma_d$, unstable.

Conditional instability is often responsible for the formation of cumulus clouds (e.g., Figure 9.1; note rain on horizon at left).

An interesting variant of conditional instability has been observed in the polar oceans (Foldvik and Kvinge, 1974; Jordan et al., 2015). The process depends on the formation of tiny ice crystals called *frazil*, which happens when water is supercooled (cooled below its freezing point but not yet frozen). A parcel of supercooled water can be thought of as a mixture of liquid water and frazil. As the frazil content increases, the parcel becomes less dense (since ice is lighter than water), and therefore rises. The resulting pressure drop increases the freezing point of water, so that the water is increasingly supercooled and frazil formation is accelerated.

9.2 Double Diffusive Instabilities

9.2.1 Some History

The class of processes known as double diffusion operates in a wide range of fluid environments, from the Earth's mantle to stellar interiors (Schmitt, 1983). It depends on two conditions: (i) that the buoyancy is a function of two different scalars (hence the "double"), and (ii) that they diffuse at different rates (hence "diffusion"). In the oceanographic context the scalars are temperature and salinity, and salt diffuses much more slowly than heat.

The discovery of double diffusion has an interesting history. In the nineteenth century, Jevons (1857) conducted an experiment similar to that shown in Figure 9.2. He produced the structures we now recognize as salt fingers, though he was not able to explain them. Later in 1880, the formidable Lord Rayleigh reproduced Jevons' experiment, but even he was unable to deduce the mechanism. The oceanographer Vagn Walfrid Ekman (Ekman, 1906) studied internal waves by means of a layer of milk laid carefully over a layer of salt water. He observed a shower of small vortex rings. While he recognized the role of molecular diffusion in this phenomenon, he did not realize that it would happen naturally in the ocean.

An important step forward was made when a group of oceanographers (Stommel et al., 1956) were discussing how to measure pressure changes in the tropical Atlantic Ocean using a pipe that would be inserted vertically into the water column. They realized that, if the water in the pipe was displaced upwards, it would continue to flow indefinitely (Figure 9.3). Their reasoning rested on the fact that the ocean is saltiest at the surface in that region due to evaporation. The buoyancy distribution is nevertheless stable because the surface waters are warmer (Figure 9.3, profiles at left). Water entering at the bottom of the pipe is therefore relatively cool and fresh, but as it rises, it is surrounded by warmer water and is therefore warmed by the

Figure 9.2 Laboratory experiment showing the formation of salt fingers. Assistance with the experiment provided by Christoph Funke.

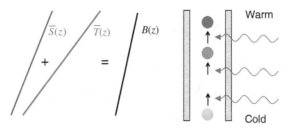

Figure 9.3 Schematic of the perpetual salt fountain, which is analogous to the growth of salt fingers when warm salty water overlies cooler, fresher water. If the water in the pipe is displaced upward, each fluid parcel moves into a warmer environment and therefore absorbs heat. It does not absorb salt, however, because the wall of the pipe prevents it. Since the water is now fresher (and thus lighter) than its surroundings, it continues to rise. Because the diffusivity of salt is negligible, the process works even without the pipe.

heat flux through the pipe. Unlike heat, salt cannot pass through the pipe, the result being that the enclosed water remains fresher, and therefore more buoyant, than its surroundings. Their now-classic paper is called "An Oceanographic Curiosity: the Perpetual Salt Fountain."

Stommel's colleague Melvin Stern soon realized that the salt fountain is more than a "curiosity" – because salt diffuses much more slowly than heat, Stommel's pipe is not actually needed! The phenomenon should occur naturally wherever warm, salty water overlays cool fresh water (Stern, 1960). We now know that this is true, and the pipeless salt fountains are called salt fingers.

9.2.2 Parameters Describing Thermohaline Stratification

When changes in density are caused by only small variations in temperature T and salinity S from reference values T_0, S_0, we use the linearized equation of state (reproducing 1.23):[1]

$$b(T, S) = g\alpha(T - T_0) - g\beta(S - S_0),\qquad(9.1)$$

where α and β are constants evaluated at $T = T_0, S = S_0$. Denoting the background profiles of T and S with overbars, the background vertical buoyancy gradient is then

$$B_z = g\alpha\bar{T}_z - g\beta\bar{S}_z.\qquad(9.2)$$

A useful and easily measured parameter for characterizing compound stratification is (minus) the ratio of thermal and saline contributions to B_z:

---

[1] More generally, the equation of state for seawater is also a function of pressure. It is determined very precisely using a 48-term polynomial equation(!), and can also involve other variables such as sediment concentration or dissolved gases.

$$R_\rho = \frac{\alpha \bar{T}_z}{\beta \bar{S}_z}.$$

The individual components of B_z can be written as

$$g\alpha \bar{T}_z = \frac{B_z R_\rho}{R_\rho - 1}, \qquad \text{and} \quad g\beta \bar{S}_z = \frac{B_z}{R_\rho - 1}. \tag{9.3}$$

We will only consider the situation where $R_\rho > 0$ and $B_z > 0$. In this case $g\alpha \bar{T}_z$ and $g\beta \bar{S}_z$ have the same sign, i.e., one makes a positive contribution to B_z while the other makes a (smaller) negative contribution. It is in this situation that double diffusive instabilities may occur. The case in which $g\alpha \bar{T}_z$ and $g\beta \bar{S}_z$ are both positive is sketched in Figure 9.3.

An important special case is that of compensating temperature and salinity. If $R_\rho = 1$, then the thermal and saline contributions to the buoyancy gradient cancel each other out. The mean density profile is then uniform despite the variations in temperature and salinity.[2]

Other useful parameters include

- the Prandtl number $Pr = \nu/\kappa_T$,
- the Schmidt number $Sc = \nu/\kappa_S$ and
- the diffusivity ratio $\tau = \kappa_S/\kappa_T$.

9.2.3 Perturbation Equations

By substituting $b = B(z) + \epsilon b'$, $T = \bar{T}(z) + \epsilon T'$, and $S = \bar{S}(z) + \epsilon S'$ into (9.1), we obtain the buoyancy perturbation

$$b' = g\alpha T' - g\beta S'.$$

This is substituted into the perturbation equation for a motionless stratified equilibrium (2.17) to give

$$\frac{\partial}{\partial t} \nabla^2 w' = g\alpha \nabla_H^2 T' - g\beta \nabla_H^2 S' + \nu \nabla^4 w'. \tag{9.4}$$

Equations for the temperature and salinity perturbations are obtained by substituting the perturbation solutions into (1.24) and linearizing:

$$\frac{\partial}{\partial t} T' = -\bar{T}_z w' + \kappa_T \nabla^2 T', \tag{9.5}$$

---

[2] It is well known that the ocean tends to arrange itself into more-or-less horizontal layers that are homogeneous in density. But because of this compensation property, those layers are not necessarily homogeneous in temperature or salinity. This can lead to *thermohaline interleaving* as discussed in section 12.2.2.

$$\frac{\partial}{\partial t} S' = -\bar{S}_z\, w' + \kappa_S \nabla^2 S'. \tag{9.6}$$

We have assumed, as for the previous case of pure convection, that the profiles of $\bar{T}(z)$ and $\bar{S}(z)$ are linear (see section 2.1.1), i.e., $\bar{T}_z$ and $\bar{S}_z$ are both constants, so that there is no change in time of the background profiles due to diffusion.

Assuming that each of the three unknowns w', T', S' has the usual normal mode form, e.g., $w' = W_r$ where

$$W \sim e^{\sigma t + \iota(kx + \ell y + mz)},$$

we can derive an algebraic equation for σ:

$$(\sigma + \nu K^2)(\sigma + \kappa_T K^2)(\sigma + \kappa_S K^2) + (\sigma + \kappa_S K^2)\frac{C^2 B_z R_\rho}{R_\rho - 1} - (\sigma + \kappa_T K^2)\frac{C^2 B_z}{R_\rho - 1} = 0, \tag{9.7}$$

where $C = \tilde{k}/K$ is the cosine of the angle of elevation (Figure 2.2).[3] Collecting powers of σ, we write (9.7) as a cubic polynomial equation

$$\sigma^3 + A_2 \sigma^2 + A_1 \sigma + A_0 = 0 \tag{9.8}$$

with coefficients

$$A_2 = (Pr + 1 + \tau)\kappa_T K^2$$
$$A_1 = (Pr + Pr\tau + \tau)\kappa_T^2 K^4 + C^2 B_z$$
$$A_0 = Pr\, \tau \kappa_T^3 K^6 + C^2 B_z \kappa_T \frac{\tau R_\rho - 1}{R_\rho - 1} K^2.$$

While (9.8) is cubic and is therefore awkward to solve analytically, it has some simplifying properties that allow us to identify regions of parameter space where different types of instability are found. First, the coefficients of (9.8) are real. Second, A_1 and A_2 are positive definite, provided only that $B_z > 0$ and $K \neq 0$. Only A_0 can take either sign. Using these properties, one may deduce that

- stationary instability exists when $A_0 < 0$, and
- oscillatory instability exists when $A_0 > A_1 A_2$.

If you enjoy this sort of thing, you may wish to complete project B.5 (appendix B), in which these regions are identified.

The requirement $A_0 < 0$ for stationary instability is fulfilled (at least for some range of K) provided that the second term in A_0 is negative, or

$$1 < R_\rho < \frac{1}{\tau}. \tag{9.9}$$

---

[3] Exercise: Eliminate salinity from (9.7) by taking $\kappa_S = 0$ and $R_\rho \to \infty$. Compare the result with (2.45). Now eliminate temperature by setting $\kappa_T = 0$ and $R_\rho = 0$ and compare again.

Similarly (albeit with more algebra), we can show that the requirement $A_0 > A_1 A_2$ for oscillatory instability is equivalent to

$$\frac{Pr + \tau}{Pr + 1} < R_\rho < 1. \tag{9.10}$$

For seawater, with $Pr = 7$ and $\tau = 0.01$, these ranges become $1 < R_\rho < 100$ for stationary modes and $0.876 < R_\rho < 1$ for oscillatory modes. As R_ρ is increased past unity, the fastest-growing instability changes from oscillatory to stationary (Figure 9.4a). Both mode types grow fastest when R_ρ is near 1, where the net stratification is weakest. The two classes of unstable modes represent distinctly different physical processes, as we describe in the upcoming subsections.

9.2.4 Salt Finger Instability

In the first instance, we will examine the basic state profiles shown in Figure 9.3, where both $\bar{T}$ and $\bar{S}$ increase with height, z. This is called the salt-finger, or just *finger regime*, for reasons that we will soon see. In this regime, $R_\rho > 1$. While salt fingering instability is mathematically possible for R_ρ as high as $\tau^{-1} \sim 100$, the phenomenon is usually observed when R_ρ is ~ 2 or less. Salt fingering instability

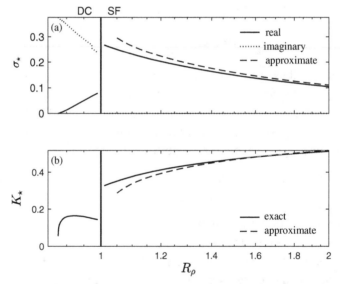

Figure 9.4 Growth rate (a; scaled by $|g\alpha T_z|^{1/2}$) and wavenumber (b; scaled by $[|g\alpha T_z|/\kappa_T^2]^{1/4}$) for double diffusive instability as described by (9.8). Results pertain to the fastest-growing instability at each R_ρ. $R_\rho > 1$: stationary salt fingering modes (section 9.2.4); $R_\rho < 1$: oscillatory diffusive convection (section 9.2.5). The Prandtl number and diffusivity ratio are set to $Pr = 7$ and $\tau = 0.01$, typical values for seawater. Dashed curves indicate the approximation (9.11).

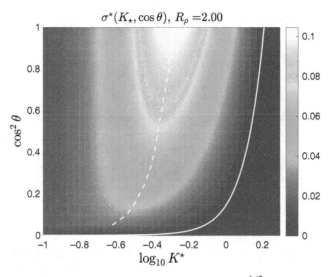

Figure 9.5 Salt fingering growth rate (scaled by $|g\alpha T_z|^{1/2}$) versus wavenumber (scaled by $[|g\alpha T_z|/\kappa_T^2]^{1/4}$) and elevation angle for salt fingering modes with $R_\rho = 2$. Dashed curve: fastest-growing mode for each θ; solid curve: stability boundary.

has a well-defined fastest-growing mode with finite wavenumber (Figure 9.5). As in ordinary convection, the optimal angle of elevation is 90 degrees, i.e., the motions are purely vertical.

Stern (1975) noticed that σ is usually $\ll \nu K^2$ but $\gg \kappa_S K^2$. Accordingly, he simplified (9.7) by replacing $\sigma - \nu K^2$ with νK^2 and $\sigma - \kappa_S K^2$ with σ, reducing the order of the equation to quadratic. The resulting approximation to the fastest-growing mode is

$$\sigma \approx \sqrt{\frac{B_z}{Pr}} \left(\sqrt{\frac{R_\rho}{R_\rho - 1}} - 1 \right) ; \qquad K^4 \approx \frac{B_z}{\kappa_T^2 Pr}. \qquad (9.11)$$

Shown by dashed curves on Figure 9.4, this approximation gives growth rates accurate to within 10 percent when $1.2 < R_\rho < 5$, the most important range for the ocean. Because K is proportional only to the fourth root of B_z, the thickness of salt fingers does not vary by much in the Earth's oceans – it is typically a few centimeters.

Laboratory Experiment

It is easy to produce the salt-finger instability in your kitchen at home, or in the classroom. All that is needed is a glass and spoon, some hot and cold water (about half a glass of each), food coloring, and salt or sugar. First, add the food coloring, and a very slight amount of sugar or salt (usually just a dozen grains or so is enough) to the hot water (about the temperature of a cup of tea), and mix it thoroughly. This will form the warm, salty upper layer. Note that the purpose

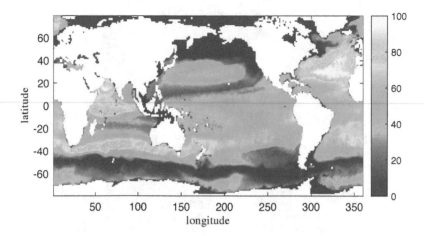

Figure 9.6 Regions of the world's oceans where R_ρ is favorable for salt fin-ger convection. Plotted is the percentage of the uppermost 2000 m in which $1 < R_\rho < 100$. These data were collected and made freely available by the International Argo Program and the national programs that contribute to it. (http://www.argo.ucsd.edu, http://argo.jcommops.org). The Argo Program is part of the Global Ocean Observing System. The BOA-Argo dataset used for this calculation was produced at China Argo Real-time Data Center (Li et al., 2017).

of the food coloring is only to visualize the instability, as in Figure 9.2. It may not be necessary to add the salt or sugar since the food coloring may already have sugar added.

The experiment begins when this colored mixture is added to half a glass of room temperature water from the tap. The most important part of the experiment is to add the warm, salty layer very carefully, so as not to mix too much with the tap water. This is why a spoon is recommended to break the fall of the warm, salty water as it is added. After about a minute, the fingers should form as shown in Figure 9.2.

Geography

The stratification required for the finger regime of double-diffusive convection is found in widespread areas of the subtropical oceans, especially in the Atlantic Ocean and the Mediterranean Sea. The reason is that both surface heating and evaporation are strong, tending to produce warm and salty surface waters. Figure 9.6 shows regions where R_ρ values are low enough for salt fingers to be likely.

9.2.5 Oscillatory Diffusive Convection

We turn now to the stability properties of profiles where both T and S *decrease* with height, the opposite situation to salt fingering. This is called the *diffusive* regime of double diffusion, and leads to an instability that is very different from salt fingering.

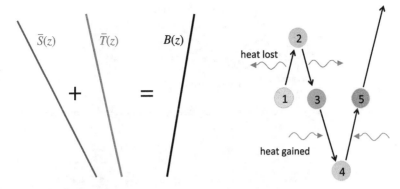

Figure 9.7 Schematic of diffusive convection instability. Cold, fresh water overlies warm, salty water, such that the net buoyancy gradient is positive. A water
parcel (1), displaced upward (2), sinks due to buoyancy (3). During this excursion
it loses heat to its surroundings, but not salt, so it now has decreased buoyancy.
The parcel continues downward, becoming lighter than its surroundings (4), then
rises again (5), all the while losing heat. Larger vertical excursions lead to larger
heat exchanges and thus positive feedback.

Referring to Figure 9.7, suppose that a fluid parcel is displaced upward from its
equilibrium position. It will find itself surrounded by fluid that is both cooler and
fresher. Since $\kappa_T \gg \kappa_S$, heat will be transferred from the parcel to the surroundings
much more effectively than salt. The parcel will therefore become denser than its
surroundings and sink. In the warmer environment below, the parcel will lose heat,
but retain its salinity, therefore becoming lighter than its surroundings, after which
it will rise once again. In this way the water parcel performs exponentially growing
oscillations, as our linear stability analysis predicts.

As in salt fingering (and ordinary convection), diffusive convection is most efficient when motions are purely vertical (Figure 9.8). The imaginary part of the
growth rate generally exceeds the real part by an order of magnitude, i.e., the instability oscillates rapidly but grows slowly. This is true throughout the range of R_ρ
where diffusive convection exists (Figure 9.4).

As per (9.9), the oscillatory mode of double diffusion occurs only over a
restricted range of R_ρ, $0.876 < R_\rho < 1$. Yet, evidence of instability that is
observed in the oceans is seen for a much wider range of R_ρ. A way out of
this conundrum has been suggested by Radko (2016), who found that when even
small amounts of shear are included in the stability analysis of diffusive convection, instability is found over a much larger range of R_ρ. In addition, for
conditions found in the oceans the dominant instability is not, in fact, the oscillatory mode described above, but is instead a type of convection that depends on
the action of diffusion on the *mean profiles* of temperature and salinity (described
below).

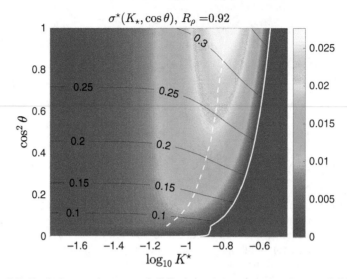

Figure 9.8 Scaled growth rates of diffusive convection for $R_\rho = 0.92$. Contours show the imaginary part. Scaling is as in Figure 9.5. Dashed curve: fastest-growing mode for each θ; solid curve: stability boundary.

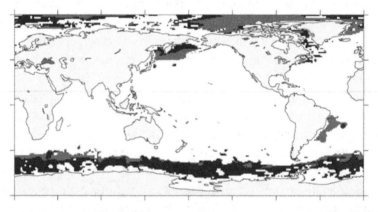

Figure 9.9 Regions of the world oceans where the stratification of the diffusive regime of double diffusion is found with $1/3 < R_\rho < 1$ in dark gray, and $1/10 < R_\rho < 1/3$ in light gray. Reprinted from Kelley et al. (2003), with permission from Elsevier.

Geography

The compound stratification needed for diffusive convection is found most often in the polar regions (Figure 9.9), where ice melt and cold river runoff create a cold, fresh surface layer.

9.2.6 Thermohaline Staircases in Fingering and Diffusive Regimes

If we look more closely at profiles from double diffusive regions, we see that the profiles of T, S, and B are often comprised of a series of mixed layers separated

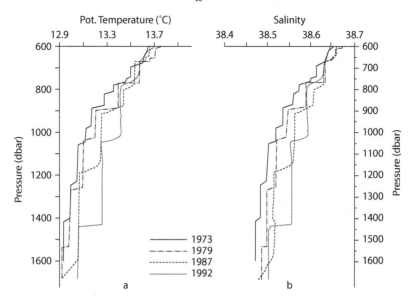

Figure 9.10 Staircase profiles measured in the Tyrrhenian Basin of the Mediterranean Sea over a 19-year period. The vertical axis can be interpreted as depth with 100 dbar roughly 100 m. Reprinted from Zodiatis and Gasparini (1996), with permission from Elsevier.

by sharp interfaces. These so-called thermohaline staircases may occur either in fingering regimes (e.g., the Mediterranean Sea; Figure 9.10) or diffusive regimes (e.g., the Arctic Ocean Figure 9.11). The mechanism of their formation has been a subject of intense discussion since they were first observed in the 1960s and remains so today.

In the diffusive regime, it appears that a convective instability is acting at the edges of each interface. This can be understood by considering the thickening of the $\bar{T}$ and $\bar{S}$ interfaces due to molecular diffusion. Since $\kappa_T > \kappa_S$, the $\bar{T}$ interface tends to become thicker than the $\bar{S}$ interface (Figure 9.12). This results in layers above and below each interface where the gravitationally unstable thermal stratification is not stabilized by salinity. Once the relevant Rayleigh number exceeds the critical value (section 2.4.2), the layers become unstable and produce turbulent convection rolls above and below each interface. The layers are thereby maintained at roughly uniform, but different, temperatures and salinities while the interfaces remain sharp (Carpenter et al., 2012). In the salt fingering regime, staircases may be related to the turbulence-induced instabilities discussed later in section 12.2.

In our discussion of double diffusion we have focused on oceanic observations, but there are many other systems that exhibit double diffusive phenomena. A number of brackish lakes exhibit spectacular thermohaline staircases. The best example is Lake Kivu, located in east Africa on the border between Rwanda and the Democratic Republic of Congo. There, in about 400 m of water, more than

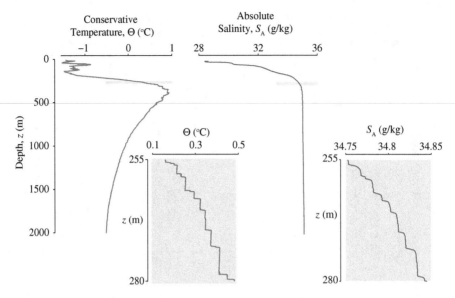

Figure 9.11 Profiles of T and S from the thermohaline staircase in the Arctic Ocean. The large-scale profiles are shown above, with more detailed plots of the small-scale staircase structure taken from the gray regions. Data for this plot was provided from the Beaufort Gyre Observing System (www.whoi.edu/beaufortgyre/).

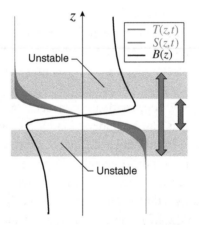

Figure 9.12 Schematic of an interface in a thermohaline staircase of the diffusive regime. Sequential profiles show the thickening of the T interface. Arrows represent the final thickness of the T and S interfaces. The final buoyancy profile is also plotted, with gray regions roughly indicating the gravitationally unstable layers.

300 individual mixed layers can be identified. This diffusive convection system is also interesting in that it has a significant buoyancy contribution from dissolved carbon dioxide and methane gases. In fact, there is so much methane gas present

in the deep waters of Lake Kivu that it is currently being extracted and burned to produce electricity. This extraction is also a safety measure: sudden de-gassing events in smaller lakes have resulted in dense CO_2 gas clouds that suffocated the surrounding inhabitants (Schmid et al., 2004).

Radko (2013) is an excellent source for further information on double diffusive instabilities.

9.3 Bioconvection

The reader may thus far have noticed that instability through convection has a common theme: it is a mechanism that releases the potential energy of a fluid column and converts it to kinetic energy. Normally this potential energy comes from some distribution of a scalar quantity such as temperature, salinity, or water vapor, but more exotic cases can be found. One such case is when the scalar quantity is the concentration of microorganisms. This curious *bioconvection* was noticed by biologists in the laboratory when populations of swimming creatures spontaneously formed hexagonal cell patterns if either the thickness of the fluid layer was large enough, or the concentration of microorganisms was high enough (see Figure 9.13). The potential energy needed for convection exists because (i) the microorganisms are slightly denser than water, and (ii) they instinctively swim upwards in search of sunlight, and are therefore concentrated near the surface.

9.3.1 Equations of Motion

We can understand the onset of bioconvection as a normal mode instability. To this end, we couple the standard Boussinesq equations of motion, reproducing (2.1, 2.3),

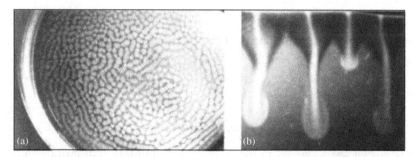

Figure 9.13 Bioconvection in laboratory suspensions of swimming microorganisms. (a) Top view of bioconvection cells in a container of 4 cm diameter and a depth of 6.8 mm. (b) Side view of bioconvection plumes in a deep, narrow container. Photos from Pedley and Kessler (1992), where more information can be found.

$$\vec{\nabla} \cdot \vec{u} = 0 \tag{9.12}$$

$$\frac{D\vec{u}}{Dt} = -\vec{\nabla}\pi + b\hat{e}^{(z)} + \nu\nabla^2\vec{u} \tag{9.13}$$

with an equation that describes the buoyancy in terms of the volumetric concentration of the biota, c, and an equation that describes the evolution of c. We assume that the density of the biota, ρ_1, is a constant slightly greater than the ambient fluid density, ρ_0, which is also assumed to be constant. The total density of the mixture is then given by $\rho = \rho_1 c + \rho_0(1 - c)$, which we can write in terms of buoyancy as

$$b = -\tilde{g}c, \tag{9.14}$$

where $\tilde{g} \equiv g(\rho_1 - \rho_0)/\rho_0$ is referred to as the reduced gravity.

Now let the conservation equation for c take the form

$$\frac{Dc}{Dt} = -\vec{\nabla} \cdot \vec{F}, \tag{9.15}$$

where $\vec{F}$ is a microorganism flux. The flux can be thought to consist of two parts: (i) the flux due to swimming, with speed and direction given by a vector $\vec{V}$, and (ii) the diffusion of c, also due to swimming but on a random trajectory meant to avoid crowds. Thus,

$$\vec{F} = c\vec{V} - \kappa\vec{\nabla}c. \tag{9.16}$$

(In more sophisticated models, the diffusivity is a tensor that incorporates swimming behavior in more detail.) Substituting (9.16) into (9.15) and assuming that the swimming is directed upwards at uniform velocity, $\vec{V} = V_s\hat{e}^{(z)}$, yields a conservation equation for c:

$$\frac{Dc}{Dt} = -V_s c_z + \kappa\nabla^2 c. \tag{9.17}$$

Equations (9.12, 9.13, 9.14) and (9.17) form a complete set. Boundaries are located at $z = -H, 0$, and assumed to be impermeable ($w' = 0$) and either rigid ($w'_z = 0$) or frictionless ($w'_{zz} = 0$). Impermeability also requires that the flux of biota not penetrate the boundaries, viz. $F^{(z)} = 0$.

9.3.2 Equilibrium Solution

We now search for solutions in the form of a motionless equilibrium with small perturbations imposed on it, i.e., $c(\vec{x}, t) = C(z) + \epsilon c'(\vec{x}, t)$, $\vec{u} = \epsilon\vec{u}'$, and $\pi = \Pi(z) + \epsilon\pi'(\vec{x}, t)$. The conservation equation (9.15) requires that $F^{(z)}$ be independent of depth, and since $F^{(z)}$ is zero at the boundaries, it must be zero everywhere. Hence,

$$V_s C_z = \kappa C_{zz} \quad \Rightarrow \quad C(z) = C_0 e^{\frac{V_s}{\kappa} z}, \tag{9.18}$$

where the constant C_0 is the biota density at the surface.

9.3.3 The Perturbation Equations

We now proceed to the order ϵ and look at the evolution of small perturbations. Following the usual steps to eliminate the pressure and the horizontal velocity components (e.g., section 3.1.2), the equations are reduced to:

$$\frac{\partial}{\partial t} \nabla^2 w' = -\tilde{g} \nabla_H^2 c' + \nu \nabla^4 w', \tag{9.19}$$

$$\frac{\partial}{\partial t} c' = -C_z w' - V_s c'_z + \kappa \nabla^2 c'. \tag{9.20}$$

Note that (9.19) and (9.20) are analogous to the perturbation equations for ordinary convection (2.17, 2.18) except for the second term on the right-hand side of (9.20), which arises from the microorganisms' swimming. A second important difference is that the coefficient C_z is not constant but rather depends on z, necessitating a numerical solution.

We now make the usual normal mode assumption $w' = W_r$; $W = \hat{w}(z) e^{\sigma t + \iota(kx + \ell y)}$ and similar for c', leading to

$$\sigma \nabla^2 \hat{w} = \tilde{g} \tilde{k}^2 \hat{c} + \nu \nabla^4 \hat{w} \tag{9.21}$$

$$\sigma \hat{c} = -C_z \hat{w} - V_s \hat{c}_z + \kappa \nabla^2 \hat{c}, \tag{9.22}$$

where $\nabla^2 = d^2/dz^2 - \tilde{k}^2$. together with

$$C_z = C_0 \frac{V_s}{\kappa} \exp\left(\frac{V_s}{\kappa} z\right) \tag{9.23}$$

and the boundary conditions

$$\hat{w} = 0 ; \quad \hat{c}_z = \frac{V_s}{\kappa} \hat{c} \quad \text{on } z = -H, 0.$$

In the usual fashion, this is all expressed as a generalized eigenvalue problem and solved using the numerical methods developed in previous chapters. The resulting solution may be written as

$$[\sigma, \hat{w}, \hat{c}] = \mathcal{F}(z, \tilde{g}, \tilde{k}, \nu, \kappa, C_0, V_s, H).$$

9.3.4 Scaling

It is convenient to scale the problem using the length H, the time H^2/κ, and the concentration C_0, resulting in

$$\sigma_\star \nabla_\star^2 \hat{w}^\star = Ra \, Pr \, \tilde{k}_\star^2 \hat{c}^\star + Pr \, \nabla_\star^4 \hat{w}^\star \tag{9.24}$$

$$\sigma_\star \hat{c}^\star = -\lambda \exp(\lambda z^\star) \hat{w}^\star - \hat{c}_{z\star}^\star + \nabla_\star^2 \hat{c}^\star, \tag{9.25}$$

with boundary conditions

$$\hat{w}^\star = 0, \quad \hat{c}_{z\star}^\star = \lambda \hat{c}^\star, \quad \text{and either } \hat{w}_{z\star}^\star = 0 \text{ or } \hat{w}_{z\star z\star}^\star = 0$$

on $z^\star = -1, 0$. The quantity Ra appearing in (9.24) plays a role similar to the Rayleigh number in ordinary convection, and is defined as

$$Ra = \frac{\tilde{g} C_0 H^3}{\kappa \nu}.$$

The parameter Pr is not the usual Prandtl number; it is the ratio of the viscosity of the water to the diffusivity of the biota. The analogy is close, though, since it is the biota that determine the buoyancy via (9.14). A typical value for Pr is 2 (see below). The parameter λ can be thought of as a nondimensional ratio of the swimming flux to the diffusive flux:

$$\lambda = \frac{V_z H}{\kappa}.$$

The solution procedure is now

$$[\sigma_\star, \hat{w}_\star, \hat{c}^\star] = \mathcal{F}(z^\star, Ra\,Pr, \tilde{k}_\star, Pr, 1, 1, \lambda, 1),$$

i.e., the solution is determined by the horizontal wavenumber $\tilde{k}^\star$ and the parameters Ra, Pr, and λ. As in ordinary convection, the direction of the wave-vector (k, ℓ) is immaterial.

9.3.5 The Stability Boundary

Figure 9.14 shows the (purely real) growth rate as a function of $k^\star$ and Ra for the particular case $Pr = 2, \lambda = 3$. The bottom boundary is rigid; the top is frictionless (to mimic a lab experiment). As in the Rayleigh-Benard problem, there is for each wavenumber a critical value of Ra, Ra_c, above which $\sigma^\star = 0$ grows monotonically from zero. The minimum critical Rayleigh number for this case is 488, occurring at $k_\star = 1.54$. For every $Ra > Ra_c$, there is a unique fastest-growing mode which occurs at monotonically growing wavenumbers.

On the stability boundary of a stationary mode, $\sigma^\star = 0$, so that the left-hand sides of (9.24) and (9.25) vanish. If we then divide (9.24) by Pr, Pr is removed from the problem. This tells us that both the critical Rayleigh number and the wavenumber at which it occurs are independent of Pr, as in Rayleigh-Benard convection (section 2.4.2). However, both parameters depend on the flux ratio λ as shown in Figures 9.15(a,b), respectively.

You can explore bioconvection further yourself in project B.3.

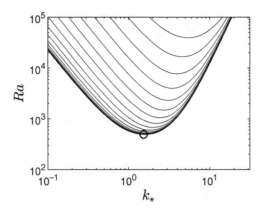

Figure 9.14 Scaled growth rate of bioconvection versus wavenumber and Rayleigh number. The circle indicates the critical state. Logarithmic contour spacing represents a factor of 1.8; maximum growth rate is 291.

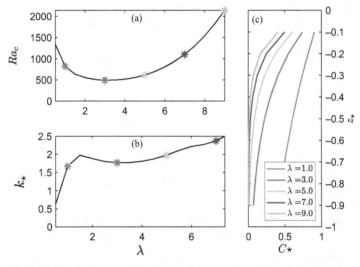

Figure 9.15 Critical values of the Rayleigh number (a) and the corresponding scaled wavenumber (b) as functions of λ. (c) Equilibrium concentration profiles for selected λ, color-coded to match (a) and (b).

9.4 CO₂ Sequestration

The burning of fossil fuels for energy production has led to a potentially harmful increase in the concentration of carbon dioxide gas in the atmosphere (CO_2 is a byproduct of the reaction). One of the techniques proposed to combat this increase is to store CO_2 in deep (around $800-3,000$ m) underground aquifers. The aquifers consist of a porous rock medium that is filled with saline groundwater, and

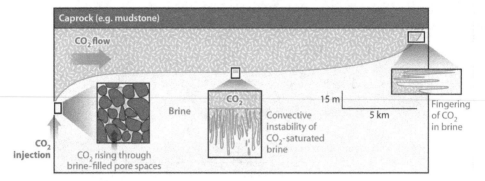

Figure 9.16 Injection and flow of liquid CO_2 along a capping rock in a saline underground aquifer (Huppert and Neufeld, 2014).

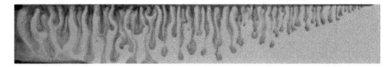

Figure 9.17 Laboratory experiment performed to study the instability of buoyant CO_2 current (dark colors). Analog fluids are water (which takes the role of the CO_2) and propylene glycol (which takes the role of the groundwater). The dark CO_2-like fluid was injected on the left and sheds convective plumes as it propagates to the right along the top of the photo [see MacMinn and Juanes (2013) for more details]. Photo courtesy of Christopher MacMinn.

is capped above by an impermeable layer that is usually gently sloping. The idea is to inject CO_2 into the porous medium below the capping layer where it will be stored. Since the liquid CO_2 at these pressures and temperatures is more buoyant than the saline groundwater, it will rise to the capping rock and spread laterally (Figure 9.16). It is possible for the CO_2 to flow laterally for tens or hundreds of kilometers, whence it may escape the coverage of the capping rock and find its way back to the atmosphere. However, there are ways of trapping the CO_2, one of which consists of dissolving it into the underlying groundwater.

This dissolution and trapping of the CO_2 into the ambient groundwater is aided by a convection-type instability that arises due to the equation of state. Despite the liquid CO_2 being more buoyant than the ambient water, the *mixture* of CO_2 and water is actually denser than either of the two fluids when pure. This has the effect of creating a negatively buoyant layer at the interface of the two fluids that can become unstable when conditions are right. The situation is similar to that described for the maintenance of thermohaline staircases in the diffusive regime (see the end of section 9.2.5, and Figure 9.12). The instability can be seen in the laboratory experiment shown in Figure 9.17, where the buoyant gravity current forms convective plumes as it propagates from left to right along the top boundary. It is important to understand the increased dissolution of CO_2 that is caused by this instability, as it is an important mechanism in setting the possible travel distance of the current, as well as influencing the amount of CO_2 that is trapped in the aquifer.

10

Summary

10.1 Equilibrium States

Equilibrium states are exact solutions of the equations of motion with all occurrences of the time derivative $\partial/\partial t$ set to zero. Such solutions are normally possible only after extreme simplification of the flow geometry, although for generality this is desirable (i.e., we want to focus on features common to a broad class of flows while ignoring the details that distinguish individual instances). Intrepid analysts often apply normal modes to states that are not exactly in equilibrium – the "frozen flow" hypothesis. The validity of this hypothesis must be checked after the fact to ensure that the instability grows faster than the background flow changes.

10.1.1 Mass Conservation

In most cases we have assumed that, if the equilibrium state involves a nonzero current, that current will be directed in one and only one of the coordinate directions.[1] Such a unidirectional current can be incompressible (see 1.17) only if its speed does not vary in the direction of flow. In Cartesian coordinates, this invariance of the equilibrium flow implies that the nonlinear self-advection term in the momentum equation (1.16, 1.14) must vanish:

$$[\vec{u} \cdot \vec{\nabla}]\vec{u} = 0, \tag{10.1}$$

a major simplification. The single exception to (10.1) is circular flow in cylindrical coordinates (Chapter 7), where the self-advection term is not zero but instead contributes the centrifugal force.

---

[1] We have found ways to accommodate some other classes of flow, e.g., veering flows (section 4.12) and flows that are not quite in equilibrium (sections 5.2 and 6.1.3).

Table 10.1 *Summary of equilibrium states. In each case the named force balances the pressure gradient force.*

Force	Equilibrium	Chapter(s)
none		3
gravity	hydrostatic	2, 4, 8, 9
viscosity	frictional	5, 6
Coriolis	geostrophic	8
Coriolis+gravity	thermal wind	8
centrifugal	cyclostrophic	7

10.1.2 Force Balances

In equilibrium, the momentum equation (1.19) reduces to a statement that the sum of forces must be everywhere zero. Each of the force terms on the right-hand side of the momentum equation can be neglected under certain plausible assumptions, with the exception of the pressure gradient force. The gravitational term can be zero either in a zero-gravity environment where $g = 0$ or in a fluid with uniform density such that $b = 0$. The viscous term can be zero if the fluid is assumed to be inviscid, $v = 0$, or if the flow is such that the Laplacian of the velocity field is everywhere zero. The same is true of the diffusion term in the buoyancy equation. The Coriolis term can be zero in a non-rotating environment, $f = 0$, or in a state of no motion. The centrifugal force vanishes in a parallel flow.

In each of the equilibria considered here, most (or all) of these "optional" force terms are assumed to be zero, and that the pressure field is arranged so as to balance whichever force terms remain. We can therefore classify equilibria in terms of the force that the pressure gradient must balance (see Table 10.1).

10.2 Instabilities

10.2.1 Mechanisms

The mechanism of convective instability is intuitively simple: gravity drives vertical accelerations that must overcome the damping effects of viscosity and diffusion. An analogous mechanism was identified for centrifugal, inertial, and symmetric instabilities.

Shear and baroclinic instabilities are understood in terms of wave resonances: vortical waves in the case of parallel shear flow (section 3.12); Eady waves in the case of baroclinic instability. In a stratified environment, gravity waves can also take part in a resonant interaction. Barotropic instability of a circular vortex also falls into this category.

Table 10.2 *Rules of thumb. Double asterisks ** indicate that the wavelength pertains to the critical state. Otherwise it is the wavelength of the fastest-growing mode.*

instability	wavelength	growth rate	criterion	chapter
convection, $B_z < 0$	$2.8H^{**}$		$Ra > 657.5$	2
shear layer, $U = u_0 \tanh \dfrac{z}{h}$	$7 \times 2h$	$0.2 \dfrac{u_0}{h}$	inflection point	3
jet (sinuous mode), $U = u_0 \operatorname{sech}^2 \dfrac{z}{h}$	$3.5 \times 2h$	$0.16 \dfrac{u_0}{h}$	inflection point	3
stratified shear flows (all)			$Ri_{min} < 1/4$	4
vortex (barotropic)			inflection point	7
vortex (axisymmetric)			$\min_r 2\Omega Q < 0$	7
plane Poiseuille flow, $U = 4u_0\dfrac{z}{h}\left(1 - \dfrac{z}{h}\right)$	$3h^{**}$		$Re > 11600$	5
inertial, $U = U_y y$	none	$\sqrt{-f(f - U_y)}$	$f(f - U_y) < 0$	8
symmetric, $U = U_y y + U_z z$	none	$\|f\|\sqrt{\dfrac{1}{Ri} - \dfrac{f_a}{f}}$	$Ri < \dfrac{f_a}{f}$	8
baroclinic (Eady), $U = U_z z$	$\dfrac{3.9}{P}H$	$0.3\dfrac{\|f\|}{\sqrt{Ri}}$		8
salt fingering	$2\pi\sqrt{\dfrac{B_{Tz}}{\nu\kappa_T}}$		$1 < R_\rho < \tau^{-1}$	9

10.2.2 Rules of Thumb

Table 10.2 includes a (non-exhaustive) list of properties of stability boundaries, critical states and fastest-growing instabilities in the simplest more or less accurate form. See the chapter listed for details.

Part II
The View Ahead

11

Beyond Normal Modes

Throughout Part 1, we employed the classical normal mode description of instability growth. All perturbations, $\vec{X}'$, from the equilibrium state were represented by expressions like $\vec{X}' \propto e^{\sigma t + i(kx + \ell y)}$. The goal was to find complex values of σ that signal growth, decay, oscillations/waves, or a combination of these. We also focused on the fastest-growing mode, which is expected to dominate the solution in the long time limit. However, this focus can be too restrictive; other modes can be an important part of the solution. In this chapter we will introduce a different approach to stability analysis that allows for more general temporal structures. We will find that, over limited times, disturbances can grow considerably faster than the fastest-growing normal mode. This has especially important implications for the emergence of turbulence in geophysical flows.

11.1 Instability as an Initial Value Problem

Consider the evolution of a *specific* initial perturbation of some equilibrium state. This is equivalent to an initial value problem where the perturbation is specified at some initial time, $\vec{X}_0 = \vec{X}(t_0)$, and we seek its state at all subsequent times, $\vec{X}(t)$. Previously, we sidestepped consideration of the initial condition by focusing on the fastest-growing mode, which should dominate the solution at late times. At earlier times, however, solutions can exhibit non-intuitive transient behavior, including rapid initial growth. If this initial growth is large enough, it may trigger nonlinear effects such as the transition to turbulence before the fastest-growing normal mode can be established. Also, when unstable conditions last only for a limited time, it is important to know which disturbances will grow the most over that interval.

To fix these ideas, consider a system of coupled, linear, first-order ordinary differential equations,

$$\frac{d\vec{X}}{dt} = A\vec{X}, \tag{11.1}$$

which reduces to an eigenvalue problem when the time dependence $\vec{X} \propto e^{\sigma t}$ is assumed. We have often reduced our system of equations to such a form, and it may help the reader to review (2.17, 2.18) from the convection chapter or (3.12) describing shear instability. In the case of convection $\vec{X}(t)$ could represent a concatenation of the vertical velocity and buoyancy discretized in the vertical, or possibly the coefficients of these variables in a Fourier series (Chapter 13). After solving for the eigenvalues, σ_j, and the associated eigenvectors, $\vec{\zeta}_j$, we can write the general solution as

$$\vec{X}(t) = B_1 \vec{\zeta}_1 e^{\sigma_1 t} + B_2 \vec{\zeta}_2 e^{\sigma_2 t} + \cdots \tag{11.2}$$

The coefficients $B_1, B_2, \ldots$ are then determined by the initial condition:

$$\vec{X}_0 = B_1 \vec{\zeta}_1 + B_2 \vec{\zeta}_2 + \cdots \tag{11.3}$$

or, more compactly,

$$\vec{X}_0 = Z\vec{B}, \tag{11.4}$$

where Z is the matrix whose columns are the eigenvectors of A:

$$Z = [\, \vec{\zeta}_1 \mid \vec{\zeta}_2 \mid \ldots \,],$$

and $\vec{B}$ is a vector of coefficients

$$\vec{B} = \begin{bmatrix} B_1 \\ B_2 \\ \vdots \end{bmatrix}.$$

If we let σ_1 denote the eigenvalue with the maximum real part (which could be negative), then we can see from (11.2) that, as $t \to \infty$, the solution will become dominated by that mode:

$$\vec{X}(t) \to B_1 \vec{\zeta}_1 e^{\sigma_1 t}.$$

This simplification is the reason we have focused on the fastest-growing mode in previous sections.

But what if we're interested in growth over a finite time interval? We will see that, in geophysical stability problems, the fastest-growing mode alone is often not sufficient to describe the growth of perturbations over finite times. In fact, over a finite time, a problem like (11.1) can have *all eigenvalues decaying* (i.e., $\sigma_r < 0$), and still exhibit growth! An example is the plane Couette flow that we discussed in section 5.1. That flow has no growing eigenmodes, yet is known to become unstable at sufficiently high Reynolds number (Orszag and Kells, 1980).

11.2 Transient Growth in Simple Linear Systems

As a simple illustration of transient growth, consider a system like (11.1) consisting of only two components, $\vec{X} = (X_1, X_2)$. As a specific example, we choose

$$A = \begin{bmatrix} -0.1 & 1 \\ 0 & -0.2 \end{bmatrix}. \tag{11.5}$$

Both eigenvalues of A are real and represent exponential decay: $\sigma_1 = -0.1$ and $\sigma_2 = -0.2$. As is conventional, we order the eigenvalues so that $\sigma_1 > \sigma_2$. The eigenvectors, $\vec{\zeta}_1 = (1, 0)$ and $\vec{\zeta}_2 = (-0.9950, 0.0995)$, are not orthogonal; in fact, they are nearly parallel.

To complete the problem, we choose an initial condition that is arbitrary except that it contains substantial contributions from both eigenvectors, as we can tell by solving (11.4) for $\vec{B}$:

$$\vec{X}_0 = (0.1, 1); \quad (B_1, B_2) = (10.1, 10.05).$$

The solution is then

$$\vec{X}(t) = 10.1\vec{\zeta}_1 e^{-0.1t} + 10.05\vec{\zeta}_2 e^{-0.2t}$$

The evolution of this system is shown in Figure 11.1. The left-pointing eigenvector $\vec{\zeta}_2$ decays rapidly, and with it, its contribution to $\vec{X}$. As a result, $\vec{X}$ points

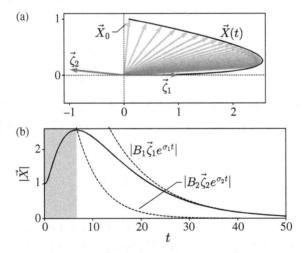

Figure 11.1 (a) Evolution of $\vec{X}$ from its initial value $\vec{X}_0$, with the eigenvectors shown as red arrows (after Schmid, 2007). After an initial period of growth, the evolution of $\vec{X}$ follows that predicted by the eigenvector $\vec{\zeta}_1$ and eigenvalue ($\sigma_1 < 0$) with the largest growth rate. This is also seen in (b), where the amplitude of $|\vec{X}|$ is shown over time (solid line), along with the two different eigenvector terms, $|B_1\vec{\zeta}_1 e^{\sigma_1 t}|$ and $|B_2\vec{\zeta}_2 e^{\sigma_2 t}|$ (dashed lines). The initial growth interval is indicated by gray shading.

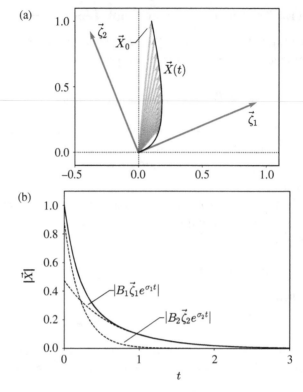

Figure 11.2 As in Figure 11.1, but for the case of orthogonal eigenvectors.

progressively more to the right, and its magnitude increases. At late times, the contribution from $\vec{\zeta}_2$ is negligible. $\vec{X}$ is then nearly parallel to $\vec{\zeta}_1$, and decays with rate σ_1.

As you might anticipate from the last example, a necessary condition for large initial growth is that the eigenvectors be non-orthogonal. In fact, when the eigenvectors are orthogonal, transient growth is not observed. To demonstrate this, let

$$A = \begin{bmatrix} -2 & 1 \\ 1 & -4 \end{bmatrix}. \tag{11.6}$$

The symmetry of A guarantees that the eigenvectors can be chosen to be orthogonal: $\vec{\zeta}_2 = (-0.38, 0.92)$, and $\vec{\zeta}_1 = (0.92, 0.38)$. The corresponding eigenvalues are again both negative: $\sigma_1 = -1.59$ and $\sigma_2 = -4.41$.

The evolution begins with the same initial condition as in the previous case: $\vec{X}_0 = (0.1, 1)$. The resulting evolution of $\vec{X}(t)$ is shown in Figure 11.2. Once again the contribution from $\vec{\zeta}_2$ decays rapidly so that $\vec{X}$ ultimately becomes parallel to $\vec{\zeta}_1$, but the length of $\vec{X}$ decreases monotonically. This is what we would expect based on our experience with normal modes.

Since transient growth requires that the eigenvectors of the system are non-orthogonal, it would be helpful to have a general rule to assess whether this is the case for a given matrix A. It is a general property of linear algebra that *normal* matrices contain only eigenvectors that are orthogonal,[1] with the converse also holding, i.e., any matrix with orthogonal eigenvectors is normal. A normal matrix, A, has the property that $AA^\dagger = A^\dagger A$, with the dagger indicating the Hermitian transpose (the transpose of the complex conjugate, also called the adjoint) of A.

11.3 Computing the Optimal Initial Condition

We've seen how an arbitrarily chosen initial state can exhibit transient growth regardless of the signs of the eigenvalues. Under what conditions is this likely to happen? For a given system, can we identify an initial state that is "optimal," in some sense, for transient growth? The meaning of "optimal" can vary depending on the particular behavior we're interested in, but the methods of matrix calculus allow us to explore a wide range of those behaviors.

We begin by writing the general solution to the initial value problem (11.1) as

$$\vec{X}(t) = e^{At}\vec{X}_0, \tag{11.7}$$

where e^{At} is the matrix exponential function. In case you're unfamiliar, e^{At} is a matrix defined by

$$e^{At} = I + \sum_{n=1}^{\infty} \frac{A^n t^n}{n!}, \tag{11.8}$$

(similar to the infinite series representation of the scalar function e^{at}), where the symbol I is the identity matrix. It has two properties that we'll need:

$$\frac{d}{dt}e^{At} = Ae^{At} \quad \text{and} \quad e^{At}|_{t=0} = I,$$

both of which you can (and should) derive using (11.8). With these you can confirm that $e^{At}\vec{X}_0$ is indeed the solution of (11.1).

To generalize the concept of growth beyond the exponential kind, we define an amplification factor (sometimes called the gain) G, at any given time, as

$$G(t, \vec{X}_0) = \frac{\langle \vec{X}(t), \vec{X}(t) \rangle}{\langle \vec{X}_0, \vec{X}_0 \rangle}. \tag{11.9}$$

This is the factor by which the squared amplitude of the solution $\vec{X}(t)$ exceeds that of the initial condition, $\vec{X}_0$.

[1] Or can be chosen to be orthogonal.

Two aspects of (11.9) should be noted:

- We have used $\langle \cdot, \cdot \rangle$ to denote the inner product, which could take different forms. Here we will consider only the Euclidean inner product – equivalent to the vector dot product, $\langle \vec{a}, \vec{b} \rangle = \vec{a}^* \cdot \vec{b}$ for complex vectors $\vec{a}$ and $\vec{b}$ (more on this in section 11.8).
- G depends on (i) the initial state $\vec{X}_0$, (ii) the time t over which the system has evolved, and (iii) the matrix A that controls the evolution of $\vec{X}$. The first two are listed explicitly in (11.9), the third is implicit.

Given the solution (11.7), the amplification factor can be written as

$$G(t, \vec{X}_0) = \frac{\langle e^{At}\vec{X}_0, e^{At}\vec{X}_0 \rangle}{\langle \vec{X}_0, \vec{X}_0 \rangle} = \frac{\langle e^{A^\dagger t}e^{At}\vec{X}_0, \vec{X}_0 \rangle}{\langle \vec{X}_0, \vec{X}_0 \rangle}. \tag{11.10}$$

The last equality uses the properties

$$\langle \vec{X}, A\vec{y} \rangle = \langle A^\dagger \vec{X}, \vec{y} \rangle,$$

and

$$(e^{At})^\dagger = e^{A^\dagger t}.$$

This shows that it is the matrix $e^{A^\dagger t}e^{At}$ that determines the amplification at time t.

Now let us ask, what is the initial condition that optimizes growth over a given time interval $0 \le t \le T$? To answer this, we must maximize $G(T, \vec{X}_0)$ with respect to the initial condition $\vec{X}_0$. For tidiness, we define the matrix

$$E(T) = e^{A^\dagger T}e^{AT}$$

and write the elements of $\vec{X}_0$ as $x_1, x_2, \ldots, x_N$, so that the gain becomes

$$G = \frac{E_{ij}x_i^* x_j}{x_k^* x_k}. \tag{11.11}$$

We now take the derivative of G with respect to a generic element x_ℓ^*, resulting in

$$\frac{\partial G}{\partial x_\ell^*} = \frac{E_{\ell j}x_j}{x_k^* x_k} - \frac{E_{ij}x_i^* x_j}{(x_k^* x_k)^2}x_\ell = \frac{1}{x_k^* x_k}\left[E_{\ell j}x_j - Gx_\ell\right] = 0. \tag{11.12}$$

(Variations with respect to x_ℓ and x_ℓ^* are independent. Differentiating with respect to x_ℓ gives the complex conjugate of 11.12.)

The quantity in square brackets must be zero, or

$$E\vec{X}_0 = G\vec{X}_0. \tag{11.13}$$

This tells us that the optimal gain G is the largest eigenvalue of E, and $\vec{X}_0$ is the corresponding eigenvector. (Because E is Hermitian, the gain G is guaranteed to be real.)

To sum up, here is the procedure for calculating the maximum growth over time *T*.

(i) Compute the stability matrix A.
(ii) For a target time *T*, compute e^{AT}, e.g., in Matlab using the built-in function `expm(A*T)`.
(iii) Multiply e^{AT} by its Hermitian conjugate to get E.
(iv) Compute the largest eigenvalue of E and its eigenvector $\vec{X}_0$.
(v) If desired, compute the evolution $\vec{X}(t) = e^{At}\vec{X}_0$.

The interested student may want to investigate the method of *singular value decomposition*, described in the appendix to this chapter, which facilitates certain aspects of this calculation.

11.4 Optimizing Growth at $t = 0^+$

As we have seen, the optimal initial condition is a function of the target time *T*. An important special case is the limit $T \to 0$, i.e., the initial condition that grows most rapidly immediately after $t = 0$. This can be found by Taylor expanding $e^{A^\dagger t}e^{At}$ about $t = 0$, i.e.,

$$e^{A^\dagger t}e^{At} = (I + A^\dagger t + \cdots)(I + At + \cdots) = I + (A + A^\dagger)t + O(t^2) \quad (11.14)$$

Substituting this into the amplification factor gives

$$G(t, \vec{X}_0) = 1 + \frac{\langle(A + A^\dagger)\vec{X}_0, \vec{X}_0\rangle}{\langle\vec{X}_0, \vec{X}_0\rangle}t + O(t^2). \quad (11.15)$$

Recalling the definition (11.9), we express the initial exponential growth rate of $|\vec{X}|$ in terms of *G*:

$$\frac{1}{2}\frac{dG}{dt} = \frac{|\vec{X}|}{|\vec{X}_0|^2}\frac{d}{dt}|\vec{X}| \to \frac{1}{|\vec{X}|}\frac{d}{dt}|\vec{X}| \quad (11.16)$$

as $t \to 0$, with $|\vec{X}|^2 = \langle\vec{X}, \vec{X}\rangle$. This shows that the initial growth rate is given by $0.5 dG/dt$. Therefore, taking the derivative of (11.15) we can write

$$\text{Initial growth rate} = \frac{1}{2}\frac{dG}{dt} = \frac{\langle 0.5(A + A^\dagger)\vec{X}_0, \vec{X}_0\rangle}{\langle\vec{X}_0, \vec{X}_0\rangle}. \quad (11.17)$$

Comparing (11.17) with (11.10), the definition of *G*, we see that the problem of maximizing dG/dt at $t = 0$ is isomorphic to the maximization of *G* in section 11.3. The same calculation therefore shows that the maximum initial growth rate is the largest eigenvalue of $(A + A^\dagger)/2$, and the initial condition $\vec{X}_0$ that achieves that growth rate is the corresponding eigenvector.

11.5 Growth at Short and Long Times: a Simple Example

To demonstrate the different choices for optimal amplification of a perturbation we look at the system described by the matrix

$$A = \begin{bmatrix} -1 & -31.8 \\ 0 & -2 \end{bmatrix}. \tag{11.18}$$

This matrix has real decaying eigenvalues $\sigma_{1,2} = -1, -2$, but is non-normal and exhibits transient growth, as shown in Figure 11.3. The evolution of the amplification factor, $G(t)$, is shown for

   (i) the optimal perturbation for initial growth (section 11.4),
  (ii) the optimal perturbation for the target time $T = 0.1$ (section 11.3), and
 (iii) the optimal $G(t)$ possible *at each time*, denoted by $G_{\text{opt}}(t)$, which acts as an upper bound on $G(t)$.

As expected from the preceding discussion, the optimal *initial* growth curve (blue curve on Figure 11.3) is steep initially, exceeding the growth of the optimal for $T = 0.1$ (red curve; see the closeup in the right-hand frame). By $t = 0.03$, however, the optimal curve for $T = 0.1$ has caught up and grows more rapidly from then on.

The overall optimum growth G_{opt} is close to the blue curve near $t = 0$ and close to the red curve near $T = 0.1$. This is not surprising since the initial states for the blue and red curves are optimized for maximum growth over those times.

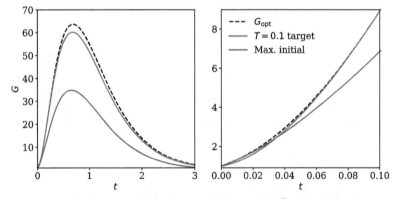

Figure 11.3 Evolution of the amplification factor, $G(t, \vec{X}_0)$, for two different initial conditions in time, corresponding to optimal initial growth (blue), and the optimal amplification at the target time 0.1 (red). The dashed black curve represents the maximum amplification possible at each time, $G_{\text{opt}}(t)$. The right panel is a closeup of the evolution for the time interval $0 \le t \le T$, with $T = 0.1$.

The final case to consider is the maximum possible amplification over all times. This is referred to as the global optimal, and can be seen to occur at $t = 0.7$ (Figure 11.3, left frame). This initial condition is a good candidate to reach large amplitude and possibly trigger a transition to turbulence.

11.6 Example: The Piecewise Shear Layer

A geophysical example that we are able to solve analytically is the piecewise-linear shear layer of section 3.3. Recall that the solution for the vertical velocity eigenfunction can be written as

$$\hat{w}(z) = B_1 e^{-k|z-h|} + B_2 e^{-k|z+h|} \tag{11.19}$$

with the coefficients B_1, B_2 determined from (3.31). The latter can be reformulated as an eigenvalue problem $\sigma \vec{B} = A\vec{B}$, where

$$A_\star = -\iota k_\star \begin{bmatrix} 1 - \frac{1}{2k_\star} & -\frac{e^{-2k_\star}}{2k_\star} \\ \frac{e^{-2k_\star}}{2k_\star} & -1 + \frac{1}{2k_\star} \end{bmatrix}, \tag{11.20}$$

and the shear scaling

$$\hat{w}_\star \equiv \hat{w}/u_0\,, \quad k_\star \equiv kh\,, \quad \sigma_\star \equiv \sigma h/u_0$$

has been applied.

Note that the dispersion relation for the dimensionless eigenvalue problem, $\det(A_\star - \sigma_\star I) = 0$, returns equation (3.32) in dimensionless form, and the results presented in Figure 3.7. The normal modes of the shear layer are unstable for long waves with $0 < k_\star < 0.64$, and are neutral propagating vorticity waves for $k_\star \geqslant 0.64$. Here, we investigate the example $k_\star = 0.2$, for which the eigenvalues are $\sigma_\star = \pm 0.149$. Using the method described in section 11.4 we determine the initial state that grows fastest at $t = 0^+$, then follow its evolution to later times.

The result is shown by the solid curve on Figure 11.4. The dashed lines represent the optimal initial growth rate (steeper line) and the maximum eigenvalue of A (less steep). This exponential growth appears linear because the amplitude is plotted on a log scale. As expected, the perturbation grows at the optimal rate near $t = 0$, then converges to the fastest-growing normal mode as $t \to \infty$.

Note that the optimal perturbation attains the same amplitude as the eigenmode despite its initial amplitude being smaller by about a factor of 2. Conversely, if the two were initialized with the same amplitude, the optimal perturbation would end up bigger by a factor of 2. Along with the change in growth rate comes a change in the structure of the disturbance, which converges over time to the fastest-growing normal mode.

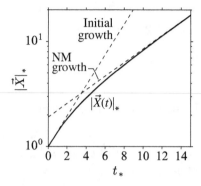

Figure 11.4 Example of large initial growth of a perturbation on the piecewise shear layer with $k_\star = 0.2$. The dashed lines show two curves with different growth rates corresponding to the optimal initial growth and that of the fastest-growing normal mode (NM growth).

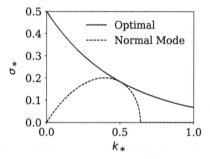

Figure 11.5 Optimal initial growths for the piecewise shear layer (black line). The normal mode growth rates are shown by the dashed line together for comparison. Adapted from Heifetz and Methven (2005).

11.7 Mechanics of Transient Growth in a Shear Layer

Using the method of section 11.4, we may calculate the optimal initial growth for each value of the dimensionless wavenumber, $k_\star$. The result is simply

$$\frac{1}{2}\frac{dG}{dt} = \frac{1}{2}e^{-2k_\star}, \tag{11.21}$$

the right-hand side being the positive eigenvalue of $(\mathbf{A} + \mathbf{A}^\dagger)/2$ with $\mathbf{A}$ given by (11.20). As shown in Figure 11.5, this optimal initial growth rate is a monotonically decreasing function of $k_\star$, and exceeds the growth rate of the fastest-growing normal mode for all $k_\star$ except $k_\star = 0.5$.

In section 3.12 we learned of the wave resonance mechanism of shear layer instability. These same ideas can be used to understand the enhanced initial growth found here. In particular, we learned that there is an optimum phase relationship between the upper and lower vorticity waves to cause mutual amplification, $\Delta\theta = \pi/2$ (as sketched in Figure 3.23). This optimal phase difference is only found for the special case $k_\star = 0.5$, where the phase speeds of the vorticity waves in isolation from each other are equal. For all other unstable $k_\star$, different $\Delta\theta$ are required so that the waves will phase-lock (i.e., bring each other to a stationary state) and thereby undergo sustained growth.[2]

In the present case, however, we are not concerned about sustained growth; we seek only to maximize *instantaneous* growth at $t = 0$. Therefore, phase-locking is not required. In our discussion of wave resonance we derived the dimensionless growth rate (3.88) in terms of $\Delta\theta$:

$$\sigma_\star = \frac{1}{2}\sin(\Delta\theta)e^{-2k_\star}. \tag{11.22}$$

The factor $e^{-2k_\star}$ quantifies the strength of the interaction between the two waves and is greater when $k_\star$ is small. If $\Delta\theta$ is not constrained by the requirement of phase-locking, we can simply set it to $\pi/2$ for all $k_\star$, in which case (11.22) becomes the optimal *initial* growth rate (11.21). Once again, enhanced instantaneous growth is possible for all $k_\star \neq 0.5$. The initial growth rate is largest in the limit $k_\star \to 0$.

Take note – this result poses a challenge to our "rule of thumb" that says instability on a shear layer has wavelength ~ 8 times the layer thickness (section 3.3.3). Since naturally occurring shear instabilities do not have infinite time to grow, (11.21) suggests that longer wavelengths are more likely to attain visible amplitude. Is this true? We don't know; go find out!

Exercise: Review the other rules of thumb that we derived in Chapter 3 based on the fastest-growing normal mode. How do you think these would change if we considered the optimal initial disturbance instead? Is there an analog of Squire's theorem for optimals? How might three-dimensional disturbances grow?

11.8 Generalizing the Inner Product

Throughout this chapter we have used the simple Euclidean inner product to calculate the magnitude of our solution vector $|\vec{X}(t)|$. However, we have either looked only at arbitrary abstract systems or avoided stating exactly what this quantity corresponds to. Here we mention some different choices that are common, and note that this choice affects the results of the transient growth analysis.

In our featured example of a shear layer, we used the magnitude of the coefficient vector $\vec{X} = (B_1, B_2)$, with the B_j given in (11.19). In this case, our inner

---

[2] The exact relationship $\Delta\theta(k_\star)$ is plotted in Figure 3.26.

product $\langle \vec{X}, \vec{X} \rangle = |\vec{X}|^2$ corresponds to the vertical part of the kinetic energy,[3] given explicitly by

$$\int \hat{w}(z, t)^2 dz = \int \hat{w}^* \hat{w} dz \tag{11.24}$$

$$= \int \left[|B_1|^2 \delta(z - h) + |B_2|^2 \delta(z + h) \right] dz \tag{11.25}$$

$$= |B_1|^2 + |B_2|^2. \tag{11.26}$$

Other choices of norms are possible, and these are usually based on physical considerations for each particular problem. Another common choice is an energy norm, which measures the "size" of the perturbation in terms of its total energy. Regardless of the choice of norm, we can switch by a simple transformation so that all of the results in this chapter remain valid.

11.9 Summary

In this chapter we have seen that a more general linear stability analysis is possible when taking into account the evolution of normal modes other than the fastest-growing mode. The results can be summarized as follows.

- Transient growth exceeding the fastest-growing normal mode is possible for systems described by non-normal matrices, which give rise to non-orthogonal eigenvectors. (For normal systems, in contrast, maximum growth is given by the fastest-growing, or least-decaying, normal mode.)
- For any desired target time T, eigen-analysis of the matrix $e^{A^\dagger T} e^{AT}$ gives the optimal amplification factor and the initial condition corresponding to it.
- The fastest-growing initial condition and growth rate can be found from an eigen-analysis of the matrix $(A + A^\dagger)/2$.
- Transient growth in the piecewise shear layer can be understood intuitively from the wave interaction perspective: the requirement of phase-locking is removed.

---

[3] As noted previously, these B_j are proportional to vorticity anomalies associated with the vorticity waves on the edges of the shear layer, and can be written as $B_j \propto \hat{q}_j = -\Delta Q_j \hat{\eta}_j$. Therefore, the amplitude factor $G(t) \propto |\hat{q}_1|^2 + |\hat{q}_2|^2$ corresponds to a quantity called *enstrophy*, which is generally defined as

$$\mathcal{E}(t) = \int q'(z, t)^2 dz. \tag{11.23}$$

In this case we say that we are using the *enstrophy norm*. This is a common alternative to energy when examining the transient growth of shear flows.

11.10 Appendix: Singular Value Decomposition

The procedure for calculating the transient growth of initial conditions can be streamlined through the use of the singular value decomposition (SVD). In fact, this is such a standard method in matrix algebra that it can be accomplished in Matlab with a single command: `svd(A)`. To better understand what is inside the "black box" of the SVD, let us think how we might construct an ideal solution to the problem of optimal growth of $\vec{X}(t)$ without prior knowledge of the SVD.

It is important to keep in mind that we already know a full solution to our problem of determining the time evolution of $\vec{X}(t)$. From (11.2) it is expressed as a linear combination of the eigenvectors, $\vec{\zeta}_j$, of A, along with their amplification factors $e^{\sigma_j t}$, and the coefficients, B_j, required to produce the initial condition $\vec{X}_0$, i.e.,

$$\vec{X}(t) = B_1 \vec{\zeta}_1 e^{\sigma_1 t} + B_2 \vec{\zeta}_2 e^{\sigma_2 t} + \cdots \tag{11.27}$$

This can all be written compactly in matrix form

$$\vec{X}(t) = Z e^{Dt} Z^{-1} \vec{X}_0, \tag{11.28}$$

with the coefficients determined from $\vec{B} = Z^{-1} \vec{X}_0$.

We have learned that the problem with this representation of the solution, for non-normal A, is that the $\vec{\zeta}_j$ are not orthogonal. This leads to amplification factors, $e^{\sigma_j t}$, that are not representative of the growth of the perturbation. Can we find a different representation of $e^{At} = Z e^{Dt} Z^{-1}$, so that the eigenvector columns of Z are orthogonal, and therefore the entries in the diagonal matrix are representative of the perturbation growth? It turns out we can, and this representation is the SVD.

To see this, we first note that the different "representation" of e^{At} that we seek, is in fact a different set of basis vectors for $\vec{X}(t)$ than the eigenvectors of A. This new set of basis vectors will have the desirable property that they will all be orthogonal to each other – exactly what is missing in the $\vec{\zeta}_j$. Such a representation of e^{At} will necessarily have the form of

$$e^{At} = U\Sigma V^{-1}, \tag{11.29}$$

where the columns, $\vec{u}_j$, of U are the basis vectors we are seeking, Σ is a diagonal matrix with amplification factors, Σ_j, of each basis vector $\vec{u}_j$, and V^{-1} has the job of converting $\vec{X}_0$ to the coordinates of our new basis, i.e., $V^{-1}\vec{X}_0$ will play an equivalent role as $\vec{B}$.

Since the $\vec{u}_j$ are orthogonal (and normalized to have unit amplitude) we can write

$$\langle \vec{u}_i, \vec{u}_j \rangle = \vec{u}_i^\dagger \cdot \vec{u}_j = \begin{cases} 0, & i \neq j \\ 1, & i = j \end{cases}. \tag{11.30}$$

In the language of linear algebra we say that U is a *unitary* matrix. An important property of a unitary matrix is that $U^{-1} = U^{\dagger}$. It can be shown that V is also a unitary matrix, so that we can write (11.29) exactly as the SVD, i.e.,

$$e^{At} = U\Sigma V^{\dagger}. \tag{11.31}$$

A useful property of the SVD is that we may order the entries such that the largest amplification factor is first, and given by Σ_1, together with its corresponding basis vector $\vec{u}_1$. It is this combination that we are ultimately after, since we know it provides the maximum amplification (analogous to the fastest-growing mode for a normal matrix). Given what we already know of the SVD, we are now able to come up with a simple recipe to determine all of U, Σ, and V, with the following steps.

(i) Take the right product of (11.31) with V and consider only the first column to give

$$e^{At}\vec{v}_1 = \Sigma_1\vec{u}_1. \tag{11.32}$$

(ii) Then find a similar relationship by performing the left product of (11.31) with $U^{\dagger}$ and then taking the conjugate transpose to give

$$e^{A^{\dagger}t}\vec{u}_1 = \Sigma_1\vec{v}_1. \tag{11.33}$$

(iii) A formula for $\vec{v}_1$ and Σ_1 can then be found by taking the left product of (11.32) with $e^{A^{\dagger}t}$ to give

$$e^{A^{\dagger}t}e^{At}\vec{v}_1 = \Sigma_1 e^{A^{\dagger}t}\vec{u}_1 = \Sigma_1^2\vec{v}_1, \tag{11.34}$$

where the last step comes from using (11.33). This shows that Σ_1^2 is the largest eigenvalue of the matrix $e^{A^{\dagger}t}e^{At}$, and $\vec{v}_1$ is its eigenvector.

(iv) Similarly, we can find $\vec{u}_1$ from taking the left product of (11.33) with e^{At}, and use (11.32) to find

$$e^{At}e^{A^{\dagger}t}\vec{u}_1 = \Sigma_1 e^{At}\vec{v}_1 = \Sigma_1^2\vec{u}_1. \tag{11.35}$$

Our recipe to construct the SVD is complete, and we have the ideal form for expressing the solution $\vec{X}(t)$, in terms of orthogonal basis vectors with ordered, real, amplification factors. This also demonstrates that the SVD can be found directly from the eigen-properties of the matrices $e^{A^{\dagger}t}e^{At}$ and $e^{At}e^{A^{\dagger}t}$. Note that an identical statement was arrived at when we considered a formula for $G(t)$ in (11.10).

11.11 Further Reading

Nice overview papers on transient and optimal growth, which this material has been based on, are Farrell (1996), Trefethen et al. (1993), and Schmid (2007). A full treatment of transient growth and wave interactions in the piecewise shear layer is discussed in Heifetz and Methven (2005). More recent advances are described by Kerswell et al. (2014), Kaminski et al. (2014), and Luchini and Bottaro (2014).

12

Instability and Turbulence

Instability is of interest largely because it creates turbulence. But the relationship between instability and turbulence is complex. To take advantage of the relative simplicity of linear instability theory, we must understand that relationship much better than we do today.

One approach is to relax the assumption of small-amplitude perturbations by numerically solving the fully nonlinear equations. This reveals a sequence of instabilities that lead to turbulence (section 12.1).

Though we think of instability as the "cause" of turbulence, that causality can actually be reversed. In section 12.2 we will explore two mechanisms by which turbulence can create instability.

In observational science, we find that nature rarely cooperates in creating the simple, idealized scenarios that our equations and theorems can describe. Instead, we must take a step back and look at turbulence from a less-rigorous, yet intuitively appealing, perspective such as that described in section 12.3.

12.1 Secondary Instabilities and the Transition to Turbulence

What happens after an unstable mode begins to grow? At some point it becomes large enough that our linearized equations become invalid. Exponential growth is damped. Beyond this time, the disturbance may simply decay, or it may exhibit a secondary instability. Analysis of this evolution is extremely difficult, and is best done using a direct numerical simulation (DNS) of the fully nonlinear equations of motion.

12.1.1 Nonlinear Development

Figure 12.1 shows a sequence of snapshots from DNS of a stratified shear layer. Initially, the minimum Richardson number is 0.08, so the Miles-Howard instability

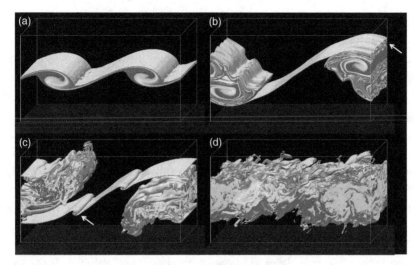

Figure 12.1 Snapshots of the buoyancy field from a nonlinear simulation of an unstable shear layer as it becomes turbulent. (a) Kelvin-Helmholtz instability. (b) Subharmonic pairing instability, convective instability (arrow). (c) Secondary Kelvin-Helmholtz instability (arrow). (d) Turbulence. Parameter values $Ri_b = 0.08$, $Re = 800$, $Pr = 7$. Horizontal boundaries are periodic. Colors range from $-0.6b_0$ to $+0.6b_0$; values outside this range are transparent. (Smyth and Thorpe, 2012)

criterion is easily satisfied. The horizontal boundaries are periodic, with domain length chosen to accommodate two wavelengths of the fastest-growing mode.[1] In the first snapshot (Figure 12.1a), the fastest-growing mode has grown to form a sequence of billows (compare with Figure 4.4, for example). Within each billow is a region where dense fluid overlies buoyant fluid, inviting convective instability.

The next stage is the growth of a subharmonic instability, which causes adjacent billows to pair. The result appears as a single large billow straddling the edge of the domain (Figure 12.1b). This instability is essentially a manifestation of the primary instability with wavenumber equal to half that of the fastest-growing

---

[1] The molecular properties of the modeled fluid are consistent with thermally inhomogeneous water. The kinematic viscosity ν is $1.0 \times 10^{-6} \mathrm{m^2 s^{-1}}$ and the diffusivity κ is $1.4 \times 10^{-7} \mathrm{m^2 s^{-1}}$, so that the Prandtl number is 7. Domain dimensions are $4.19 \times 0.87 \times 2.18$ m. The domain is designed to accommodate two wavelengths of the primary Kelvin-Helmholtz instability and four of the secondary convective instability. Boundary conditions are periodic in the horizontal directions, and free-slip and insulating in the vertical. Initial mean profiles of streamwise velocity and density are chosen to represent a stratified shear layer:

$$\frac{U(z)}{u_0} = -\frac{B(z)}{b_0} = \tanh\frac{z}{h}$$

with half-changes $b_0 = 1.6 \times 10^{-5}$ m s^{-2} in buoyancy and $u_0 = 5.3 \times 10^{-3}$ m s^{-1} in velocity, and half-thickness $h = 0.15$ m. A small perturbation is added to excite primary and secondary instabilities. The minimum Richardson number is Ri_b is 0.08 and the initial Reynolds number Re is 800.

mode. Although its growth rate is relatively small, its large wavelength allows it to grow more or less independently of the fastest-growing mode, and when it reaches large amplitude it simply swallows up the original billows. If Ri is small enough, this process can occur repeatedly as successively larger modes reach finite amplitude. This is an example of an upscale energy cascade, in which kinetic energy is transferred to successively larger-scale motions.

Within the merging billows, we see a row of four convection cells resulting from the convective overturning noted earlier (arrow on figure 12.1b). Note that the cells are oriented in the direction of the mean flow, as predicted by stability theory (homework problem 19, Appendix A). This secondary instability grows very quickly, leading to the development of turbulence within the core of the merged billow (Figure 12.1c).

In the center of the domain, the strain between the large billows compresses the transition layer to form a sharp gradient of buoyancy and velocity. If the gradient Richardson number in that thin layer is small enough, secondary Kelvin-Helmholtz billows develop. The secondary convective and Kelvin-Helmholtz instabilities are examples of a downscale energy cascade, in which energy is transferred to successively smaller motions. The ultimate result of the cascade is that the entire structure breaks down into turbulence (Figure 12.1d).

This is only one possible evolution of a stratified shear layer. With different values of Ri_b, Re, and Pr, different sequences of secondary instabilities are found. In nature, billows often bypass the subharmonic pairing instability and break down directly into smaller-scale motions.

12.1.2 Return to Stability

Instability can be thought of as the means by which an unstable flow adjusts to a stable state. In Figure 12.1d, for example, the transition layer is much thicker than it was initially (Figure 12.1a). If we take horizontal averages of velocity and buoyancy and compute the bulk Richardson number, it ends up greater than its initial value by a similar factor (Figure 12.2). That final value is greater than the critical value 1/4. In other words, the end state consists of a *stable* stratified shear layer plus decaying turbulence. This state of gradual decay persists indefinitely, unless some external force re-accelerates the shear layer. That possibility is the subject of section 12.3.

12.2 Turbulence-Driven Instabilities

In contrast to instability causing turbulence, we now look at two cases where turbulence causes instability.

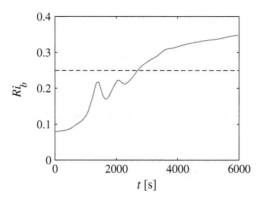

Figure 12.2 Evolution of the bulk Richardson number in the DNS shown in Figure 12.1.

12.2.1 Phillips' Layering Instability

While molecular viscosity and diffusivity can usually be regarded as constant, turbulent viscosity and diffusivity vary in space and time. Because viscosity and diffusivity change in response to changes in the flow, there is a potential for positive feedback and hence instability.

Consider a highly idealized fluid in which

- buoyancy $b(z, t)$ varies with height and (possibly) time, and
- the only motion is in the form of small-scale turbulence, which alters b via a turbulent diffusivity $K(z, t)$.

The evolution of b is governed by the diffusion equation:

$$\frac{\partial b}{\partial t} = -\frac{\partial F}{\partial z}.$$

where F is the vertical buoyancy flux due to turbulence:

$$F = -K\frac{\partial b}{\partial z}.$$

Differentiating with respect to z yields an equation for the buoyancy gradient b_z:

$$\frac{\partial b_z}{\partial t} = \frac{\partial^2}{\partial z^2}(Kb_z), \tag{12.1}$$

where the subscript z indicates the partial derivative, just like $\partial/\partial z$.

Now suppose that the turbulent diffusivity K is determined by b_z, so that $K = K(b_z)$. For example, it is plausible to think that turbulence would be weaker in strongly stratified regions, in which case K would be a decreasing function of

b_z. This creates the possibility of a positive feedback between stratification and turbulence.

Imagine now that b_z consists of a constant plus a small perturbation:

$$b_z = b_{z0} + b_z'(z, t),$$

and approximate $K(b_z)$ by a first-order Taylor series expansion

$$K = K_0 + K^* b_z',$$

where the constant K^* is defined as

$$K^* = \left(\frac{\partial K}{\partial b_z}\right)_{b_z = b_{z0}}. \tag{12.2}$$

If, as we have imagined, turbulence is reduced in regions of stronger stratification, then $K^* < 0$.

We now write the buoyancy flux as

$$F = -K b_z = -(K_0 + K^* b_z')(b_{z0} + b_z') = -K_0 b_{z0} - K_0 b_z' - b_{z0} K^* b_z'.$$

The quadratic term has been neglected as usual. Substituting this into (12.1), we have

$$\frac{\partial b_z'}{\partial t} = \frac{\partial^2}{\partial z^2}\left[K_0 b_{z0} + (K_0 + b_{z0} K^*)b_z'\right] = (K_0 + b_{z0} K^*)\frac{\partial^2 b_z'}{\partial z^2}.$$

The perturbation b_z' is therefore governed by a standard diffusion equation in which the "diffusivity" is $K_0 + b_{z0} K^*$. If K^* is not only negative but less than $-K_0/b_{z0}$, then this effective diffusivity is negative. Buoyancy gradients, rather than being smoothed over time, become sharper.[2] Negative diffusivity can lead to the formation of layers of uniform buoyancy ($b_z = 0$) separated by thin layers in which buoyancy changes rapidly. Step-like features observed in the ocean (e.g., Figure 12.3) may be accounted for by instabilities like this.

12.2.2 Thermohaline Interleaving

Steppy profiles of temperature and salinity like those shown in Figures 9.10 and 9.11 are ubiquitous in the ocean. Here we'll look at a second possible explanation for these features. In the previous section we supposed that the buoyancy flux between adjacent layers was carried by some generic brand of turbulence. A slightly more complicated variant of this instability involves gradients of both temperature and salinity and fluxes driven by turbulent salt fingering (section 9.2.4). As

---

[2] This is like the diffusion of money: the rich get richer, and the poor get poorer.

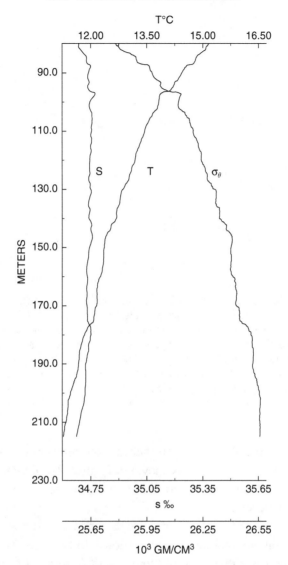

Figure 12.3 Profiles of salinity, temperature, and density measured in the California current (Gregg, 1975). Salinity S is in parts per thousand by mass, temperature T is in degree Celsius. Density is represented as σ_θ = potential density minus 1000 kg/m^3.

salt fingers grow, they eventually break down and become turbulent, but they continue to carry with them an upward flux of buoyancy. As in the previous subsection, variations in that buoyancy flux can create a new instability.

Imagine a fluid stratified by both temperature and salinity so that warm, salty water overlies cool, fresh water, the configuration suitable for salt fingering

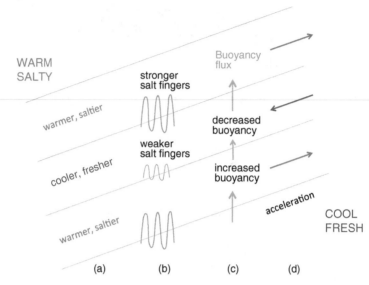

Figure 12.4 Schematic of thermohaline interleaving instability. Overall stratification favors salt fingering, i.e., warm, salty fluid over cool, fresh fluid, but the
gradients also have a horizontal component. Temperature and salinity perturbations (a) have the form of tilted layers. At the layer boundaries, perturbations
alternately enhance and diminish the tendency to form salt fingers (b), along with
the upward buoyancy flux the fingers carry (c). As a result, alternating layers
become more and less buoyant, causing them to slide (d). This motion advects the
temperature and buoyancy fields so as to increase the perturbations.

instability (section 9.2.4). Imagine further that the temperature and salinity gradients both tilt upward to the left as shown in Figure 12.4. For simplicity, we will
also assume that the temperature and salinity gradients balance exactly so that the
density is the same everywhere.

Now imagine a normal mode disturbance that induces in-phase perturbations of
temperature and salinity on tilted layers as shown. In between two such layers, the
stratification becomes either slightly more favorable to salt fingering, or slightly
less so. Like all convective motions, salt fingers carry an upward buoyancy flux,
and that flux is stronger when the salt fingering is stronger.

As a result of the varying buoyancy flux, alternating layers become lighter and
denser, so that gravity pulls them upward to the right or downward to the left. These
motions advect the original temperature and salinity gradients so as to reinforce the
perturbation. This positive feedback leads to exponential growth of the disturbance.

Interleaving is common wherever currents of water with different properties
meet. For example, warm, salty water from the North Atlantic flows into the cold,
fresh Arctic Ocean through Fram Strait (Figure 12.5, red and green curves). The
purple curve shows inflow from the Barents Sea, where the temperature/salinity

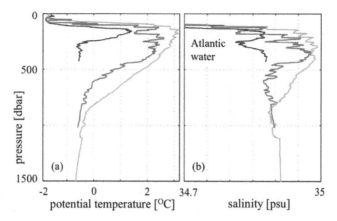

Figure 12.5 Potential temperature (a) and salinity (b) profiles from the eastern Kara Sea slope. Pressure [dbar] is equivalent to depth in meters. Red and green curves show Atlantic water interleaving with Arctic water. Adapted from Rudels et al. (2009).

contrast is reduced and interleaving is less prominent. Interleaving layers in the Arctic are the largest in the world ocean, extending over ~1000 km.

Interleaving can also be driven by diffusive convection (section 9.2.5). In Figure 12.5, interleaving on the upper (lower) edge of the Atlantic layer is driven by diffusive convection (salt fingering). Shear instability can develop on the boundaries between the oppositely moving layers. If the temperature and salinity gradients do not exactly balance, the resulting lateral density gradient can be balanced by the Coriolis force as in Chapter 8 so that the motion takes on some aspects of baroclinic instability.

12.3 Cyclic Instability

In instability theory, we assume that at some initial time the flow takes a very simple, laminar, equilibrium form. The equilibrium is then perturbed, and we compute the initial growth of the perturbation. In real geophysical fluids this specific chain of events is rare. Instabilities observed in nature grow in environments that are already turbulent, thanks to previous instabilities. At large amplitude, instabilities break and generate turbulence whose ultimate effect is to return the mean flow to a stable state.

The hypothesis of cyclic instability suggests that a flow can remain in a state of approximate equilibrium through the combined action of external forcing, which tends to destabilize the flow, and sporadic instabilities, which restore stability. A nice analogy was proposed by Bak et al. (1987): a sandpile is continually steepened by the addition of more sand, but if it becomes too steep, an avalanche will form, reducing the slope to a stable value (Figure 12.6). Through the action of many,

Figure 12.6 A marginally stable, forced-dissipative system. The angle of repose is maintained, on average, by random, small avalanches. Source: iStock/Getty Images

sporadic avalanches, the slope is maintained, on average, at the *angle of repose*. That angle depends on the grade of the sand.

Called self-organized criticality in the physics community, this behavior is characteristic of many forced-dissipative systems. Other examples include earthquakes, forest fires, and solar flares (Aschwanden, 2016). In the case of geophysical turbulence, we imagine that some external force such as the wind acts continually to destabilize the fluid while sporadic turbulent events relieve the instability.

Figure 12.7 shows an example from the eastern equatorial Pacific Ocean. The trade winds blow the surface water to the west, generating the south equatorial current (SEC, Figure 12.7a). These pile up against the Asian coastlines, and the weight of that extra water increases the subsurface pressure. The resulting pressure gradient drives a return flow at around 100 m depth, the equatorial undercurrent (EUC). Shear and stratification both increase to maxima just above the EUC core (Figure 12.7b). Their ratio, Ri (Figure 12.7c), is remarkably uniform over much of this depth range, and is conspicuously close to Miles and Howard's critical value 1/4 (section 4.7). We'll have more to say about this shortly.

A useful measure of turbulence intensity is the rate at which turbulent kinetic energy is dissipated by viscosity, ϵ.[3] In our example from the EUC, turbulence remains strong down to 80 m depth (Figure 12.7d). Because this turbulence coexists with strong stratification, the turbulence is doing a lot of work against gravity by exchanging warm, buoyant water from the surface mixed layer with cold water from the thermocline. This is one reason that the eastern equatorial Pacific is the global maximum of ocean heat uptake.

Now as for the Richardson number being "conspicuously close" to 1/4, what is happening (Figure 12.8) is that forcing from the trade winds acts continuously to increase the shear, and therefore to decrease Ri. But whenever Ri drops below

---

[3] Equation (5.35) gives the linearized form.

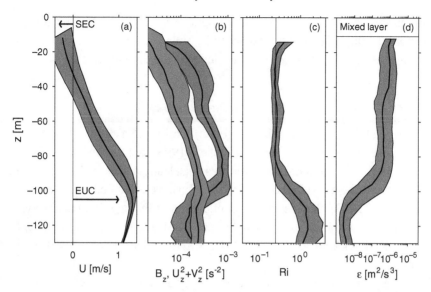

Figure 12.7 Mean flow and turbulence in the upper equatorial Pacific. Two weeks of shipboard observations were taken on the equator at 140 W in October, 2008. Profiles show the median and the quartile range at each depth. (a) Zonal velocity, showing the South Equatorial Current and the Equatorial Undercurrent. (b) Buoyancy gradient B_z (blue) and squared shear (red). (c) Gradient Richardson number with the vertical line at $Ri = 1/4$. (d) Turbulent kinetic energy dissipation rate. See Smyth and Moum (2013) for further details.

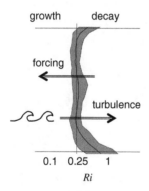

Figure 12.8 Schematic of cyclic instability. The red area shows a Richardson number profile from the upper equatorial Pacific (cf. Figure 12.7c).

1/4, instability generates turbulence, which mixes out the shear, thereby increasing Ri to values greater than $1/4$ (as in section 12.1.2). The ultimate result is that Ri fluctuates more or less randomly around $1/4$.

This is the oceanic analog of the sandpile described earlier. The trade winds correspond to the sand source and sporadic shear instabilities to the avalanches. The "angle of repose" for stratified turbulence is the critical state $Ri = 1/4$.

12.4 Further Reading

- The original discussion of self-induced criticality is Bak et al. (1987). Comprehensive overviews may be found in Jensen (1998) and Aschwanden (2016).
- The sequence of secondary instabilities that leads a Kelvin-Helmholtz billow train to the turbulent state is described in Mashayek and Peltier (2012a,b).
- Posmentier (1977) provides a nice explanation of the layering instability.
- For more information on thermohaline interleaving, try Ruddick and Kerr (2003), Ruddick and Richards (2003), Smyth and Ruddick (2010), or Radko (2013).
- For more on marginally unstable states and cyclic instability, see Thorpe and Liu (2009), Smyth and Moum (2013), or Smyth et al. (2017).

13

Refining the Numerical Methods

The matrix method of numerical stability analysis can become cumbersome when working with large datasets or when very fine resolution is needed. The reason is that the processor time needed to find the eigenvalues of an $N \times N$ matrix increases rapidly with increasing N. Here, we'll explore a few more-refined methods aimed at improving resolution without increasing processing time.

13.1 Higher-Order Finite Differences

The most obvious refinement of our second-order matrix method is to increase the accuracy of the derivative matrices. In section 1.4.3, you derived the second-order, second-derivative matrix with Dirichlet boundary conditions $f_0 = f_{N+1} = 0$:

$$
D^{(2)} = \frac{1}{\Delta^2}
\begin{bmatrix}
-2 & 1 & 0 & 0 & \cdots \\
1 & -2 & 1 & 0 & \cdots \\
 & & \ddots & & \\
\cdots & 0 & 1 & -2 & 1 \\
\cdots & & & 1 & -2
\end{bmatrix}
\tag{13.1}
$$

If we repeat the derivation of this matrix, keeping two more terms in the Taylor series approximations (e.g., 1.7), the result is accurate to fourth order in Δ:

$$
D^{(2)} = \frac{1}{12\Delta^2}
\begin{bmatrix}
-15 & -4 & 14 & -6 & 1 & 0 & \cdots \\
16 & -30 & 16 & -1 & 0 & 0 & \cdots \\
-1 & 16 & -30 & 16 & -1 & 0 & \cdots \\
 & & & \ddots & & & \cdots \\
\cdots & 0 & -1 & 16 & -30 & 16 & -1 \\
\cdots & 0 & 0 & -1 & 16 & -30 & 16 \\
\cdots & 0 & 1 & -6 & 14 & -4 & -15
\end{bmatrix}
\tag{13.2}
$$

273

In principle, one can continue from here to calculate finite difference approximations of arbitrarily high order, but the resulting formulae become very complicated
and results can be contaminated by roundoff error.

It is important to recognize that higher-order approximations are designed to be
more accurate *in the limit as* $\Delta \to 0$. In practice, we make Δ as small as we can
with the available resources, but we often find ourselves using the method when
Δ is not as small as we would like (i.e., compared with the flow features we are
interested in). In that case, *there is no guarantee that a higher-order method will
be more accurate.*

How do these methods compare when used for stability analysis? Consider the
Rayleigh equation (3.16, 3.17), reproduced here for convenience:

$$\sigma \nabla^2 \hat{w} = -\imath k U \nabla^2 \hat{w} + \imath k \frac{d^2 U}{dz^2} \hat{w}; \quad \nabla^2 = \frac{d^2}{dz^2} - \tilde{k}^2. \tag{13.3}$$

We choose $U = \tanh z$ and impose impermeable boundary conditions at $z_1 = -4$
and $z_2 = +4$. The fastest-growing instability is two-dimensional, with wavenumber $k = \tilde{k} = 0.47$, and has (real) growth rate $\sigma = 0.1768316166$. (This very
precise value was determined using the shooting method to be described later in
section 13.4.) The solution method is exactly as described in section 3.5, except
that the second-derivative matrix is now the fourth-order version, (13.2).

In some applications, our aim is to approximate the solution of (13.3) as accu-
rately as possible, while in others we try to attain acceptable accuracy while
minimizing processing time and memory. Therefore, we will compare methods
based on two criteria:

- How quickly does the error go to zero as $N \to \infty$?
- What is the smallest N needed to achieve 1 percent accuracy?

Figure 13.1 shows the relative error in the growth rate for the second- and fourth-
order finite difference methods. The number of grid points ranges from 9 to 513
(illustrated in Figure 13.1a). Not surprisingly, both methods become more accurate
as N is increased. At large N, the error in the second-order method (blue) decreases
like N^{-2}, while the error in the fourth-order method is proportional to N^{-4}. At low
N, however, the order makes much less difference.

To achieve 1 percent accuracy in the growth rate requires $N = 23$ with the
second-order method and only $N = 18$ with the fourth-order method. This may
not seem like much of a difference, but the time needed to compute the eigenvalues is reduced by 40 percent.[1] In the analysis of large geophysical datasets,
the computation must be repeated thousands of times, taking several weeks (on a

---

[1] The time needed to calculate eigenvalues in Matlab is approximately proportional to N^2, and
$(18/23)^2 = 0.6$.

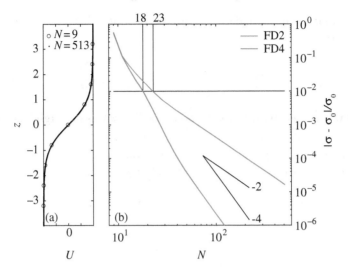

Figure 13.1 (a) Velocity profile $U = \tanh z$ spanned by 9 and 513 grid points. (b) Relative error in the growth rate for second- (blue) and fourth-order (red) finite difference approximations. The problem is specified by the Rayleigh equation (3.19), $U = \tanh z$, $k = 0.47$, and $\hat{w}(-4) = \hat{w}(4) = 0$.

2016-vintage workstation). Therefore, the minor task of upgrading the derivative matrices to fourth order is well justified.

13.2 Finite Differences on an Adaptive Grid

It stands to reason that accuracy could be improved efficiently by focusing resolution where it is needed most, i.e., by letting the grid spacing Δ vary with z such that the spacing is fine in regions where the solution varies most rapidly. In a shear layer, for example, one might make Δ smaller at the center of the layer and larger outside the layer. To do this requires two things: derivative matrices that allow for non-uniform Δ, and a plan for distributing the grid points efficiently.

The first requirement is straightforward. The derivation in section 1.4.2 is easily generalized to allow for variable Δ, though the algebra is a bit more complicated. The second-derivative becomes

$$\tilde{f}_i'' = \frac{2f_{i-1}}{\Delta_{i-1}(\Delta_{i-1} + \Delta_i)} - \frac{2f_i}{\Delta_{i-1}\Delta_i} + \frac{2f_{i+1}}{\Delta_i(\Delta_{i-1} + \Delta_i)}, \tag{13.4}$$

where

$$\Delta_i = z_{i+1} - z_i. \tag{13.5}$$

Check that, if the Δ_i are all the same, this reduces to the familiar approximation for the second-derivative that you derived in homework problem 2.

Incorporation of the boundary conditions is straightforward. In the case $f_0 = f_{N+1} = 0$, the result is

$$\tilde{f}_1'' = -\frac{2f_1}{\Delta_0\Delta_1} + \frac{2f_2}{\Delta_1(\Delta_0 + \Delta_1)}; \quad \tilde{f}_N'' = \frac{2f_{N-1}}{\Delta_{N-1}(\Delta_{N-1} + \Delta_N)} - \frac{2f_N}{\Delta_{N-1}\Delta_N}.$$

(13.6)

Now, how shall we distribute the grid points? There is no "right" answer to this; we are required to guess in advance where the solution will vary most rapidly. Let's suppose that this will happen in regions where the shear is strongest.

We begin with a set of points spaced evenly between the boundaries z_B and z_T:

$$z_i = z_B + i\Delta, \quad \text{where } i = 0, 1, 2, \ldots, N+1$$

and

$$\Delta = \frac{z_T - z_B}{N - 1} > 0.$$

Note this list *includes* the top and bottom points: $z_0 = z_B$ and $z_{N+1} = z_T$.

We now define the absolute shear

$$s_i = \frac{|U_{i+1} - U_i|}{z_{i+1} - z_i}, \quad \text{for } i = 0, 1, 2, \ldots, N$$

(13.7)

and a transformed version

$$\zeta_i = \frac{\ln(s_i/s_{min})}{\ln(s_{max}/s_{min})}.$$

(13.8)

The latter variable approaches 0 and 1 when the absolute shear is at its minimum and maximum values, s_{min} and s_{max}, respectively.[2] We now define the new grid increment

$$\delta_i = \frac{z_T - z_B}{\sum\limits_{j=0}^{N}(1 - a\zeta_j)}(1 - a\zeta_i), \quad i = 0, 1, 2, \ldots, N,$$

where a is a constant such that $0 \le a < 1$. If $a = 0$, $\delta_i = \Delta$, i.e., the original uniform grid is recovered (cf. 13.2). As $a \to 1$, the stretching becomes extreme.

The new grid points are

$$\xi_0 = z_B; \quad \xi_j = z_B + \sum\limits_{i=0}^{j-1}\delta_i; \quad j = 1, 2, \ldots, N+1.$$

Results for the hyperbolic tangent shear layer are shown in Figure 13.2. With $a = 0$, the second-order result from Figure 13.1 is reproduced (blue curve). With

---

[2] When coding this, it is advisable to increment s_{min} with a tiny value, e.g., 10^{-16}, to avoid the possibility of dividing by zero.

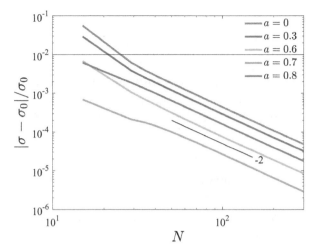

Figure 13.2 Relative error in the growth rate for second-order finite difference approximations using various degrees of grid stretching. The problem is specified by the Rayleigh equation (3.19), $U^\star = \tanh z^\star$, $k^\star = 0.47$, and $\hat{w}(-4) = \hat{w}(4) = 0$. Calculation courtesy of Qiang Lian.

nonzero a, the error decreases, the most accurate choice being $a = 0.7$ (yellow curve). Increasing a further leads to reduced accuracy as resolution becomes inadequate at the edges of the shear layer. As $N \to \infty$, the error decreases as N^{-2} in all cases.[3]

Note that our choice of the grid adaptation algorithm (13.7, 13.8) is based on the assumption that shear is the primary factor determining the resolution requirement at a given location. In more general situations where properties other than shear become important, some creative trial-and-error may be called for.

13.3 Galerkin Methods

Finite difference methods are *local*, in the sense that the derivative at each point is approximated using nearby points only. In a *global* method, derivatives are approximated using functions that span the entire domain. Global methods are intrinsically more complicated, but the extra effort is often repaid in fast, accurate results. An example is the Galerkin methods, two of which are described in the next two subsections.

In the Galerkin technique, we expand the solution in terms of a set of $\mathcal{N}$ orthogonal basis functions, i.e.,

---

[3] This result is somewhat surprising because (13.4) is formally accurate only to first-order. See Ferziger and Peric (1999) for further discussion.

$$\hat{w}(z) = \sum_{n=1}^{\mathcal{N}} w_n F_n(z).$$

(13.9)

If the basis functions F_n are chosen properly,[4] the result becomes exact in the limit $\mathcal{N} \to \infty$. A common example is the Fourier sine series:

$$F_n(z) = \sin\frac{n\pi}{H}z ; \quad n = 1, 2, \cdots \mathcal{N}.$$

(13.10)

In practice, we choose the largest value of $\mathcal{N}$ that our resources allow and hope that the result will be accurate enough for the purpose. The examples given below show that this approach can be spectacularly successful if $\mathcal{N}$ is sufficiently large.

In general the basis functions $F_n(z)$ must satisfy three criteria.

(i) Each basis function must obey the boundary conditions. For example, the set (13.10) obeys the conditions $\hat{w} = 0$ at $z = 0$ and $z = H$.

(ii) The basis functions must obey an *orthogonality relation*, meaning

$$\int_{z_1}^{z_2} W(z) F_m(z) F_n(z) dz = \delta_{mn},$$

(13.11)

where $z = z_1$ and $z = z_2$ are the boundaries (which may be at infinity) and $W(z)$ is a weighting function. In the Fourier example (13.10), $z_1 = 0$, $z_2 = H$, and the weighting function is a constant, $W = 2/H$.

(iii) The basis functions must form a *complete set* with respect to the chosen boundary conditions, meaning that any smooth function satisfying the boundary conditions can be approximated to arbitrary precision by making $\mathcal{N}$ sufficiently large.

There are many alternative choices, e.g., cosine series, Hermite polynomials, Chebyshev functions, ..., which can be found in most applied math references (e.g., Spiegel, 1968). One may also invent basis functions for a particular problem. An example is the *hydrostatic normal modes* for a particular buoyancy profile (Gill, 1982; McWilliams, 2006).

Having chosen our basis functions, we substitute (13.9) into the equation we want to solve. Finally, we multiply through by $W(z)F_m(z)$ and integrate over the domain. If we've done everything right, the result is an algebraic eigenvalue problem for the coefficients w_m.

---

[4] Specifically, the basis functions must form a *complete set* over the space of functions satisfying the boundary conditions.

As an example, we'll solve the Rayleigh equation (13.3) subject to $\hat{w}(0) = \hat{w}(H) = 0$ using the Fourier basis functions (13.10). This is called the Fourier-Galerkin (or just Fourier) method. Substituting (13.9) into (13.3) gives

$$\sigma \sum_{n=1}^{\mathcal{N}} w_n \nabla^2 F_n = -\iota k U \sum_{n=1}^{\mathcal{N}} w_n \nabla^2 F_n + \iota k U'' \sum_{n=1}^{\mathcal{N}} w_n F_n. \tag{13.12}$$

where primes indicate differentiation with respect to z. For the Fourier basis functions (13.10), we can simplify using the fact that $F_n'' = -(n\pi/H)^2 F_n$, and therefore the Laplacian $\nabla^2 F_n$ becomes $D_n F_n$, where

$$D_n = -\left(\frac{n\pi}{H}\right)^2 - \tilde{k}^2. \tag{13.13}$$

We now multiply through by $(2/H)F_m$ and integrate, remembering (13.11):

$$\sigma \sum_{n=1}^{\mathcal{N}} w_n D_n \underbrace{\frac{2}{H} \int_0^H F_m F_n dz}_{=\delta_{mn}} = -\iota k \sum_{n=1}^{\mathcal{N}} w_n D_n \frac{2}{H} \int_0^H F_m U F_n dz$$

$$+ \iota k \sum_{n=1}^{\mathcal{N}} w_n \frac{2}{H} \int_0^H F_m U'' F_n dz.$$

The summation on the left-hand side can be done explicitly, resulting in $w_m D_m$. Dividing through by D_m then gives the algebraic eigenvalue problem:

$$\sigma w_m = A_{mn} w_n, \tag{13.14}$$

where

$$A_{mn} = -\iota k \frac{D_n}{D_m} \int_0^H F_m U F_n dz + \iota k \frac{1}{D_m} \int_0^H F_m U'' F_n dz. \tag{13.15}$$

As in previous methods, the eigenvalue is the growth rate σ, but the eigenvector is now composed of the coefficients w_n in the expansion (13.9). When coding the Galerkin method, note that the two integrals appearing in (13.15) are independent of the wavenumber. When looping over many values of the wavenumber, those integrals only have to be computed once.

We have some freedom in choosing the number of basis functions, $\mathcal{N}$. In the Fourier case, the smallest wavelength resolvable with grid spacing Δ is 2Δ (the Nyquist wavelength, e.g., Harris and Stocker, 1998), so $\mathcal{N}$ should be at least $N/2$. Here, we choose $\mathcal{N} = N$.

In Figure 13.3, we compare the Fourier-Galerkin technique with the second- and fourth-order finite difference techniques for two idealized velocity profiles. The first case shown is the hyperbolic tangent shear layer (Figure 13.3a). The finite

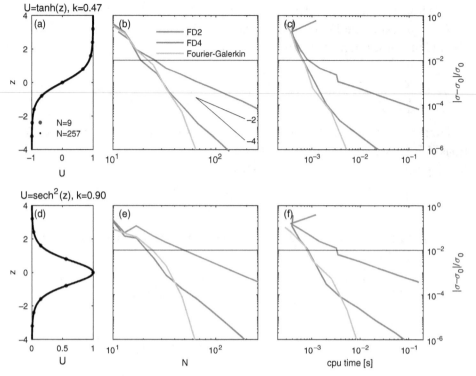

Figure 13.3 Comparisons of Fourier-Galerkin and finite difference solutions of the Rayleigh equation for the hyperbolic tangent shear layer with $k = 0.47$ (a,b) and the Bickley jet with $k = 0.9$ (c,d). Left panels show the velocity profile with the minimum and maximum grid spacings tested. Middle panels show the relative error in the growth rate versus the number of grid points. The right panels show relative error versus processing time (numerical values are specific to the computer used, but relative values should be general). The "exact" growth rate σ_0 is computed using a high-precision shooting method (section 13.4).

difference results (blue and red curves on Figure 13.3b) reproduce Figure 13.1. The Fourier-Galerkin result is shown in yellow.

For N greater than about 40, the Fourier method converges much more rapidly than the finite difference methods. Based on this, the Fourier method is preferred if we seek the most accurate results possible and computation time is not a consideration. If we only need 1 percent accuracy, the required number of grid points is intermediate between the two finite difference methods. The second case tested is the Bickley jet (Figure 13.3d,e,f). The resolution requirement is somewhat more stringent than in the tanh case because the profile is more sharply curved. The 1 percent error tolerance requires $N = 46$ for the second-order method, about $N = 20$ for either the fourth-order or the Fourier method. Either of the more sophisticated methods reduces the computation time by about a factor of four. In both cases, the extra coding required to implement the Fourier method is justified only if one requires relative errors less than about 10^{-4}.

13.4 The Shooting Method

The shooting method is fundamentally different from the matrix methods considered so far. Its main advantage is precision: it can deliver a much more accurate result for a given amount of processing time. The main disadvantage is reliability: there is no guarantee that all modes, or even the fastest-growing mode, will be found. Considerable ingenuity is needed to manage this problem, and the resulting codes tend to be quite complicated. The latter issue is the reason we focus on matrix methods in this introductory-level book.

Aptly named, the shooting method is analogous to an aiming method used by artillery gunners to hit a distant target. They begin with an initial "test" shot and note how much they miss by. With this information, they adjust their aim and try again, repeating the process until the target is hit.

To solve a differential eigenvalue problem via the shooting method, you begin with an initial guess for the eigenvalue. You then use the boundary conditions to specify the solution at one boundary (call it the near boundary), and integrate the differential equation from there to the other (far) boundary. In general, the boundary condition at the far boundary will not be satisfied, but the size of the mismatch tells you how to adjust your estimate of the eigenvalue. You then try again, repeating the procedure until the far boundary condition is satisfied.

13.4.1 A Simple Illustration

Consider the Rayleigh equation for 2D modes:

$$\hat{w}_{zz} = \left(\frac{U_{zz}}{U - c} + k^2 \right) \hat{w}, \qquad (13.16)$$

where the subscript z indicates differentiation. We choose $U = \tanh z$, $k = 0.47$, and impermeable boundary conditions at $z_1 = -4$ and $z_2 = +4$. It remains only to find the value (or values) of c that permit satisfaction of both boundary conditions. The results shown in Figure 13.4 were obtained via the following sequence of steps.

(i) Start with an initial guess for c, which we'll call c_1. Cheating shamelessly, we use results from section 3.9.1 to "assume" that c is purely imaginary, and choose $c = 0.44\iota$.

(ii) Begin the integration at the near boundary, $z_1 = -4$. Because the differential equation is second order, we must specify both $\hat{w}(z_1)$ and $\hat{w}_z(z_1)$, in accordance with the Dirichlet boundary conditions, $\hat{w}(z_1) = 0$.

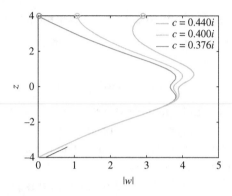

Figure 13.4 Three example integrations of (13.16), with $U = \tanh z$ and $k = 0.47$ and different estimates of c. For simplicity we assume that c is purely imaginary. Integration begins at the lower boundary with $\hat{w} = 0$ and $\hat{w}_z = 1$ (indicated by short black line). Successive attempts converge to a solution that satisfies the upper boundary condition.

(iii) How do we specify $\hat{w}_z(z_1)$? We know that it can't be zero, or the integration would yield the null solution $\hat{w}(z) = 0$. What value should we use? It actually doesn't matter. Because the equation is homogeneous, the solution is defined only up to a multiplicative constant. Therefore, set $\hat{w}_z(z_1)$ to some arbitrary (nonzero) value, e.g., $\hat{w}_z(z_1) = 1$.

(iv) Integrate the equation from z_1 to z_2 using an initial value solver such as Matlab's ode45. The result is the green curve on Figure 13.4.

(v) Unless we're very lucky, the value of $\hat{w}$ at the far boundary z_2 will not be zero. We'll call that value

$$M(c_1) = \hat{w}(z_2). \tag{13.17}$$

In this case $M(c_1) = 2.9$ (green circle).

(vi) Now try again with a different value of c, say c_2, integrate again, and calculate $M(c_2)$. Here we choose $c_2 = 0.40\iota$, giving us the yellow curve and $M(c_2) = 1.1$. We're headed in the right direction!

(vii) Based on this information, continue refining the guess until $M = 0$ to within some predefined tolerance. For the example shown, trial and error leads us to the final choice $c = 0.376\iota$, which gives a value of M very close to zero (blue curve).

13.4.2 The General Method

For the example above, we allowed ourselves the huge advantage of assuming in advance that c is purely imaginary. In general, c is complex, and so is $M(c)$. A

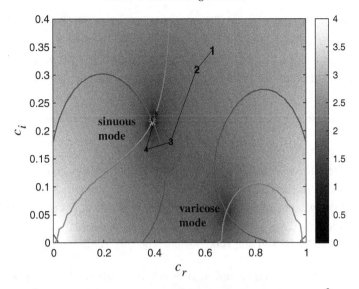

Figure 13.5 The function $M(c)$ for the Bickley jet profile $U = \text{sech}^2 z$ with $k = 0.7$ and $\hat{w} = 0$ at $z = \pm 5$. Shading indicates the absolute value $|M|$. The contours $M_r = 0$ and $M_i = 0$ are shown in blue and red, respectively. Numerals indicate a sequence of guesses for the eigenvalue c leading to $c = 0.3932 + 0.2164\iota$ (green asterisk).

solution is valid only if both the real and imaginary parts of M are zero. The most straightforward approach is to make a contour plot such as that shown in Figure 13.5. In this case we have used the Bickley jet profile $U = \text{sech}^2 z$ with impermeable boundaries at $z = \pm 5$. The wavenumber k is set to 0.7.

The complex function M is represented in two ways. Gray shading indicates $|M|$, with dark regions showing near-zero values. The real part of M is zero on the blue contour; the imaginary part is zero on the red contour. The requirement $M = 0$ is satisfied wherever the red and blue contours cross.

There are two such intersections, corresponding to the sinuous and varicose modes of the Bickley jet (see section 3.9.2). If you have time, you can simply read the values of c_r and c_i from the graph. It is usually preferable to automate that procedure, though. The simplest method is linear extrapolation:

(i) Make two initial guesses. In this case, we have chosen $c_1 = 0.6311 + 0.3437\iota$ and $c_2 = 0.5680 + 0.3094\iota$, shown on Figure 13.5 by the numerals 1 and 2. The first value was chosen at random; the second is 0.9 times the first.

(ii) Solve the equation for each of these values and denote the results M_1 and M_2.

(iii) Compute the coefficients of a linear function passing through (c_1, M_1) and (c_2, M_2), then find the value of c at which the linear function is zero:

$$c = c_1 - \frac{c_2 - c_1}{M_2 - M_1} M_1. \tag{13.18}$$

Use this as the next guess. The result is shown by the numeral 3 on Figure 13.5.

(iv) Iterate the procedure, each time using the most recent two results to define the linear function, until $|M|$ converges. In this case the convergence criterion was

$$\left| \frac{M_2 - M_1}{M_1} \right| < 10^{-6}. \tag{13.19}$$

This criterion was satisfied at the point shown by the green asterisk, which corresponds to the sinuous mode.[5]

(v) To find the varicose mode, we would repeat this process until a distinct, second mode was identified. (The varicose mode is visible on Figure 13.5 as the intersection of the red and blue curves near $c = 0.68 + 0.05\iota$.)

The shooting method is an example of *nonlinear root-finding*, a notoriously difficult class of problems. Everything depends on the initial guess, and there is usually no way to guarantee that all roots have been found. For the complicated background flows that one encounters in nature, the function $M(c)$ can be much more complicated than that shown in Figure 13.5. The absolute value $|M|$ can have numerous peaks with valleys meandering fractally around them. Occasionally, the bottom of one of those valleys will touch zero, and there lies an eigenvalue. In this case, success depends entirely on the accuracy of the initial guess. It may be necessary to try many initial guesses to be reasonably confident of finding all unstable modes, and even then we cannot be certain that every mode, or even the fastest-growing mode, has been found.[6]

A few fine points:

- It helps to put some thought into the initial guesses. In the example above, our two initial guesses were guided by Howard's semicircle theorem (section 4.8).

---

[5] To compute the highly accurate results used earlier to test the various matrix methods, this convergence value was reduced to 10^{-12}.

[6] A blind man sets off to climb Mount Everest. "But how," a friend asks him, "will you find your way?" "Easy", he replies. "The summit of Mount Everest is the highest place on Earth, so I'll just keep walking uphill until I get there!"

The flaw in this plan, of course, is that walking continually uphill will get him to the top of the nearest hill, which is probably not Mount Everest. The shooting method works in the same way: you start with an initial guess for the eigenvalue, then use some algorithm to refine it until it attains some chosen level of accuracy. You might end up at the fastest-growing mode, but you might end up at some other mode, or even no mode at all – it all depends on the accuracy of your initial guess.

Now suppose that, instead of one blind man, 1,000 blind men set off to climb Mount Everest, each starting from some random location on Earth's surface. The odds of one of them reaching the goal would be improved (though still not very good). This is what we do with the shooting method. We make many initial guesses, refine each one, pick the ones most relevant to our problem (e.g., the largest growth rate), and hope for the best.

Because $U(z)$ varies between 0 and 1, unstable modes must have $0 < c_r < 1$ and $0 < c_i < 0.5$ or, more precisely, $(c_r - 1/2)^2 + c_i^2 < 1/4$; $c_i > 0$.

- The linear extrapolation described in step 3 can be improved by raising the order: fit a quadratic function to the most recent three estimates and find the nearest root (there will be two).
- A common problem is that several initial guesses will converge to the same root – a waste of computer time. This can be avoided by redefining the function M. Once a particular root has been found, say $c^{(1)}$, redefine $M(c)$ as $M(c)/|c-c^{(1)}|$. This effectively causes the search algorithm to avoid the known root. As further roots are found, M is redefined in the same way.
- Even a good initial value solver like Matlab's ode45 can accumulate roundoff error, which grows exponentially and renders the results meaningless. This can be avoided by means of *multiple shooting*. The integration is done in two parts: one starting from each boundary and meeting in the middle. We'll call that central point $z_m = (z_1 + z_2)/2$. The criterion for a solution is that $\hat{w}$ and its derivative be continuous at z_m. We define the matching function

$$M(c) = \hat{w}(z_m^+)\hat{w}_z(z_m^-) - \hat{w}(z_m^-)\hat{w}_z(z_m^+), \qquad (13.20)$$

where z_m^+ and z_m^- indicate the middle point approached from above and below, respectively. Remembering that the solution is defined only up to a multiplicative constant, you can persuade yourself that continuity is achieved if and only if $M = 0$. The rest of the analysis proceeds as before.
- Asymptotic boundary conditions can be applied using the multiple shooting method. At the upper boundary the asymptotic condition is $\hat{w}_z = -k\hat{w}$. So we begin the integration with the conditions $\hat{w} = 1$, $\hat{w}_z = -k$, and similarly for the integration from the lower boundary. An advantage of the shooting method is that asymptotic boundary conditions can also be used in the analysis of stratified and/or viscous flows. In those cases, the matrix method doesn't work because the asymptotic boundary condition depends on the growth rate (or the phase speed), and the problem therefore does not reduce to an algebraic eigenvalue problem. Exercise: Demonstrate this by deriving asymptotic boundary conditions for the Taylor-Goldstein equation (4.18), approximated by second-order finite differences.

13.5 Generalizations

All methods discussed here have been illustrated using the simplest nontrivial problem: inviscid, homogeneous, parallel shear flows as described by Rayleigh's equation. It is straightforward (though not necessarily easy) to generalize any or all of these methods to more complicated situations, e.g., stratified flows as described by the Taylor-Goldstein equation (5.15) or viscous flows as described by the Orr-Sommerfeld equation (4.18).

13.6 Further Reading

- A thorough discussion of finite difference methods may be found in chapter 3 of Ferziger and Peric (1999).
- The failure of higher-order finite difference methods to improve accuracy has been seen in the analysis of observed alongshore currents (Putrevu and Svendsen, 1992).
- Hazel (1972) is a classic example of the shooting method applied to simple flows.
- Sun et al. (1998) and Smyth et al. (2013) describe the application of shooting methods and finite difference matrix methods in the analysis of observational data.
- Orszag (1971) is a classic demonstration of the Galerkin-Chebyshev method applied to the plane Poiseuille problem (section 5.8.2).
- Rees and Monahan (2014) describe a modern version of the shooting method applied to the Taylor-Goldstein equation.
- As mentioned several times by now, Spiegel (1968) is an extremely useful (and economical) reference for all facets of applied mathematics.

Appendix A

Homework Exercises

1. **First-derivative matrix**

   Let

   $$f'_1 = A f_1 + B f_2 + C f_3$$
   $$f'_N = A f_N + B f_{N-1} + C f_{N-2}$$

   (where the constants A, B, and C have different values in each formula).

   (a) Find expressions for the constants in each formula so that the error is proportional to Δ^2.

   (b) Use your results from part (a) to define a matrix D such that

   $$f'_i = D_{ij} f_j ; \quad i = 1, 2, \ldots, N.$$

   (c) Type in the Matlab function ddz printed below. Verify that it corresponds to the finite difference approximation to the first-derivative that you defined in part (b). Then type in the script ddz_err (in a separate m-file). This script tests the accuracy of ddz for a given function ($f = z^5$ in this case). Run the script to demonstrate that the error is second order in Δ.

   (d) Try it with a few other functions to see if the result is generally valid. (Two is enough.) Now try it for the case $f = z^2$. Can you make sense of the result?

2. **Second-derivative matrix**

   Repeat the analyses above for the second-derivative. To begin with, assume that:

   $$f''_i = A f_{i-1} + B f_i + C f_{i+1} ; \quad i = 1, 2, \ldots, N$$
   $$f''_1 = A f_1 + B f_2 + C f_3 + D f_4$$
   $$f''_N = A f_N + B f_{N-1} + C f_{N-2} + D f_{N-3}$$

(a) Find expressions for the constants A, B ... in each formula so that the error is proportional to Δ^2 .

(b) Write a Matlab function called ddz2, similar to ddz, that computes a second-derivative matrix using your results from part (a). Modify the script ddz_test so that it computes the second-derivative using your function ddz2 and tests its accuracy. Demonstrate that your approximation is accurate to second order.

3. **Differential Eigenvalue Problem**

(a) Analytically determine the values of the constant λ for which the following boundary value problem has solutions:

$$f'' = \lambda f ; \quad f(0) = f(\pi) = 0. \tag{A.1}$$

(b) Now do the same thing numerically. Start by defining a vector of equally spaced z values z_i; $\quad i = 1, 2, \ldots, N$, such that z_0 and z_{N+1}, if they were included, would be equal to 0 and π. Use your subroutine ddz2 to compute the second-derivative matrix for z , then replace the top and bottom rows so as to be consistent with the boundary conditions $f_0 = f_{N+1} = 0$. Set $N = 10$. The eigenvalues of your matrix should now correspond to the values of λ that you found in part (a) (at least inasmuch as the finite difference derivative you derived is accurate). Check this by using the Matlab routine eig to find the eigenvalues, then the routine sort to sort them from smallest to largest. Plot the eigenvectors corresponding to the smallest and largest[1] eigenvalues. You should find that the former is a smooth, well-resolved function, whereas the latter has a lot of poorly resolved small-scale structure. Correspondingly, the smallest eigenvalues should match the analytical solution closely, whereas the largest will not.

[At the end of this assignment is a sample script that you can use as you wish.]

(c) Repeat part (b) using one-sided derivatives for the top and bottom rows instead of boundary conditions. What difference does this make to the result?

Matlab Code

```
%%%%%%%%%%%%%%%%%%%%%%%%%%%%%%%%%%%%%%%%%%
function d=ddz(z)
% First derivative matrix for independent variable z.
```

---

[1] In absolute value.

```
% 2nd order centered differences.
% Use one-sided derivatives at boundaries.

% check for equal spacing
if abs(std(diff(z))/mean(diff(z)))>.000001
    disp(['ddz: values not evenly spaced!'])
    d=NaN;
    return
end

del=z(2)-z(1);N=length(z);

d=zeros(N,N);
for n=2:N-1
    d(n,n-1)=-1.;
    d(n,n+1)=1.;
end
d(1,1)=-3;d(1,2)=4;d(1,3)=-1.;
d(N,N)=3;d(N,N-1)=-4;d(N,N-2)=1;
d=d/(2*del);
return
end

%%%%%%%%%%%%%%%%%%%%%%%%%%%%%%%%%%%%%%%
% Script ddz_err
% Example script for OC680 Hmwk #1.
% This script tests the first-derivative matrix computed in ddz.
% The result
% shows that the method is 2nd order in the time step grid
% increment.
% The assignment is to do the same for the second-derivative.
NN=[10:10:100];
% compute error at each N
for i=1:length(NN);
    N=NN(i);
    % 0<z<1
    del(i)=1/N;
    z=[0:1:N-1]'*del(i);
    % specify test function f(z) and its (exact)
    % derivative fp(z)
    f=z.^5;
    fp=5*z.^4;
```

```
    d=ddz(z);          % compute derivative matrix
    df=d*f;            % compute finite-difference approximation
                       % to the derivative
    err(i)=sqrt(mean((fp-df).^2)); % compute root-mean-squared
                                   % error
end

% plot error vs. N
figure
loglog(del,err,'*')
xlabel('\Delta')
ylabel('ERROR')
hold on

% regress to find power law and plot
p=polyfit(log(del),log(err),1)
err_th=exp(p(2))*del.^p(1);
plot(del,err_th,'-')
ttle=sprintf('ERROR = %.2f\\Delta^{ %.2f}',exp(p(2)),p(1))
title(ttle)

%%%%%%%%%%%%%%%%%%%%%%%%%%%%%%%%%%%%%%%%
% HMWK 1 Part 3
%
clear
close all

% define z values
N=10;
z=pi*[1:N]'/(N+1);

% compute derivative matrix
d=ddz2(z);
dz=z(2)-z(1);

% To use 1-sided derivatives, comment out the next two lines.
d(N,:)=0;d(N,N-1)=1/dz^2;d(N,N)=-2/dz^2;
d(1,:)=0;d(1,1)=-2/dz^2;d(1,2)=1/dz^2;

% compute eigvals & eigvecs
[v ee]=eig(d);e=diag(ee);
```

```
% sort
[~,ind] =sort(abs(e),'ascend');
e=e(ind);v=v(:,ind);

% Plot eigenvalues /i^2
% If the eigfn is well-resolved, this will be close to -1.
figure
plot([1:N],e./[1:N].^2','*','markersize',10)
xlabel('i');
ylabel('\lambda_i / i^2')
title('Is \lambda_i / i^2 = -1 ?')

% Plot first and last eigvecs.
% The first is well-resolved, and its eigval is close to -i^2.
% The last is poorly-resolved, and the eigval is not close
% to -i^2.
figure
subplot(1,2,1)
plot(v(:,1),z,'b*'); hold on
plot(v(:,1),z,'b')
ylabel('z')
title('First eigvec (smallest abs(eigval))')
subplot(1,2,2)
plot(v(:,end),z,'r*'); hold on
plot(v(:,end),z,'r')
title('Last eigvec (largest abs(eigval))')
```

4. **Benard Convection**
   (a) Given

   $$\sigma^2 + (\nu + \kappa)K^2\sigma + \nu\kappa K^4 + B_z\cos^2\theta = 0 \qquad (A.2)$$

   as derived in class, show that

   $$\frac{\partial\sigma}{\partial\cos^2\theta} > 0.$$

   [Hint: You don't have to solve the quadratic equation to do this. Just differentiate each term.] What assumptions do you have to make about σ and B_z for this to be true? Write a brief (one-sentence) justification for each assumption. At what value of $\cos^2\theta$ will σ be greatest (if all other parameters are fixed)?

   (b) Minimize the function $(\tilde{k}^{*2} + n^2\pi^2)^3/\tilde{k}^{*2}$ with respect to $\tilde{k}^{*2}$. Give both the minimum value, and the value of $\tilde{k}^{*2}$ at which the minimum occurs, as functions of n. Show that the critical Rayleigh number is 657.5.

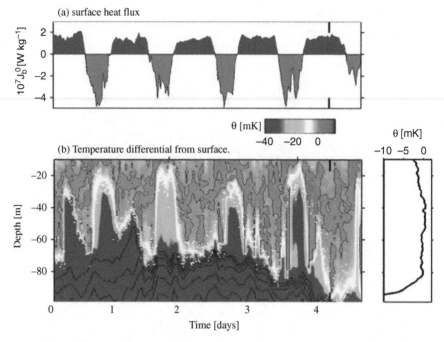

Figure A.1 The diurnal cycle of upper ocean convection (courtesy J. Moum). For reference, a temperature change $\Delta T = 2\mathrm{mK}$ is equivalent to a relative density change $\Delta\rho/\rho = 10^{-6}$.

5. **A Convective Mixed Layer**

   Suppose that nocturnal convection in the upper ocean is driven by a density difference $\Delta\rho/\rho_0 = 10^{-6}$ over the upper 40 m (as in Figure A.1). Compute the Rayleigh number, using the following values:

   $$\nu = 1.0 \times 10^{-6}\mathrm{m}^2\mathrm{s}^{-1}$$
   $$\kappa = 1.4 \times 10^{-7}\mathrm{m}^2\mathrm{s}^{-1}$$
   $$g = 9.81\mathrm{ms}^{-2}.$$

   Plot $\sigma(\tilde{k})$ for these parameter values. What is the horizontal wavelength $(2\pi/\tilde{k})$ of the fastest-growing instability? What are its growth rate and e-folding time? Give the e-folding time in hours, and compare it with the length of time over which convective conditions persist (say 12 hours). By what factor would the amplitude of this instability grow during that time?

6. **An Unstable Layer in an Inviscid Fluid**

   In a fluid with $\nu = \kappa = 0$, suppose that the mean buoyancy gradient has the following profile:

   $$B_z = B_{z0}(1 - 2\mathrm{sech}^2\alpha z). \tag{A.3}$$

Sketch this function and show that the fluid is stably stratified except for an unstable layer surrounding $z = 0$.

Solve (2.29) with boundary conditions $\hat{w} \to 0$ as $z \to \pm\infty$ for the special case $\tilde{k} = \alpha$. (Hint: try $\hat{w} = \text{sech}^2 \beta z$, where β is a constant to be determined.) In a later project you will solve this numerically for a full range of $\tilde{k}$.

Note: Hyperbolic functions provide a useful model for simple shear flows. Here are a couple of useful properties:

$$\frac{d}{dx}\tanh x = \text{sech}^2 x = 1 - \tanh^2 x.$$

7. **Numerical Analysis of Shear Instability**

   (a) Write a Matlab function to find eigenvalues σ and eigenfunctions $\hat{w}$ of the Rayleigh equation in finite difference form:

   $$\sigma A_{ij}\hat{w}_j = B_{ij}\hat{w}_j \tag{A.4}$$

   for $i, j = 1, 2, \ldots, N$ with boundary conditions

   $$\hat{w}_0 = \hat{w}_{N+1} = 0. \tag{A.5}$$

   The matrices A and B are defined by

   $$A_{ij} = D_{ij} - \tilde{k}^2 I_{ij}$$
   $$B_{ij} = -ik(U_i A_{ij} - U_i'' I_{ij}) \tag{A.6}$$

   where D is the second-derivative matrix [including the boundary conditions (A.5)], $\tilde{k}$ is the wave vector magnitude $\sqrt{k^2 + \ell^2}$, $\vec{U}$ is the background velocity profile, $\vec{U}''$ is its second-derivative, I is the identity matrix and there is no sum on the repeated index i. After defining A and B, use

   [w,e]=eig(B,A); sigma=diag(e);

   to solve the generalized eigenvalue problem (A.4). Finish by sorting the eigenvalue/eigenvector pairs in order of descending growth rate. Output the pair with the largest growth rate. Your function should accept the vectors $\vec{z}$, $\vec{U}$, and the scalars k and ℓ as inputs and deliver σ and $\hat{w}$ for the mode with the maximum growth rate as outputs.

   Hints:

   - In Matlab, a simple way to left-multiply a vector onto a matrix, $v_i A_{ij}$ (with no sum on i), is like this: `diag(v)*A`.
   - The identity matrix of size $N \times N$ is given by the built-in function `eye(N)`

- It's a good idea to use the second-derivative matrix to compute U'', but use one-sided derivatives rather than boundary conditions for the top and bottom rows, since it does not obey the same boundary conditions as $\hat{w}$.
- Make sure your z vector excludes the boundaries!
- Sort using `sort(...,'descend')`.

(b) Write a script to test the function you developed in part (a) for the following test case:

$$U^\star = \tanh(z^\star); \quad k^\star = 0.45; \quad \ell^\star = 0; \quad z^\star \in (-4, 4); \quad \Delta = 0.2.$$

The script should define the inputs for the function, call the function, then plot the outputs. The resulting plot should show $\hat{w}^\star$ versus $z^\star$, both as real and imaginary parts and in polar form (magnitude and phase versus $z^\star$), and should include an annotation that gives the growth rate, e.g.,

```
title(sprintf('\\sigma*=\%.3f',your_value_of_sigma)).
```

You should get $\sigma^\star = 0.175$.

[Hint: Remember that your vector of $z^\star$ values should exclude the boundaries.]

8. **The Piecewise-Linear Shear Layer: Numerical Solution**

Here you will solve the shear layer problem numerically for comparison with the analytical solution.

(a) Repeat the derivation of $\sigma^\star = \sigma^\star(k^\star)$ for the piecewise-linear shear layer with all of the algebra included.

(b) Test your result from (a) using the numerical function developed in project 7. Use the scaled variables, so that the velocity profile is:

$$U^\star = \begin{cases} 1, & z^\star > 1 \\ z^\star, & -1 \le z^\star \le 1 \\ -1, & z^\star < -1 \end{cases}$$

Compare plots of $\sigma^\star(k^\star)$ as well as eigenfunctions and growth rates of the fastest-growing mode for both the analytical and numerical solutions. Try a few different ranges for z, e.g. z=[−3 3]; z=[−6 6]; z=[−10 10], and plot $\sigma^\star(k^\star)$ for each. You should find that only when z=[−10 10] or larger is the analytical form of (3.34) reproduced. That is because the boundary conditions are different. In the analytical solution, we assumed that the vertical domain is infinite, so that $\hat{w} \to e^{-\tilde{k}|z|}$ as $|z| \to \infty$.

[Hint: When comparing eigenfunctions, remember that they are only defined up to a multiplicative constant, which may be complex. As a result, eigenfunctions that should be the same can look totally different. The solution is to normalize. The easiest way is to divide the eigenfunction through by its value at some fixed height, e.g., $z = 0$.]

(c) Resolve the discrepancy in boundary conditions between (a) and (b) by deriving and implementing an asymptotic boundary condition in your code. For consistency with $\hat{w} \to e^{\mp \tilde{k}z}$, require that $\hat{w}' = +\tilde{k}\hat{w}$ and $\hat{w}' = -\tilde{k}\hat{w}$ at $z = z_1$ and $z = z_N$, respectively. At $z = z_1$:

$$\hat{w}'_1 = \frac{\hat{w}_2 - \hat{w}_0}{2\Delta} = \tilde{k}\hat{w}_1 \quad \Rightarrow \quad \hat{w}_0 = \hat{w}_2 - 2\tilde{k}\Delta\hat{w}_1.$$

Now substitute into the finite difference expression for the second-derivative:

$$\hat{w}''_1 = \frac{\hat{w}_0 - 2\hat{w}_1 + \hat{w}_2}{\Delta^2} = \frac{2\hat{w}_2 - 2(1 + \tilde{k}\Delta)\hat{w}_1}{\Delta^2}$$

After a similar process at the upper boundary, you should have:

$$D_{1,1} = -2(1 + \tilde{k}\Delta)/\Delta^2 \; ; \; D_{1,2} = 2/\Delta^2 \; ; \; D_{1,j} = 0 \text{ otherwise}$$
$$D_{N,N} = -2(1 + \tilde{k}\Delta)/\Delta^2 \; ; \; D_{N,N-1} = 2/\Delta^2 \; ; \; D_{N,j} = 0 \text{ otherwise.}$$

After making this replacement in the derivative matrix, show that you can match the analytical result with a much smaller domain.

9. **Transforming the Rayleigh Equation**

The vertical displacement η can be defined in terms of vertical velocity: $w = D\eta/Dt$.

(a) Linearize this equation by assuming small perturbations about a parallel shear flow $U(z)\hat{e}^{(x)}$. Assuming a normal mode solution, show that

$$\hat{w} = \iota k(U - c)\hat{\eta}.$$

(b) Now show how the Rayleigh equation:

$$\hat{w}_{zz} = \left(\frac{U_{zz}}{U - c} + \tilde{k}^2\right)\hat{w}$$

can be transformed into an equation for the vertical displacement eigenfunction:

$$[(U - c)^2\hat{\eta}_z]_z = \tilde{k}^2(U - c)^2\hat{\eta}.$$

10. **Energy Analysis for a Shear Layer**

Using the function developed in problem 7, compute $\sigma^*(k^*)$ for the hyperbolic tangent shear layer $U^* = \tanh(z^*)$. Your plot should cover the range $0 < k^* < 1$. You should find that the growth rate and wavenumber of the fastest growing mode are close to the "test case" suggested in homework problem 7b.

Using the value of $k^\star$ you identified as the fastest-growing mode, do the following.

(From here on, stars indicating scaled quantities are dropped.)

(a) Plot profiles of

    (1) $\hat{w}$ (amplitude and phase)

    (2) $\overline{u'w'}$

    (3) $\overline{\pi'w'}$

(b) Plot profiles of

    (1) the kinetic energy $K'(z) = \frac{1}{2}(\overline{u'u'} + \overline{w'w'})$,

    (2) the shear production rate $SP(z) = -\overline{u'w'}dU/dz$, and

    (3) the flux convergence $FC(z) = -d\overline{\pi'w'}/dz$.

(c) Describe the pattern of energy transfer in words, i.e., where it's created, where it's fluxed from and to.

(d) Plot profiles of $2\sigma_r K'(z)$ and $SP(z) + FC(z)$ on the same axes. Check that they are equal to within, say, a few percent. If not, debug and recheck your results for (b) and (c).

(e) Add the x dependence to the eigenfunction: $w'(x, z) = \{\hat{w}(z)e^{ikx}\}_r$ for $x \in [0, \lambda]$ and $\lambda = 2\pi/k$. The result will be a matrix, $w'_{ij} = w'(x_i, z_j)$. Make a contour or image plot of w'. Is the tilt consistent with positive shear production? [Hint: In Matlab, good choices for plotting functions of two variables are `contourf` and `pcolor`.]

11. **The Bickley Jet**

(a) Using the techniques developed above, investigate the stability of the Bickley jet:

$$U^\star = \text{sech}^2(z^\star).$$

To encompass the domain of instability, you'll need to scan wavenumbers in the range 0–2.

    (1) For each wavenumber, plot not only the fastest-growing but also the second-fastest. You'll find that there are two families of modes, the *sinuous* and *varicose* modes.

    (2) For the fastest growing mode of each family, repeat the analyses of project 10.

    (3) Write 1–2 paragraphs describing and comparing the properties of these modes. Which grows fastest? What are their spatial scales? Where are the critical levels? Where are the inflection points? Where is kinetic energy created? Do these locations coincide with critical levels, or with inflection points, or neither? You might get some ideas from Smyth & Moum (2002, section 3.1, figures 3 and 4).

Figure A.2 Cloud patterns over Guadalupe Island, near the Baja peninsula (NASA). For use with project 11.

(b) Predict the ratio of wavelength of the fastest-growing mode to the jet width. Obtain the same ratio graphically from the NASA satellite photo below, and compare your results. (The comparison will of course be approximate. Be happy if you get agreement to within a factor of 2. Sketch on the satellite photo to indicate the lengths you used in your estimate.)

12. **Sinusoidal Flow**

Investigate the stability of the sinusoidal velocity profile

$$U^\star = \sin(z^\star)$$

with impermeable boundaries at $z^* = \pm H^*$.

(a) Choose the boundaries such that $H^* = \pi$. Solve the Rayleigh equation numerically for $0 < k^* < 1$ and plot $\sigma^*(k^*)$.
(b) Repeat the procedure for several smaller values of H^*. What is the effect of reducing H^*? At what H^* does the instability vanish?
(c) What aspect of the inflection point theorem does this illustrate?

13. **The Fourth-Derivative Matrix**

(a) Derive a second-order finite difference approximation to the fourth derivative having the following form:

$$f_i^{(4)} = C f_{i-2} + B f_{i-1} + A f_i + B f_{i+1} + C f_{i+2}. \qquad (A.7)$$

Hint: To simplify the algebra, write (A.7) in this form:

$$f_i^{(4)} = A f_i + B(f_{i-1} + f_{i+1}) + C(f_{i-2} + f_{i+2}). \qquad \text{(A.8)}$$

In the Taylor series expansions for the pairs of terms in parentheses, every second term will cancel.

(b) In (A.7), the expressions for f_1, f_2, f_{N-1}, and f_N involve "ghost points" (at which f is not specified). Explain how these expressions can be evaluated using each of the following boundary conditions (in finite difference form):

- Rigid boundaries: $f_0 = f_{N+1} = 0$; $f_0' = f_{N+1}' = 0$.
- Frictionless boundaries: $f_0 = f_{N+1} = 0$; $f_0'' = f_{N+1}'' = 0$.

Note: It is sufficient to express the boundary conditions to second-order accuracy, e.g.,

$$f_0' = \frac{f_1 - f_{-1}}{2\Delta} = 0.$$

14. **Matrix Solution of the Orr-Sommerfeld Equation**

(a) Write a Matlab function to find eigenvalues and eigenvectors for the discretized Orr-Sommerfeld equation. Your function should accept as inputs a column vector of z values, the corresponding background velocity vector $U(z)$, the viscosity v, the wavenumbers k, ℓ, and a choice of rigid or frictionless boundary conditions at each boundary. It should deliver as output the growth rate and vertical velocity eigenfunction for the fastest-growing mode. The function should compute U_{zz} internally.

You will need to write a subroutine **ddz4(z)** to compute the fourth derivative using your results from project 2. That routine need not include one-sided derivatives at the boundaries (because you will not actually use it to compute the fourth-derivative of anything).

Try your code for the following test case:

$$0 < z^\star < 1; \quad \Delta z^\star = 0.005; \quad U^\star = 4z^\star(1 - z^\star); \quad v^\star = 1/1e5;$$
$$k^\star = 1.55; \quad \ell^\star = 0.$$

Refer to your notes on scaling to make sure you understand what the starred variables mean, how they are input to your subroutine, and how to interpret the output. Use rigid boundary conditions. We get $\sigma^\star = 0.015 - 0.243\iota$.

(b) Suppose you needed to apply this result to a particular channel flow, with width 15 m and maximum flow speed 2 m/s. Give the wavelength in meters and the e-folding time in seconds (or minutes if that seems more sensible).

Figure A.3 Schematic velocity profile for a triangular jet.

15. **Wave Resonance in a Jet**

    The triangular jet profile shown in Figure A.3 has three kinks where vortical waves can propagate. Sketch the three waves such that each adjacent pair satisfies the criteria for resonance:

    (a) The vertical velocity perturbations of each wave amplify the crests and troughs of the other.

    (b) The propagation velocities allow for the waves to be stationary relative to each other.

    Make your own sketch if you prefer.

    Comparing with your analysis of the Bickley jet in homework 11 does your sketch represent the sinuous or the varicose mode?

16. **A Convectively Unstable Layer in an Inviscid Fluid, Revisited**

    In an earlier project you developed a code to solve the Rayleigh equation. Adapt this code to solve (2.29), the equation for convection in a stationary, inviscid fluid with an arbitrary buoyancy profile. You will now use this code to address the unstable layer project 6 problem 2 more thoroughly.

    (a) Using the B_z profile (A.3), reproduced below for convenience, compute and plot the growth rate for a full range of $\tilde{k}$. For simplicity choose $\alpha = 1$.

    $$B_z = B_{z0}(1 - 2\text{sech}^2\alpha z). \tag{A.9}$$

    (b) For the special case $\tilde{k} = \alpha$, do your numerical results for σ and $\hat{w}$ match the analytical solution? Make a plot to illustrate the comparison.

    (c) Does this case represent the fastest-growing mode?

(d) If $\tilde{k} = \alpha$ is not the fastest-growing mode, compute $\tilde{k}$, σ, and $\hat{w}$ for the fastest-growing mode and discuss any differences you observe. Make a plot to illustrate the comparison.

(e) Is your result consistent with the upper bound on the growth rate given in (2.34)?

17. Instability of a Separating Boundary Layer

A bottom boundary layer flowing over an obstacle tends to separate on the downstream side. Figure A.4 is an aerial photo of Knight Inlet, a fjord on the coast of British Columbia. Tidal flow in the fjord must cross a shallow sill, which is the site of strong instability and turbulence. In this assignment, you will do stability analyses of this flow.

Here is a sketch of flow over the sill. It shows regions of instability: a stratified shear layer above, and a separating boundary layer lower down.

Below is an echosounder image of the flow.

We'll examine the stratified shear flow instabilities soon. Here, we'll look at the instabilities of the separating boundary layer, where stratification is not important. The velocity profile

$$U^\star = \begin{cases} z^{\star 2}\left(6 - 8z^\star + 3z^{\star 2}\right), & 0 \leq z^\star < 1 \\ 1, & z^\star \geq 1 \end{cases}$$

is a model of a boundary layer about to separate.

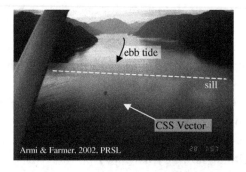

Figure A.4 Aerial photo of Knight Inlet, after Armi and Farmer (2002).

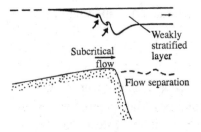

Figure A.5 Sketch of instabilities observed in sill flow (Armi and Farmer, 2002).

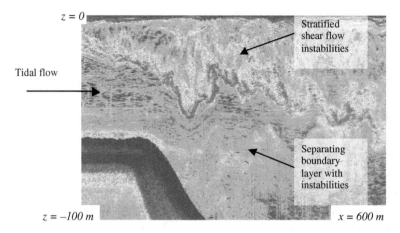

z = 0

Tidal flow

Stratified
shear flow
instabilities

Separating
boundary
layer with
instabilities

z = –100 m

x = 600 m

Figure A.6 Echosounder image of instabilities observed in sill flow. (Adapted from Armi and Farmer, 2002).

Before doing any stability analysis, what would you guess about the stability of this flow?

Use your Orr-Sommerfeld code (written as part of project 14) to investigate the stability of this profile. Place the upper boundary at $z^\star = 3$ and use a grid spacing $\Delta^\star = .02$. Use a rigid boundary condition at the bottom and a frictionless boundary at the top. Make a contour plot (using contourf or pcolor) of the growth rate versus wavenumber and Reynolds number for $0 \leq k^\star \leq 2.1$; $10 \leq Re \leq 10^6$. Assume $\ell = 0$.

What is the minimum Reynolds number for which there is instability? (We get 280.) Is the instability stabilized or destabilized by viscosity? At this Reynolds number, is the frozen flow hypothesis valid?

In inviscid shear instability, the growth rate is approximately proportional to the maximum absolute value of the background shear,

$$S_{max} = \max_z |U_z|.$$

Find the maximum growth rate for $Re = 10^6$. Express this growth rate as a fraction of the maximum background shear.
Find the wavenumber of the fastest-growing mode for $Re = 10^6$. Express the corresponding wavelength as a multiple of the original boundary layer depth, 1. Now look at the instabilities on the separating boundary layer in the echosounder image (Figure A.6). Estimate the wavelength of those instabilities as a multiple of the original thickness of the boundary layer. (Don't forget to account for the aspect ratio of the image!) How does this result compare with the result from your stability analysis?

How do you interpret the result?

- The stability theory for a separating boundary layer describes the observation perfectly.
- We're in the ballpark, but the data is imprecise and there may be other phenomena involved.
- This is not a shear instability of a separating boundary layer.
- ... ?

18. **Instabilities in a Plunging Downslope Flow**

Starting from your Orr-Sommerfeld function you wrote for project 14, write a function to find eigenvalues and eigenvectors for a stratified shear flow in a viscous, diffusive fluid. Your function should accept as inputs a column vector of z values, the corresponding background velocity $U(z)$ and buoyancy gradient $B_z(z)$, the viscosity ν, the diffusivity κ, and the wavenumbers (k, ℓ). Boundary conditions should be frictionless and fixed-buoyancy. (Later, you'll upgrade the function to include other choices.) The function should deliver as output the growth rate σ and the vertical velocity and buoyancy eigenfunctions for the fastest-growing mode.

The following test cases crudely model the stratified shear flow instabilities in the Knight Inlet observations. In each case, report the growth rate and wavenumber of the FGM, and assess the validity of the frozen flow approximation.

(a) $U^\star = -\tanh(z^\star)$; $B^\star_{z^\star} = 0$; $z^\star \in [-4, 4]$; $\Delta^\star = 0.2$; $Re = 10^6$; $Pr = 1$. Boundaries are frictionless and fixed-buoyancy. (I get $\sigma^\star = 0.1753$ at $k^\star = 0.47$.)

Compute the growth rate of the fastest-growing mode as a fraction of the maximum background shear. Compare with the corresponding result from project #1 of the this homework, the separating bottom boundary layer. How do these two results compare? Can you guess the reason for the difference?

(b) $U^\star = -\tanh(z^\star)$ $B^\star_{z^\star} = 0.1 \, \text{sech}^2(z^\star) \, z^\star \in [-4, 4]$; $\Delta^\star = 0.2$; $Re = 10^6$; $Pr = 1$. Boundaries are frictionless and fixed-buoyancy. (I get $\sigma^\star = 0.1146$ at $k^\star = 0.44$.)

(c) Same as (b), but $Re = 100$. (I get $\sigma^\star = 0.0977$ at $k^\star = 0.43$.)

(d) Same as (b), but $Re = 20$.

In cases (a–c), the wavenumber of the fastest-growing mode remains more or less consistent. Use a typical value of this wavenumber, along with the scaling relations discussed in class, to estimate the wavelength of the instability as a multiple of the vertical scale over which the background velocity varies. Compare this result visually with the shear instabilities in the echosounder image below. How does this ratio of scales of the computed instabilities compare with what you see in the observations?

19. **Sheared Convection**

Here you will use your existing stratified shear flow function to investigate the effect of background shear on convective instability. Sheared convection is an important aspect of the dynamics of thunderstorms and of upper ocean response to intense surface forcing (e.g., hurricanes). You'll begin by testing the code by reproducing the results we obtained analytically in the first week of the course. You'll then repeat the analysis with a background shear flow added, and find something interesting.

(a) **Test your software.**

Using the Matlab script given below along with your subroutine for stratified shear flow, plot growth rate versus wavenumber and Rayleigh number for a flow with

$$0 < z^\star < 1; \quad \Delta^\star = 0.05; \quad U^\star = 0; \quad v^\star = Pr = 7; \quad \kappa^\star = 1;$$

$$B^\star = -Ra\,Pr\beta; \quad \beta = z^\star; \quad B^\star_{z^\star} = -Ra\,Pr \cdot 1$$

Compute with both frictionless and rigid boundaries. In both cases, buoyancy can obey fixed-buoyancy conditions. Check to make sure that the critical values of Ra and k delivered by your code match the theoretical values.

The following script will guide you through most of this, but you must insert the call to your function for stability analysis of stratified shear flow with the appropriate inputs and outputs. If you prefer to write your own script, that is fine.

```
% Hmwk 7, project 1A
clear
close all
fs=18;
lw=1.6;

% set parameters
Pr=7;        % Prandtl number
irigid=0;    % boundary conditions (0=frictionless, 1=rigid)

% define z values
del=.05;
z_st=[del:del:1-del]';
```

```
% define profiles
beta=z_st;
beta_zst=ones(size(z_st));
U_st=zeros(size(z_st));

% analytical results for critical Rayleigh number
% and wavenumber
if irigid==0
    Ra_c=(27/4)*pi^4; % frictionless boundaries (as derived
                      % in class)
    k_c=pi/sqrt(2);
    bcw='ff'
elseif irigid==1;
    Ra_c=1708;   % rigid boundaries (Kundu 11.3)
    k_c=3.12;
    bcw='rr';
end
bcb='cc';

% ranges for loops over k and Ra
ks=[0:.2:12];nks=length(ks)
Ras=10.^[2:.2:5];nRa=length(Ras);
l_st=0;      % 2D modes only

% loop over k and Ra
for i=1:nks
    k_st=ks(i);
    for j=1:nRa
        Ra=Ras(j);
        Bz_st=-Ra*Pr*beta_zst;

        %%%%%%%%%%%%%%%%%%%%%%%%%%%%%%%%%%%%%%%
        % put call to your stratified shear flow routine here
        [s_st] = ...
        %%%%%%%%%%%%%%%%%%%%%%%%%%%%%%%%%%%%%%%

        sig(i,j)=real(s_st);
    end
    disp([num2str(i/nks) ' done'])      % indicate progress
end

% plot growth rates along with analytical values for the
% critical Ra and k
```

```
figure
contourf(ks,Ras,sig',max(sig(:))*[0:.05:.95]);shading flat
hold on
plot([k_c-1 k_c+1],Ra_c*[1 1],'k','linewidth',lw)
plot(k_c*[1 1],Ra_c*[1/1.5 1.5],'k','linewidth',lw)
set(gca,'yscale','log')
colorbar
ylabel('Ra','fontsize',fs,'fontangle','italic',
'fontweight','bold')
xlabel('k*','fontsize',fs,'fontangle','italic',
'fontweight','bold')
title('Benard convection scaled growth rate', 'fontsize',
fs,'fontweight','normal')
set(gca,'fontsize',fs-2)

print('-djpeg','hmwk7_1a')
```

(b) **The sheared case.**

First, investigate the possibility that oblique modes (modes whose wave vector points in a different direction than the sheared flow) are the most unstable by doing the following. Modify the script you used for part A so that the Rayleigh number is fixed at 1000, and loop instead over both $k^\star$ and $\ell^\star$. The range $0 \leq k^\star, \ell^\star \leq 4$ is sufficient. Assume frictionless boundaries. How do growth rates vary as the angle of obliquity is increased?

Now apply a uniformly sheared background flow in the x direction, $U^\star = RePr\, z^\star$, setting $RePr = 20$. Run the script again. Now how do growth rates of oblique modes compare with those of 2D modes?

Consider convective modes with wave vector parallel to the sheared flow, i.e., $\ell^\star = 0$. What effect does the presence of shear have on these modes? Does it increase the growth rates, decrease them, or leave them unchanged? Now consider the whole range of $k^\star$ and $\ell^\star$. What is the angle of obliquity for the fastest-growing mode? Knowing the effect of shear on convective modes, could you have *predicted* the angle of obliquity for the fastest-growing mode? [Hint: review section 4.3.2.]

20. **Instabilities of the Eady Model**

Based on section 8.8, implement a solution procedure

$$[\sigma, \hat{q}, \hat{w}, \hat{b}] = \mathcal{F}(z, U_z, B_z, f, k, l).$$

(a) **Baroclinic modes**

Test your code by computing growth rates as a function of k with $\ell = 0$ and $Ri = 100$. Compare the growth rate and wavenumber of the fastest-growing mode with (8.60) and (8.61) or (8.62).

(b) **Symmetric modes**

Computing growth rates as a function of ℓ with $k = 0$ and $Ri = 0.75$. Is the growth rate independent of ℓ as in the analytical solution for symmetric instability? If not, why not?

(c) **Comparison**

What is the critical value of Ri above which the baroclinic mode grows faster than the symmetric mode?

Appendix B

Projects

B.1 Shallow Flow Over Topography

B.1.1 Background

In this project you will figure out how the instability of a parallel shear flow is modified by topography. Examples include the flow of a river over a submerged bar or island (B.1) and an alongshore ocean current encountering a seamount or a sand bar. We'll start by developing the basic theory; read carefully and fill in any algebraic steps that aren't obvious.

Consider an inviscid, homogeneous fluid flowing over a non-horizontal bottom $z = -H(x, y)$ and allowing a free surface deflection $z = \eta(x, y, t)$. In making the shallow water approximation, we assume that

- the pressure is entirely hydrostatic:

$$\pi = g(\eta - z)$$

- the horizontal velocity is independent of depth:

$$\frac{\partial u}{\partial z} = \frac{\partial v}{\partial z} = 0.$$

With these assumptions, the horizontal momentum equations are

$$\frac{\partial u}{\partial t} + u\frac{\partial u}{\partial x} + v\frac{\partial u}{\partial y} = -g\frac{\partial \eta}{\partial x}; \quad \frac{\partial v}{\partial t} + u\frac{\partial v}{\partial x} + v\frac{\partial v}{\partial y} = -g\frac{\partial \eta}{\partial y} \qquad \text{(B.1)}$$

To complete the set, we integrate the divergence equation (1.17) over the vertical extent of the fluid, $-H(x, y) \leq z \leq \eta(x, y, t)$. After combination with (B.1), the result is

$$\frac{\partial \eta}{\partial t} + \frac{\partial}{\partial x}[(H + \eta)u] + \frac{\partial}{\partial y}[(H + \eta)v] = 0. \qquad \text{(B.2)}$$

Figure B.1 Braided drainage pattern near the confluence of the Yukon River and Koyukuk River, Alaska. The Koyukuk River (dark) joins the silt-laden Yukon River (lighter) at the right (Image by U.S. Army Air Corps, 1941).

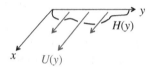

Figure B.2 Shallow flow over variable topography.

We have assumed that the background flow has no vertical shear, but it can be horizontally sheared. We therefore look for perturbations about a background flow $\vec{u} = U(y)\hat{e}^{(x)}$, $\eta = 0$. This flow represents an equilibrium state provided that the further condition

$$H = H(y) \tag{B.3}$$

is satisfied.

Accordingly, we seek a perturbation solution of the form

$$u = U(y) + \epsilon u'(x, y, t);$$
$$v = \epsilon v'(x, y, t);$$
$$\eta = \epsilon \eta'(x, y, t). \tag{B.4}$$

Linearizing as usual, we obtain

$$\frac{\partial u'}{\partial t} + U \frac{\partial u'}{\partial x} = -v' \frac{dU}{dy} - g \frac{\partial \eta'}{\partial x}; \tag{B.5}$$

$$\frac{\partial v'}{\partial t} + U \frac{\partial v'}{\partial x} = -g \frac{\partial \eta'}{\partial y}; \tag{B.6}$$

$$\frac{\partial \eta'}{\partial t} + U \frac{\partial \eta'}{\partial x} = -\frac{\partial}{\partial x} [Hu'] - \frac{\partial}{\partial y} [Hv']. \tag{B.7}$$

The perturbation vertical vorticity is given by

$$\xi' = \frac{\partial v'}{\partial x} - \frac{\partial u'}{\partial y} \tag{B.8}$$

We can now derive an evolution equation for ξ' from the x and y components of the perturbation momentum equation. Subtracting the y derivative of (B.5) from the x derivative of (B.6) yields

$$\left(\frac{\partial}{\partial t} + U \frac{\partial}{\partial x} \right) \xi' = \underbrace{U_{yy} v'}_{advection} + \underbrace{U_y \left(\frac{\partial u'}{\partial x} + \frac{\partial v'}{\partial y} \right)}_{stretching}. \tag{B.9}$$

Note that the total derivative U_y is (minus) the vertical vorticity of the background flow. Therefore, the first term on the right-hand side represents advection of that background vorticity by the cross-stream velocity perturbation. The second represents vortex stretching: changes in fluid depth accompanying convergence (divergence) of the horizontal flow can add to (subtract from) the background vorticity.

Here we'll look at the special case of *zero surface deflection*, $\eta' = 0$, for which the linearized equations collapse to a single equation. With $\eta' = 0$, the mass conservation equation becomes

$$\frac{\partial}{\partial x} [Hu'] + \frac{\partial}{\partial y} [Hv'] = 0. \tag{B.10}$$

This condition can be satisfied by defining a streamfunction ψ' such that

$$Hu' = -\frac{\partial \psi'}{\partial y}; \quad Hv' = \frac{\partial \psi'}{\partial x}. \tag{B.11}$$

In terms of this streamfunction, the perturbation vorticity becomes

$$\xi' = \frac{\partial}{\partial x} \left(\frac{1}{H} \frac{\partial \psi'}{\partial x} \right) + \frac{\partial}{\partial y} \left(\frac{1}{H} \frac{\partial \psi'}{\partial y} \right) \tag{B.12}$$

and its evolution equation can be written as

$$\left(\frac{\partial}{\partial t} + U \frac{\partial}{\partial x} \right) \left\{ \frac{\partial}{\partial x} \left(\frac{1}{H} \frac{\partial \psi'}{\partial x} \right) + \frac{\partial}{\partial y} \left(\frac{1}{H} \frac{\partial \psi'}{\partial y} \right) \right\} = \left(\frac{U_y}{H} \right)_y \frac{\partial \psi'}{\partial x}. \tag{B.13}$$

B.1.2 Project

You are now ready to explore instability of a parallel shear flow that is modified by topography. You'll do this in six sub-projects, two analytical and four numerical.

(A) In (B.13), assume a normal mode solution of the form

$$\psi' = \hat{\psi}(y)e^{ik(x-ct)}. \tag{B.14}$$

and substitute to obtain an ordinary differential equation.

Test your understanding: How do we know that $\hat{\psi}$ has to be a function of y?

(B) Following the proof of Rayleigh's theorem in section 3.15.1, show that a necessary condition for instability is

$$\left(\frac{U_y}{H}\right)_y = 0$$

for some y. You can assume that the side walls are steep, so that the impermeable boundary condition can be written as $v' = 0$, or

$$\hat{\psi} = 0. \tag{B.15}$$

Does instability require an inflection point? Suppose that U_y is constant. Show that instability is possible only if $H(y)$ has a local maximum or minimum.

(C) Write a subroutine that accepts as input the vectors y, U, and H and the scalar k, and gives as output σ (the growth rate of the FGM) and $\hat{\psi}$ (the corresponding eigenfunction). For this exercise, you will solve the equation in scaled form. The length scale will be the half-width of the channel, so that y is within the range $0 < y < 2$. The velocity will be given by a parabolic profile symmetric about the river center, scaled by the maximum velocity:

$$U = y(2 - y). \tag{B.16}$$

The water depth will have a uniform value H_0 except for a bump representing the submerged bar:

$$H = H_0 \left(1 - a \operatorname{sech}^2 \frac{y - y_0}{W}\right)$$

where $a < 1$ is the amplitude of the bar, y_0 is the location of the bar crest, and W is a width scale. These profiles are shown below for a typical parameter set.

For this parameter set, discretize using 200 points and compute the fastest growth rate for $k = [0.2 : 0.2 : 10]$. Plot the result. State the wavelength of the FGM as a multiple of the half-width of the river. State the e-folding time as a multiple of the time taken by the maximum current to traverse a distance equal to the half-width.

(D) In this part of the project, you will see how the instability varies with the cross-stream position of the bar. Repeat the stability analysis for $y_0 =$

1, 1.2, 1.4, ..., 2 and plot the growth rate, phase speed, and wavenumber of the FGM as a function of y_0.

(E) Now we will see how instability depends on the amplitude of the bar. Based on our earlier discussion of shear instability, do you think there will be instability when $a = 0$? Why or why not? From part (C), identify the value of y_0 at which the growth rate is a maximum. Setting y_0 to this value, vary the amplitude a from 0 to 1, and plot the growth rate, phase speed, and wavenumber of the FGM as functions of a. Is there a minimum amplitude at which instability occurs?

(F) How do the instability characteristics change if $a < 0$ (i.e., for flow over a depression)?

B.2 Barotropic Instability on the β-plane

Consider an inviscid, homogeneous fluid in a rotating environment ($v = 0$, $b = 0$, $f \neq 0$). The equations of motion are

$$\vec{\nabla} \cdot \vec{u} = 0 \tag{B.17}$$

$$\frac{D\vec{u}}{Dt} = -\vec{\nabla}\pi + \vec{u} \times f\hat{e}^{(z)}. \tag{B.18}$$

We now make three additional assumptions:

(i) The flow is purely horizontal: $\vec{u} = \{u, v\}$.
(ii) Nothing varies in the vertical: $\partial/\partial z = 0$.
(iii) Rather than setting f to a constant (the f-plane approximation), we use a first-order Taylor series approximation for the meridional dependence: $f = f_0 + \beta y$.[1]

The equations can now be written as:

$$\frac{\partial u}{\partial x} + \frac{\partial v}{\partial y} = 0 \tag{B.19}$$

$$\frac{Du}{Dt} - fv = -\frac{\partial\pi}{\partial x} \tag{B.20}$$

$$\frac{Dv}{Dt} + fu = -\frac{\partial\pi}{\partial y}, \tag{B.21}$$

where

---

[1] More exactly, $f = 2\Omega \sin\varphi$ where Ω is the angular velocity of the planet, $\varphi = y/R$ is the latitude, and R is the planet's radius. For Earth, $\Omega = 7.2921 \times 10^5 \mathrm{s}^{-1}$ and $R = 6371$ km. Near a given latitude φ_0, $f_0 = 2\Omega \sin\varphi_0$ and $\beta = 2\Omega R \cos\varphi_0$. On Earth, $2\Omega R = 2.3 \times 10^{-11} \mathrm{m}^{-1}\mathrm{s}^{-1}$. At the equator ($\varphi_0 = 0$), f_0 vanishes and β is a maximum. At the poles ($\varphi_0 = \pm\pi/2$), the opposite is true. Therefore the f-plane approximation is most accurate in polar regions, while the β-plane approximation is valid in any sufficiently small band of latitudes.

$$\frac{D}{Dt} = \frac{\partial}{\partial t} + u\frac{\partial}{\partial x} + v\frac{\partial}{\partial y}. \tag{B.22}$$

We now consider perturbations to a parallel shear flow that is purely zonal and varies only in the meridional:

$$u = U(y) + \epsilon u'; \quad v = \epsilon v'; \quad \pi = \Pi + \epsilon \pi'. \tag{B.23}$$

(**A**) Substitute this into the equations of motion and show that

- At $O(\epsilon^0)$, the background velocity and pressure fields must be in geostrophic balance: $\partial\Pi/\partial y = -fU$.
- At $O(\epsilon^1)$, obtain the perturbation velocity equations

$$u'_t + Uu'_x + (U_y - f)v' = -\pi'_x \tag{B.24}$$
$$v'_t + Uv'_x + fu' = -\pi'_y. \tag{B.25}$$

(**B**) Cross-differentiate (B.24, B.25) to obtain the equation for the perturbation vertical vorticity $\zeta' = v'_x - u'_y$:

$$\zeta'_t + U\zeta'_x = -(f_0 - U_y)_y v' = -(\beta - U_{yy})v'. \tag{B.26}$$

The right-hand side represents advection of the background vorticity gradient by the meridional velocity perturbation. The background vorticity has two parts: f_0 is due to the rotation of the planet and $-U_y$ is due to the parallel shear flow.

(**C**) Because the flow is 2D, we can represent it in terms of a streamfunction:

$$u' = \psi'_y; \quad v' = \psi'_x; \quad \zeta' = \nabla^2\psi'. \tag{B.27}$$

Substitute this into (B.26). Now assume that ζ' has the normal mode form

$$\zeta' = \hat{\zeta}(y)e^{ik(x-ct)}, \tag{B.28}$$

where k is the zonal wavenumber and c the zonal phase speed. With that assumption, derive:

$$\hat{\psi}_{yy} = \left(\frac{\beta - U_{yy}}{U - c} - k^2\right)\hat{\psi}. \tag{B.29}$$

Note that this is just the 2D form of the Rayleigh equation (3.19) with an extra term to account for planetary rotation.

(**D**) Setting $U = 0$, derive the dispersion relation for barotropic Rossby waves:

$$c = \frac{-\beta}{k^2 + \ell^2}. \tag{B.30}$$

(Be sure to state any additional assumptions you make.) Rossby waves are unique in that they travel only to the west, as you can see from the fact that c is negative definite. Rossby waves are closely analogous to the vortical waves described in section 3.12.2, with planetary rotation playing the same role as the kinks in the velocity profile.

(E) Following the classical proof of Rayleigh's theorem (section 3.15.1), show that (B.29) can have unstable normal mode solutions only if the background vorticity $f - U_y$ changes sign somewhere in the range of y.

B.3 Bioconvection

Review the introduction to bioconvection in section 9.3. Complete the derivation of the equilibrium state and the perturbation equations (9.21) and (9.22). Convert these into in normal mode form, then write a subroutine to solve them using the method of your choice. Scale the equations based on the layer thickness V_s and the diffusivity κ. Show that the solution depends only on the Prandtl number ν/κ, the scaled swimming velocity $\alpha = V_s H/\kappa$, and the Rayleigh number $Ra = g' H^3 \bar{n}/(\nu \kappa)$.

Reproduce the results shown in Figure 9.14.

(i) How does the stability boundary depend on the Prandtl number? Could you have anticipated this?

(ii) As α increases from zero, the critical Rayleigh number drops rapidly. Is this what you would expect? Why?

(iii) Identify the value of α at which the critical Ra smallest. As α increases above that value, the critical Rayleigh number increases. Is that what you would expect? Why?

B.4 Hydrostatic Normal Modes of a Density Profile.

Hydrostatic normal modes are described by the Taylor-Goldstein equation (4.18) with no mean flow ($U = 0$) and in the long-wave limit $k \to 0$:

$$\hat{w}_{zz} + \frac{B_z}{c^2}\hat{w} = 0.$$

(a) Find or invent a stable density profile typical of the ocean or atmosphere and investigate its normal modes, both the phase speeds and the vertical structures.

Use the numerical method described in section 6.2 and coded in homework problem 18. If you set ν and κ to very small values, say $10^{-6}\mathrm{m}^2/\mathrm{s}$, the viscous and diffusive terms will improve your resolution without significantly affecting the results.

(b) Find or invent a realistic mean velocity profile $U(z)$ to go with your density profile. Repeat the calculations above with $U(z)$ included and see what difference it makes.

B.5 Stability Boundaries for Double Diffusive Instability

Here you will determine the stability boundaries for salt fingering and diffusive convection by considering the algebraic properties of cubic equations.

Introduction

Consider the cubic polynomial

$$\sigma^3 + A_2\sigma^2 + A_1\sigma + A_0 = 0, \tag{B.31}$$

in which the coefficients A_2, A_1, and A_0 are real numbers. The stability equation (9.8) for double diffusive instability is a cubic of this type, with the additional properties $A_1 > 0$ and $A_2 > 0$.

Roots are in general complex: $\sigma = \sigma_r + \iota\sigma_i$. Roots with $\sigma_i = 0$ and $\sigma_i \neq 0$ describe stationary and oscillatory normal modes, respectively. A stability boundary is a curve that divides a region in which the flow is unstable (i.e., at least one root has a positive real part) from a region that is stable (all roots have negative real part).

We can simplify the analysis of (B.31) with a change of variables:

$$\boxed{s^3 + s^2 + \alpha_1 s + \alpha_0 = 0,} \tag{B.32}$$

where

$$s = \frac{\sigma}{A_2}; \quad \alpha_1 = \frac{A_1}{A_2^2}; \quad \text{and } \alpha_0 = \frac{A_0}{A_2^3}$$

Because A_2 is real and positive, this change of variables does not affect whether any of these quantities is real or complex, positive or negative. In particular, $\alpha_1 > 0$ in gravitationally stable stratification $B_z > 0$.

The cubic equation (B.32) has three roots, each of which is a function of α_0 and α_1. These can be either

- all real, or
- one real and two complex conjugates.

The three roots obey the following relations:

$$s_1 + s_2 + s_3 = -1 \tag{B.33}$$
$$s_1s_2 + s_2s_3 + s_1s_3 = \alpha_1 \tag{B.34}$$
$$s_1s_2s_3 = -\alpha_0. \tag{B.35}$$

Using this information, you will find the stability boundaries for stationary and oscillatory roots, then identify the side of each boundary on which the flow is unstable.

(A) Three real roots

Assume that all three roots of (B.32) are real.

(i) Show that, if one or more roots are zero, then α_0 must be zero. Now show the converse, i.e., if $\alpha_0 = 0$, at least one root is zero.

(ii) Now, what about the other two roots? If one of them is positive, then $\alpha_0 = 0$ is not a stability boundary. Show that this is not the case (i.e., both of the other two roots are negative). [Hint: Keep in mind that $\alpha_1 > 0$.]

(iii) On which side of the line $\alpha_0 = 0$ is the flow unstable? To answer this, differentiate (B.32) to show that $\partial s / \partial \alpha_0$, evaluated at $\alpha_0 = 0$, is $-1/\alpha_1$, and is therefore negative.

(B) one real, two complex roots

Assume that one root is real and the other two are complex conjugates.

(i) Show that the real root is zero if and only if $\alpha_0 = 0$, as in part A1.

(ii) Show that the complex conjugate roots have zero real part if and only if $\alpha_0 = \alpha_1$.

(iii) Substitute $s = s_r + \iota s_i$ into (B.32) and split the result into real and imaginary parts.

(iv) Show that, for the real root, the equation is the same as (B.32), and therefore that the argument of part A3 still holds.

(v) Now consider the complex conjugate roots. Assuming that $s_i \neq 0$, eliminate s_i to obtain

$$26s_r^3 - 16s_r^2 - 2(1 - 4\alpha_1)s_r + \alpha_1 - \alpha_0 = 0. \qquad (\text{B.36})$$

(vi) Differentiate this to show that $\partial s_r / \partial \alpha_0$, evaluated at $\alpha_0 = \alpha_1$, is $1/(8\alpha_1 - 2)$, and is therefore positive.

(C) Stability boundaries in terms of α_0 and α_1

Here, you will summarize your results from parts A and B with a simple sketch. On the $\alpha_0 - \alpha_1$ plane, sketch the lines $\alpha_0 = 0$ and $\alpha_0 = \alpha_1$. (Only the half-plane $\alpha_1 \geq 0$ matters.) Indicate the three regions in which

- real roots are positive,
- complex roots have positive real part,
- all roots have negative real part.

(D) Stability boundaries in terms of R_ρ, Pr, and τ

(i) Referring back to (9.8), use the foregoing results to show that stationary modes are unstable if $1 < R_\rho < \tau^{-1}$.

(ii) Show that oscillatory modes are unstable if $1 < R_\rho^* < (Pr + 1)/(Pr + \tau)$.

(iii) One of the above cases pertains to the case where cool, fresh water overlies warm, salty water, the other to the opposite case. Which is which?

B.6 Rayleigh-Taylor Instability in Outer Space

Consider a gravitating star surrounded by interstellar gas. Set up the equations of motion in spherical coordinates (e.g., Smyth, 2017; Kundu et al., 2016). Neglect viscosity, diffusion, and the Coriolis effect. Derive the conditions for static equilibrium and the perturbation equations.

Imagine a spherical shell, surrounding the star, across which the buoyancy $B(r)$ of the stellar gas changes by an amount b_0. This is the spherical analog of the interface in section 2.2.4. Substitute the buoyancy profile into the perturbation equations and explore the solutions.

B.7 Universality of Convection-like Instabilities.

Convective, centrifugal, and inertial instabilities are mathematically similar, and for each we have established an upper bound on the growth rate. Moreover, in each case we have seen examples where the growth rate increases with increasing wavenumber and asymptotes at the proven upper bound. In section 7.8.1 we gave a rather hand-waving argument as to why this should be so. In this project you will construct an explicit proof using Sturm-Liouville theory and results from the calculus of variations.

Groen (1948) considered internal gravity waves in an inviscid, motionless, stably stratified fluid. For a buoyancy profile that is unrestricted except that B_z is everywhere positive, he proved the following two theorems.

- Wave frequency is a monotonically increasing function of wavenumber.
- The maximum frequency, found in the limit of infinite wavenumber, is $\sqrt{\max_z B_z}$.

Work through Groen's proof, only now assume that the stratification is unstable. (2.29) is a good starting point.[2] To go a step further, see if you can prove the corresponding results for centrifugal instability, where the geometry is cylindrical.

Note that this proof works only if B_z is *everywhere* negative. What if B_z is only negative for some z? Based on our discussions so far (including sections 2.2.3 and 7.8.1 and problem 16 in Appendix A) would you expect the result to hold (i.e., the growth rate bound to be achieved) in that case? If so, can you prove it?

---

[2] If you are willing to trust Groen's mathematics (and there is no reason not to other than the habitual skepticism of a good scientist), you can prove the results easily by showing that the instability problem is isomorphic to Groen's wave problem.

References

Acheson, D. 1997. *From Calculus to Chaos: An Introduction to Dynamics*. Oxford, UK: Oxford University Press.

Adriani, A., Mura, A., Orton, G., Hansen, C., Altieri, F., Moriconi, M. L., Rogers, J., Eichstdt, G., Momary, T., Ingersoll, A. P., Filacchione, G., Sindoni, G., Tabataba-Vakili, F., Dinelli, B. M., Fabiano, F., Bolton, S. J., Connerney, J. E. P., Atreya, S. K., Lunine, J. I., Tosi, F., Migliorini, A., Grassi, D., Piccioni, G., Noschese, R., Cicchetti, A., Plainaki, C., Olivieri, A., O'Neill, M. E., Turrini, D., Stefani, S., Sordini, R., and Amoroso, M. 2018. Clusters of cyclones encircling Jupiter's poles. *Nature*, **555**.

Armi, L., and Farmer, D. 2002. Stratified flow over topography: bifurcation fronts and transition to the uncontrolled state. *Proc. Roy. Soc. London A*, **458**(2019), 513–538.

Aschwanden, M. J. 2016. 25 years of self-organized criticality: solar and astrophysics. *Space Science Rev.*, **198**, 47–166.

Baines, P. G., and Mitsudera, H. 1994. On the mechanism of shear flow instabilities. *J. Fluid Mech.*, **276**, 327–342.

Bak, P., Tang, C., and Wiesenfeld, K. 1987. Self-organized criticality: an explanation of $1/f$ noise. *Phys. Rev. Lett.*, **59**, 381–384.

Boccaletti, G., Ferrari, R., and Fox-Kemper, B. 2007. Mixed layer instabilities and restratification. *J. Phys. Oceanogr.*, **37**, 2228–2250.

Bretherton, F. P. 1966. Baroclinic instability and the short wave cut-off in terms of potential vorticity. *Q. J. R. Meteorol. Soc.*, **92**, 335–345.

Cabot, W. H., and Cook, A. W. 2006. Reynolds number effects on Rayleigh-Taylor instability with possible implications for type Ia supernovae. *Nature Physics*, **2**, 562–568.

Carpenter, J. R., Sommer, T., and Wüest, A. 2012. Stability of a double-diffusive interface in the diffusive convection regime. *J. Phys. Oceanogr.*, **42**(5), 840–854.

Carpenter, J. R., and Timmermans, M.-L. 2012. Deep mesoscale eddies in the Canada Basin, Arctic Ocean. *Geophys. Res. Lett.*, **39**.

Carpenter, J. R., Tedford, E. W., Heifetz, E., and Lawrence, G. A. 2013. Instability in stratified shear flow: review of a physical interpretation based on interacting waves. *Appl. Mech. Rev.*, **64**, 060801.

Carpenter, J. R., Tedford, E. W., Rahmani, M., and Lawrence, G. A. 2010. Holmboe wave fields in simulation and experiment. *J. Fluid Mech.*, **648**, 205–223.

Caulfield, C. P. 1994. Multiple linear instability of layered stratified shear flow. *J. Fluid Mech.*, **258**, 255–285.

Chimonas, G. 1970. The extension of the Miles-Howard theorem to compressible fluids. *J. Fluid Mech.*, **43**, 833–836.

Davies Wykes, M. S., and Dalziel, S. B. 2014. Efficient mixing in stratified flows: experimental study of a Rayleigh-Taylor unstable interface. *J. Fluid Mech.*, **756**, 1027–1057.

Drazin, P. G., and Howard, L. N. 1966. Hydrodynamic stability of parallel flow of inviscid fluid. *Advan. Appl. Math.*, **9**, 1–89.

Einaudi, F., and Finnigan, J. J. 1993. Wave-turbulence dynamics in the stably stratified boundary layer. *J. Atmos. Sci.*, **50**(13), 1841–1864.

Ekman, V. W. 1906. Beiträge zur Theorie der Meeresströmungen. *Ann. Hydrogr. Marit. Meteor.*, 1–50.

Farrell, B. 1996. Generalized stability theory. Part I. Autonomous operators. *J. Atmos. Sci.*, **53**, 2025–2040.

Ferziger, J. H., and Peric, M. 1999. *Computational Methods for Fluid Dynamics*. Cambridge University Press.

Fjortoft, R. 1950. Application of integral theorems in deriving criteria of stability of laminar flows and for the baroclinic circular vortex. *Geophys. Publ. Oslo*, **17**(6), 1–52.

Foldvik, A., and Kvinge, T. 1974. Conditional instability of seawater at the freezing point. *Deep Sea Res.*, **21**, 169–174.

Gill, A. E. 1982. *Atmosphere-Ocean Dynamics*. San Diego: Academic Press.

Gregg, M. C. 1975. Microstructure and intrusions in the California current. *J. Phys. Oceanogr.*, **5**(2), 253–278.

Groen, P. 1948. Two fundamental theorems on gravity waves in inhomogeneous incompressible fluids. *Physica*, **14**, 294–300.

Haine, T., and Marshall, J. 1998. Gravitational, symmetric and baroclinic instability of the ocean mixed layer. *J. Phys. Oceanogr.*, **28**, 634–658.

Harris, J. W., and Stocker, H. 1998. *Handbook of Mathematics and Computational Science*. New York: Springer-Verlag.

Hazel, P. 1972. Numerical studies of the stability of inviscid parallel shear flows. *J. Fluid Mech.*, **51**, 39–62.

Heifetz, E., and Methven, J. 2005. Relating optimal growth to counterpropagating Rossby waves in shear instability. *Phys. Fluids*, **17**, 064107.

Heifetz, E., Bishop, C. H., and Alpert, P. 1999. Counter-propagating Rossby waves in the barotropic Rayleigh model of shear instability. *Q. J. R. Meteorol. Soc.*, **125**, 2835–2853.

Heifetz, E., Bishop, C. H., Hoskins, B. J., and Methven, J. 2004. The counter-propagating Rossby-wave perspective on baroclinic instability. I: mathematical basis. *Q. J. R. Meteorol. Soc.*, **130**, 211–231.

Heifetz, E., Agnon, A., and Marco, S. 2005. Soft sediment deformation by Kelvin Helmholtz instability: a case from Dead Sea earthquakes. *Earth Plan. Sci. Lett.*, **236**(1), 497–504.

Holmboe, J. 1962. On the behaviour of symmetric waves in stratified shear layers. *Geophys. Publ.*, **24**, 67–113.

Howard, L. N. 1961. Note on a paper of John W. Miles. *J. Fluid Mech.*, **10**, 509–512.

Howard, L. N., and Gupta, A. S. 1962. On the hydrodynamic and hydromagnetic stability of swirling flows. *J. Fluid Mech.*, **14**(3), 463–476.

Huerre, P. 2000. Open shear flow instabilities. Pages 159–229 of: Batchelor, G. K., Moffat, H. K., and Worster, M. G. (eds.), *Perspectives in Fluid Dynamics*. Cambridge University Press, UK.

Huppert, H. E., and Neufeld, J. A. 2014. The fluid mechanics of carbon dioxide sequestration. *Ann. Rev. Fluid Mech.*, **46**(1), 255–272.

Jensen, H. J. 1998. *Self-Organized Criticality: Emergent Complex Behavior in Physical and Biological Systems*. Cambridge University Press.

Jevons, S. 1857. On the cirrus form of cloud. *London, Edinburgh and Dublin Philos. Mag. J. Sci., 4th Series*, **14**, 22–35.

Jordan, J. R., Kimura, S., Holland, P. R., Jenkins, A., and Pigott, M. D. 2015. On the conditional frazil ice instability in seawater. *J. Phys. Oceanogr.*, **45**, 1121–1138.

Kaminski, A. K., Caulfield, C. P., and Taylor, J. R. 2014. Transient growth in strongly stratified shear layers. *J. Fluid Mech.*, **758**, R4.

Kaspi, Y., Galanti, E., Hubbard, W. B., Stevenson, D. J., Bolton, S. J., Iess, L., Guillot, T., Bloxham, J., Connerney, J. E. P., Cao, H., Durante, D., Folkner, W. M., Helled, R., Ingersoll, A. P., Levin, S. M., Lunine, J. I., Miguel, Y., Militzer, B., Parisi, M., and Wahl, S. M. 2018. Jupiter's atmospheric jet streams extend thousands of kilometres deep. *Nature*, **555**.

Kelley, D. E., Fernando, H. J. S., Gargett, A. E., Tanny, J., and Özsoy, E. 2003. The diffusive regime of double-diffusive convection. *Prog. Oceanogr.*, **56**(3), 461.

Kerswell, R. R., Pringle, C. C. T., and Willis, A. P. 2014. An optimization approach for analysing nonlinear stability with transition to turbulence in fluids as an exemplar. *Rep. Prog. Physics*, **77**(8), 085901.

Kundu, P. K., Cohen, I. M., and Dowling, D. R. 2016. *Fluid Mechanics (6th ed.)*. San Diego: Academic Press.

Lee, C. Y., and Beardsley, R. C. 1974. The generation of long nonlinear internal waves in a weakly stratified shear flow. *J. Geophys. Res.*, **79**, 453–462.

Li, H., Xu, F., Zhou, W., Wang, D., Wright, J. S., Liu, Z., and Lin, Y. 2017. Development of a global gridded Argo data set with Barnes successive corrections. *J. Geophys. Res.*, **122**.

Linden, P. F. 2000. Convection in the environment. Pages 289–345 of: Batchelor, G. K., Moffatt, H. K., and Worster, M. G. (eds.), *Perspectives in Fluid Dynamics*. Cambridge University Press, UK.

Luchini, P., and Bottaro, A. 2014. Adjoint equations in stability analysis. *Annu. Rev. Fluid Mech.*, **46**(1), 493–517.

MacMinn, C. W., and Juanes, R. 2013. Buoyant currents arrested by convective dissolution. *Geophys. Res. Lett.*, **40**(10), 2017–2022.

Mashayek, A., and Peltier, W. R. 2012a. The "zoo" of secondary instabilities precursory to stratified shear flow transition, Part 1: shear aligned convection, pairing, and braid instabilities. *J. Fluid Mech.*, **708**, 5–44.

—. 2012b. The "zoo" of secondary instabilities precursory to stratified shear flow transition, Part 2: The influence of stratification. *J. Fluid Mech.*, **708**, 45–70.

McWilliams, J. C. 2006. *Fundamentals of Geophysical Fluid Dynamics*. Cambridge University Press, UK.

Miles, J. W. 1961. On the stability of heterogeneous shear flows. *J. Fluid Mech.*, **10**, 496–508.

Moum, J. N., Farmer, D. M., Smyth, W. D., Armi, L., and Vagle, S. 2003. Structure and generation of turbulence at interfaces strained by solitary waves propagating shoreward over the continental shelf. *J. Phys. Oceanogr.*, **33**(10), 2093–2112.

Moum, J. N., Nash, J. D., and Smyth, W. D. 2011. Narrowband, high-frequency oscillations in the upper equatorial ocean: part 1: intepretation as shear instabilities. *J. Phys. Oceanogr.*, **41**, 397–411.

Nash, J. D., Peters, H., Kelly, S. M., Pelegrí, J. L., Emelianov, M., and Gasser, M. 2012. Turbulence and high-frequency variability in a deep gravity current outflow. *Geophys. Res. Lett.*, **39**(18).

Orszag, S. A. 1971. Accurate solution of the Orr-Sommerfeld stability equation. *J. Fluid Mech.*, **50**, 689.

Orszag, S. A., and Kells, L. C. 1980. Transition to turbulence in plane Poiseuille and plane Couette flow. *J. Fluid Mech.*, **96**(1), 159–205.

Pedley, T. J., and Kessler, J. O. 1992. Hydrodynamic phenomena in suspensions of swimming microorganisms. *Annu. Rev. Fluid Mech.*, **24**, 313–358.

Pedlosky, J. 1987. *Geophysical Fluid Dynamics*. New York: Springer-Verlag.

Pope, S. B. 2000. *Turbulent Flows*. Cambridge, UK: Cambridge University Press.

Posmentier, E. S. 1977. The generation of salinity finestructure by vertical diffusion. *J. Phys. Oceanogr.*, **7**, 298–300.

Pujiana, K., Moum, J. N., Smyth, W. D., and Warner, S. J. 2015. Distinguishing ichthyogenic turbulence from geophysical turbulence. *J. Geophys. Res.*, **120**, 3792–3804.

Putrevu, U., and Svendsen, A. 1992. Shear instability of longshore currents: A numerical study. *J. Geophys. Res.*, **97**, 7283–7303.

Radko, T. 2013. *Double-Diffusive Convection*. Cambridge, UK: Cambridge University Press.

—. 2016. Thermohaline layering in dynamically and diffusively stable shear flows. *J. Fluid Mech.*, **805**, 147–170.

Rayleigh, Lord. 1880. On the stability, or instability, of certain fluid motions. *Proc. London Math. Soc.*, **10**, 4–13.

Rees, T., and Monahan, A. 2014. A general numerical method for analyzing the linear stability of stratified parallel shear flows. *J. Atmos. Oceanic Technol.*, **31**, 2975–2808.

Ruddick, B. D., and Kerr, O. S. 2003. Oceanic thermohaline intrusions: theory. *Prog. Oceanogr.*, **56**, 483–497.

Ruddick, B. D., and Richards, K. 2003. Oceanic thermohaline intrusions: observations. *Prog. Oceanogr.*, **56**, 499–527.

Rudels, B., Kuzmina, N., Schauer, U., Stipa, T., and Zhurbas, V. 2009. Double-diffusive convection and interleaving in the Arctic Ocean distribution and importance. *Geophysica*, **45**(12), 199–213.

Schluter, A., Lortz, D., and Busse, F. 1965. On the stability of steady finite amplitude convection. *J. Fluid Mech.*, **23**(9), 129–144.

Schmid, M., Lorke, A., Dinkel, C., Tanyileke, G., and Wüest, A. 2004. Double-diffusive convection in Lake Nyos, Cameroon. *Deep Sea Res.*, **51**(8), 1097–1111.

Schmid, P. J. 2007. Nonmodal stability theory. *Annu. Rev. Fluid Mech.*, **39**, 129–162.

Schmitt, R. W. 1983. The characteristics of salt fingers in a variety of fluid systems, including stellar interiors, liquid metals, oceans, and magmas. *Phys. Fluids A*, **26**, 2373–2377.

Smyth, W. D. 2017. *All Things Flow: Fluid Mechanics for the Natural Sciences*. OSU Open Textbook, Oregon State University, Corvallis, OR, USA. https://open.oregonstate.edu/textbooks/.

Smyth, W. D., and Moum, J. N. 2012. Ocean mixing by Kelvin-Helmholtz instability. *Oceanography*, **5**, 140–149.

—. 2013. Seasonal cycles of marginal instability and deep cycle turbulence in the eastern equatorial Pacific ocean. *Geophys. Res. Lett.*, **40**, 6181–6185.

Smyth, W. D., and Ruddick, B. R. 2010. Effects of ambient turbulence on interleaving at a baroclinic front. *J. Phys. Oceanogr.*, **40**, 685–712.

Smyth, W. D., and Thorpe, S. A. 2012. Glider measurements of overturning in a Kelvin-Helmholtz billow train. *J. Mar. Res.*, **70**, 119–140.

Smyth, W. D., and Winters, K. B. 2003. Turbulence and mixing in Holmboe waves. *J. Phys. Oceanogr.*, **33**, 694–711.

Smyth, W. D., Moum, J. N., and Nash, J. D. 2011. Narrowband, high-frequency oscillations in the upper equatorial ocean: part 1: properties of shear instability. *J. Phys. Oceanogr.*, **41**, 412–428.

Smyth, W. D., Moum, J. N., Li, L., and Thorpe, S. A. 2013. Diurnal shear instability, the descent of the surface shear layer, and the deep cycle of equatorial turbulence. *J. Phys. Oceanogr.*, **43**, 2432–2455.

Smyth, W. D., Pham, H. T., Moum, J. N., and Sarkar, S. 2017. Pulsating turbulence in a marginally unstable stratified shear flow. *J. Fluid Mech.*, **822**, 327–341.

Snow, J. T. 1978. On inertial instability as related to the multiple-vortex phenomenon. *J. Atmos. Sci.*, **35**, 1660–1677.

Spiegel, M. 1968. *Mathematical Handbook of Formulas and Tables (Schaum's Outline series).* McGraw-Hill, UK.

Stamper, M. A., and Taylor, J. R. 2016. The transition from symmetric to baroclinic instability in the Eady model. *Ocean Dyn.*, **67**, 65–80.

Stern, M. E. 1960. The salt fountain and thermohaline convection. *Tellus*, **12**, 172–175.

—. 1975. *Ocean Circulation Physics.* New York: Academic Press.

Stommel, H., Arons, A. B., and Blanchard, D. 1956. An oceanographical curiosity: the perpetual salt fountain. *Deep Sea Res.*, **3**, 152.

Stone, P. H. 1966. On non-geostrophic baroclinic stability. *J. Atmos. Sci.*, **23**, 390–400.

Sun, C., Smyth, W. D., and Moum, J. N. 1998. Dynamic instability of stratified shear flow in the upper equatorial Pacific. *J. Geophys. Res.*, **103**, 10323–10337.

Tedford, E. W., Carpenter, J. R., Pawlowicz, R., Pieters, R., and Lawrence, G. A. 2009a. Observation and analysis of shear instability in the Fraser River estuary. *J. Geophys. Res.*, **114**(C11).

Tedford, E. W., Pieters, R., and Lawrence, G. A. 2009b. Symmetric Holmboe instabilities in a laboratory exchange flow. *J. Fluid Mech.*, **636**, 137–153.

Terwey, W. D., and Montgomery, M. T. 2002. Wavenumber-2 and wavenumber-m vortex Rossby wave instabilities in a generalized three-region model. *J. Atmos. Sci.*, **59**(16), 2421–2427.

Thomas, L. N., Taylor, J. R., Ferrari, R., and Joyce, T. M. 2013. Symmetric instability in the Gulf Stream. *Deep Sea Res.*, **91**, 96–110.

Thorpe, S. A. 1971. Experiments on the instability of stratified shear flows: miscible fluids. *J. Fluid Mech.*

—. 1973. Turbulence in stably stratified fluids: a review of laboratory experiments. *Boundary Layer Meteorology*, **5**, 95–119.

—. 2005. *The Turbulent Ocean.* Cambridge, UK: Cambridge University Press.

Thorpe, S. A., and Liu, Z. 2009. Marginal instability? *J. Phys. Oceanogr.*, **39**, 2373–2381.

Thorpe, S. A., Smyth, W. D., and Li, Lin. 2013. The effect of small viscosity and diffusivity on the marginal stability of stably stratified shear flows. *J. Fluid Mech.*, **731**, 461–476.

Trefethen, L., Trefethen, A., Reddy, S., and Driscoll, T. 1993. Hydrodynamic stability without eigenvalues. *Science*, **261**, 578–584.

Van Haren, H., and Gostiaux, L. 2009. High-resolution, open-ocean temperature spectra. *J. Geophys. Res.*, **114**, C05005, doi:10.1029/2008JC004967.

Zhao, M., and Timmermans, M.-L. 2015. Vertical scales and dynamics of eddies in the Arctic Ocean's Canada Basin. *J. Geophys. Res.*, **120**, 8195–8209.

Zodiatis, G., and Gasparini, G. 1996. Thermohaline staircase formations in the Tyrrhenian Sea. *Deep Sea Res.*, **43**(5), 655–678.

Index

CPSIA information can be obtained
at www.ICGtesting.com
Printed in the USA
LVHW061802080419
613390LV00003B/13/P

9 781108 703017